20.00
33.3

DISEASE IN A MINOR CHORD

You cannot charm or interest or please
By harping on that minor chord, disease.

Ella Wheeler Wilcox

BUT YOU CAN TRY!

EDWARD A. STEINHAUS

Disease in a Minor Chord

Being a Semihistorical and Semibiographical Account of A Period in Science When One Could Be Happily Yet Seriously Concerned with the Diseases of Lowly Animals without Backbones, Especially the Insects

OHIO STATE UNIVERSITY PRESS : COLUMBUS

Chapter one is revised and expanded from E. A. Steinhaus, "Microbial Control—The Emergence of an Idea: A Brief History of Insect Pathology through the Nineteenth Century," *Hilgardia* 26 (October 1956): 107–57.

Copyright © 1975 by the Ohio State University Press
All rights reserved
Manufactured in the United States of America

Library of Congress Cataloging in Publication Data

Steinhaus, Edward Arthur, 1914–1969.
Disease in a minor chord.

Bibliography: p.
Includes index.
1. Insects—Diseases. 2. Insects—Diseases—Research—History. 3. Steinhaus, Edward Arthur, 1914–1969. I. Title. [DNLM 1. Entomology—History. 2. Insects. SB942 S822d]
SB942.S75 1975 595.7'02 75-4527
ISBN O-8142-0218-7

NMU LIBRARY

TO

PEGGY
TIM
CINDY

Contents

Foreword

Let him step to the music which he hears,
however measured or far away.

Thoreau

When Edward Arthur Steinhaus died on 20 October 1969, he left an unfinished manuscript which he had titled "Disease in a Minor Chord." It was to have been a complete history of the beginnings of insect pathology gleaned from literature and what he had experienced himself during his years at the University of California, Berkeley and Irvine. In this span of time insect pathology attained recognition as a distinct scientific discipline. He was not one to look backwards, for he felt that the past was only the beginning, but he hoped that those who follow might be interested in their roots or origin.

Most of the manuscript was written during 1968 and was in various stages of completeness when Ed died. Some sections were still in the first draft, handwritten, without additional polishing while other sections had gone through several drafts. These stages in development are evident in the book. Only chapter 6 was complete, and undoubtedly Ed would have given even this chapter additional polishing. What is now chapter 1 had been written as two separate chapters, but he decided to integrate the two and was in the process of doing this at the time of his death. Since the partially integrated material could not be published in its half-organized state, I have attempted to finish the organization as he had planned, which makes for not-so-smooth reading at times. Chapters 2, 3, 4 and 5 vary in completeness—either a planned section was not written or portions of some sections were not finished. Two additional chapters, "Insect Pathology in States Other Than California" and "Insect Pathology and Its Development in Canada," were planned but never written. There were notes relative to these two unwritten chapters plus a few others which, when possible, I have incorporated into the manuscript.

No attempt has been made to complete the manuscript. As one reads the book he will become aware why only Ed could have finished it. An effort has been made to delete repetitions and to clear up ambiguities and badly constructed sentences. In a few cases, the sequence of material has been changed, but in general the manuscript has been published much as Ed left it. Material added by either Mauro E. Martignoni or myself is noted by the initials MEM or MCS. I hope that the readers will be charitable when they encounter roughness and unevenness in this book, realizing that this was an unfinished, unpolished labor of love by Edward Steinhaus.

The bibliography has been especially difficult to assemble. Ed left only about one hundred complete references, which meant that I had to track down over seven hundred more. The extensive Steinhaus reprint collection (some 11,000 items at the time of his death) was indispensable for this chore. However, there were still many more references that had to be found, and innumerable books and periodicals were obtained through the university interlibrary loan system. To the reference librarians my many thanks for their aid. I have tried to check each reference with the manuscript to be sure that the correct one was cited. However, as I read page proof and compiled the index, I realized that there are a few references referred to in the text that I have not listed in the bibliography. My apologies to the missed authors.

Blanks, such as a name, place, or date, occurred. The necessary information to fill these has been supplied by many kind friends too numerous to mention, and I do wish to express my gratitude to each and everyone of them.

Having a scientific background, I always had and still have a very keen interest in the development of the field of insect pathology, and vicariously experienced the trials and tribulations as well as the final fruition of efforts to establish insect pathology on a sound scientific basis. However, I claim no expertise in the field, and without the wise guidance of Mauro E. Martignoni and Y. Tanada, former associates of Ed's at the University of California, Berkeley, I never would have attempted to prepare the manuscript for publication. To them for their wise counsel and tremendous assistance my heartfelt thanks.

I am most appreciative of the suggestions of Irvin M. Hall, University of California, Riverside; Kenneth M. Hughes, Forestry Sciences Laboratory, Corvallis; T. Hukuhara, University of Tokyo, Tokyo; and Gordon A. Marsh, University of California, Irvine, who read portions or all of the manuscript; the assistance of John D. Briggs, Ohio State University, Columbus, who arranged for the final typing of the manuscript and its publication; and the herculean efforts of Beatrice Weaver and Mary Lou Lauer, who typed portions of the manuscript and the total bibliography from the original reference cards.

To the following, all of the University of California, Irvine, I am indeed grateful: Elaine Hillis and Marjorie Todd, Center for Pathobiology Library, who gave me valuable aid while compiling the bibliography; Donna Krueger, Center for Pathobiology, for her kind consideration and inspiration given so many times; and Howard A. Schneiderman, School of Biological Sciences, who provided employment and made facilities available to work on this manuscript. Last but not least, I wish to thank my family and friends for their moral support when I became discouraged, and the Ohio State University Press for having faith that an unfinished manuscript had merit enough to warrant its being published.

Mabry C. Steinhaus

Prologue

It was typical of Edward Steinhaus to look far into the future of biology and of mankind: imagination and idealism were certainly among the most impressive traits of his personality. He also had a thorough knowledge of the history of civilization, including the history of biology and, especially, pathology (in its broadest meaning, the "biology of the abnormal" or pathobiology). But as much as he loved to review the past and to speculate on the future, Steinhaus was not bent on a life of meditation. He was a doer, with a constant stream of energy flowing throughout his life.

Modern studies of invertebrate and, especially, insect diseases can be said to begin with Louis Pasteur's monumental work on the pébrine and flacherie of the silkworm. However, up to the early 1940s, except for Pasteur and a mere handful of investigators, the study of insect diseases was only incidental to the solution of other problems. Insect pathology did not really begin to exist as a branch of pathobiology until publication of Steinhaus's *Principles of Insect Pathology* in 1949. By means of this classic work, his exciting lectures, and later, his *Journal of Invertebrate Pathology*, Steinhaus stimulated the development of this area of biology and gave it that coherence, form, and substance which it has today.

The distinction of Steinhaus's numerous accomplishments places him among those who made significant contributions to the history of biology. In fact, Steinhaus, as author, acknowledges his ambivalent role as historian, since he was so intimately involved with certain segments of the subject he deals with in this book. But no one can deny that the interdigitation of history and autobiography can provide a marvelous continuity of thought, just as it can offer an incomparable understanding of the actions and goals of a man within his field of endeavor. Probably what makes certain history books so utterly impossible to enjoy is the "true objectivity" observed by their authors. Fortunately, this is not the case with the present volume. Steinhaus could not, and did not, pretend to such objectivity, for he played the central role in the development of the field of invertebrate pathology.

The personality of Steinhaus emerges from the chapters that follow as a magnificent lesson to future generations of biologists. How a man, singlehandedly and against many odds, can lovingly nurse a branch of biology through its infancy and guide it through its adolescence is only part of the lesson. As the story unfolds, there is Steinhaus's concern for the problems of humanity and for the social consequences of research. The breakthroughs of research are viewed not as ends in themselves, but in the broadest spectrum of man's relationships with his fellow men and with the ecosystem. This book reveals a man of vision, of accomplishment, and of great humanity. The reader will see how he was responsive to the needs of the student, how his warmth and understanding made him an especially successful teacher. Above all, the reader will experience, as I did, Steinhaus's fascination with his studies and his deep reverence for life.

Mauro E. Martignoni

Corvallis, Oregon
February 1972

In Lieu of a Preface

I have set before you
life and death . . .

Deuteronomy 30 : 19

There is really no logical excuse for writing this book. (I use the present tense "writing" rather than "having written" because, although ironically, most prefaces are the last part of a book to be written, in the present case it is not—hence, perhaps this is not a preface at all.) Indeed, I am by no means sure that I shall end up with a book. The few passages I have already begun are somewhat vague and discursive, if not just badly written. Should I not survive to smooth out the rough passages, to delete the repetitions, to soften the overly critical and harsh statements, and to devise some mechanism of continuity for what is now but rambling thought, I hope that some kind and merciful friend will do it for me.

In any case, I shall attempt none of the usual excuses or apologies for writing about a subject that has been one of

This preface was written from notes left by EAS. No doubt it is not the preface in the final form that he would have produced, but it gives the reader some idea of the reasons why this book was written. MCS.

my life's major interests. Remembering that I am one of those who had "much to be modest about," I shall, nevertheless, enjoy recording some of my experiences in the field of invertebrate pathology—more particularly in insect pathology—while at the same time attempting to depict in semihistorical fashion the fascinating development of this specialty. The extent to which others are as fascinated as I am with the phenomenon of disease in animals without backbones is highly variable. But I shall not worry about this if only to hope for a more meaningful book, for, as someone has said, many authors fail to reach their goals because the thought of the audience gets in the way of the work itself. Then the author is no longer writing to fulfill his own needs and aspirations, but to satisfy the expectations of the outside world. It is quite impossible to please everyone, and I shall not try. As soon as a writer becomes anxious about the critics, about the public, or about the scholarly reactions to his work, he becomes a literary slave—and usually ends up gratifying none. I hope to avoid such a fate, and hope that at least my children, and perhaps some of my former students and associates, will find some interest in what I am about to write.

This account is "semihistorical and semibiographical." I must use the prefix *semi* because I have neither the talent nor the training to presume to write either history or biography. One cannot be detached enough to really write a history of things that happened in the twentieth century. Yet deep down, I am confident that I have something to say, though I make no claim to a detached impartiality when I report the things in which I was involved; but I shall try. As M. S. Nordau puts it, "Even Balaam's ass acquired speech when he had something definite to say." In the editorial words of Paul Tillich, "We only want to show you something we have seen and to tell you something we have heard." I am well aware of the problems encountered when one attempts to write contemporary history, especially when so many of the characters are still on stage. Remembering the wrath of some of James D. Watson's colleagues when *The Double Helix* was published, let me hasten to assure all concerned that, although I certainly make no claim to greatness, it is not my intention to hurt anyone or do an unkindness to any soul.

If one is to gain a proper perspective of the history of invertebrate, especially insect, pathology in North America, it is essential to have some notion of the early emergence of the ideas that gave birth to this branch of knowledge as a modern scientific discipline. There is no question but that the roots of this discipline are in Europe and Asia, even though little is known, due mostly to the language barrier, about early oriental concepts, and even less is appreciated by those of us in the Western world. Thus I shall, at least in curosry fashion, summarize the early developments in America and throughout the world from ancient times up through the nineteenth century. But the emphasis of this book is intentionally on the history of twentieth-century insect pathology, more especially since World War II, in North America. Foreign contacts, both in Europe and Asia, will be discussed as they relate to developments in this country. These associations have added immeasurable richness to my concepts and development of ideas through the years.

If only I had the skill to make the development of invertebrate pathology, or more importantly, the history of the biology of disease, "come alive" for those who seem to disdain what has happened in the past! To quote Archibold McLeish, "History, like a badly constructed concert hall, unfortunately has occasional dead spots where the music can't be heard." One thing is certain, we cannot enjoy history if we believe that we are at the end of it. So, this book will be written with the conviction that most of what is to be historical in invertebrate pathology lies in the future. The best is yet to come. And for ages yet to be, two of the things man shares with other forms of life, both animal and plant—death and disease—will be better understood if we pay heed to how they occur down toward the bottom of the evolutionary tree where man had his beginnings. As General Robert E. Lee wrote to Charles Marshall a hundred years ago, "The march of Providence is so slow and our desires so impatient; the work of progress is so immense and our means of aiding it so feeble; the life of humanity is so long, that of the individual so brief, that we often see only the ebb of the advancing wave and are thus discouraged. *It is history that teaches us to hope*" (italics mine).

Edward A. Steinhaus

DISEASE IN A MINOR CHORD

All knowledge of facts
is historical . . . one knows accurately
only what one knows historically.

Rudolf Virchow, 1847

1

Beginning with Notions and Some Facts

1. IT ALL DID NOT START WITH HIPPOCRATES

The time was 1960.[1] The day was one of those mystical autumnal days in Tokyo. An occidental could bask in the warm oriental hospitality, yet feel that somehow he was not a part of it all. The classic beauty of the Japanese garden was evident all around me, but obviously the ceremony about to be performed, for the time being at least, transcended both the gracious courtesy of the people and the delicate magic of the garden. The scene was at the Benten Shrine on a tiny island in Shinobazu in Ueno Park in the central part of the city.

As I watched, a number of Buddhist priests chanted scriptural sutras. It appeared as though a memorial of some kind was being performed. But, I wondered, to whom? I soon learned that memorial services were, indeed, being held, but not for some important Japanese personage. They were being held for dead crickets and other insects! As the priests finished the rites, a large number of living, singing crickets were released into the area of the shrine. Only later did I learn that similar services were held annually at many local temples in Tokyo to express man's gratitude to the insects for providing him with the delight of their songs during their short lifetimes. (Incidentally, early

Chinese sericulturists sometimes arranged for the proper burial of spent silkworm moths as benefactors to man.)

I left the scene humbled by the richness of the experience, but also, in some vague way, mentally associating it with my long professional concern about, and study of, the diseases of the ubiquitous and fascinating bits of life we call insects. Jonathan Swift's oft-quoted lines, which stand as an appropriate oracle to what is generally called *insect pathology*, came to mind:

> So, Nat'ralists observe, a Flea
> Hath smaller Fleas that on him prey,—
> And these have smaller Fleas to bite 'em;
> And so proceed ad infinitum.

That day I returned to the laboratory at the University of Tokyo, where I was a guest, more of a philosopher than when I had left it.

Although realizing there was no direct connection between my thoughts on diseases of insects and humans and the memorial services just witnessed other than we do hold memorial services for humans, my belief that an understanding of the phenomenon of disease that causes death is basically and scientifically related to all life was again reinforced. Insects and other invertebrates (earthworms, snails, clams, oysters, hydras, and so on) make up ninety-five to ninety-seven percent of all animal life on this planet. Whatever the relationship—insect and man, bacterium and whale, plant and animal (wild or domestic)—there can be a noble purpose in our efforts to understand disease on all levels. Disease is one of the great tragedies of living things, not only because man sees it from his own prejudiced point of view, but because all that we call living is subject to it in one form or another. To be sure, there is another side to the coin that shows us that one of the results of disease, death, is in the final analysis one of Earth's necessities and blessings, else we should long ago have been choked into oblivion. After spending a major part of a lifetime concerned with "what goes wrong" with life and living processes—especially with those of the so-called lower animals (the invertebrates or animals without backbones)—I find I have gradually come to a better understanding of the philosophies of Saint Francis of Assisi ("Canticle of Creatures") and Albert Schweitzer (*Reverence for Life*).[2] Although I acknowledge the contradictory pragmatics of killing organisms (including insects) harmful to man's interests or used by him for food, my associates have long known of my abhorrence for the unnecessary destruction of life of any kind and of my insistence in the laboratory that even excess experimental insects be accorded as respectable and "humane" a type of necessary destruction as possible. Just how one comes by such an attitude toward living things undoubtedly varies with the individual concerned. It may be borne of deep philosophical or religious convictions, or it may be nothing more than a sincere realization that after all is said and done "man cannot make a worm"—at least not yet. At any rate, since I was a young lad I could understand the feeling behind William Cooper's words, "I would not enter on my list of friends . . . the man who needlessly sets foot upon a worm."

The mysteries and hazards of disease, along with associated taboos and explanations, have come down with man through the ages. Richard Armour (1966) notwithstanding, it all did not start with Hippocrates—at least if by "it" Armour meant man's recognition of disease. Of course, Armour himself realized this because he did not arrive at a discussion of Hippocrates until chapter 5, page 27, of his book *It All Started with Hippocrates.* Although there are accounts (e.g., see Tasnádi-Kubacska, 1962) of fossil insects, most of these instances, so far as paleopathology is concerned, deal with the insects of medical importance among invertebrates. I know of no clear-cut record of paleopathology in insects and other invertebrates.

In several early papyri, ancient Egyptians referred to insects in comparisons, likening, for example, the number of men in an army or the number of dead after a battle to grasshoppers; and in one manuscript, a young man is given a field that is protected from grasshoppers by the Gods (Breasted, 1962). According to the Bible (Exod. 8:17, 24; 10:13–15), Moses caused plagues of lice, flies, and locusts to descend upon Egypt. The

Egyptians recognized disease in man, having developed fairly complicated medicines based on magic, herbs, and some not-too-palatable ingredients such as droppings of flies. Whether they recognized disease in invertebrates we do not know. One could assume that, with such large hordes of insects abounding, they were no doubt subject to diseases of various kinds. Perhaps the plagues of Moses were decimated by epizootics of some sort.

Boyd (1962) considered Haggard's (1933) three M's—mystery, magic, and medicine—in their own origins, one and the same. So it was with man's understanding of the disease of lower forms of life. However, whereas human medicine eventually became an art as well as a science, the understanding of disease in invertebrate animals became and, for obvious reasons, remains, essentially a science. This is not to say that there are not those invertebrate pathologists who seem to possess a "magic" sort of art in the way they force Nature to yield some of her secrets regarding disease in the lowliest of animals. The study of disease in vertebrate animals (and also in plants), especially man and other mammals, has served as a guidepost and has contributed greatly to the study of disease in invertebrate animals, particularly through the concepts and techniques developed through the course of history. It has been, and continues to be, a two-way street. As to which direction will see the greater flow in the future, only time will tell. As for the past, one must read the history of vertebrate medicine and make an estimate and judgment; for each scholar the evaluation will no doubt be different.

Insofar as pathology is frequently used in the restricted medical sense, early history, with a few exceptions, does not relate to invertebrate pathology. As stated by Florey (1962), the study of disease has no doubt been going on since the time man emerged as a thinking animal. Moreover, disease as a biological phenomenon certainly affected the life systems of every living creature, plant or animal, including microorganisms. Indeed, to the comparative pathologist it is some satisfaction that one of the earliest of all records, an Egyptian papyrus of about 2160 to 1788 B.C., mentioned diseases of women and cattle.

Although we cannot, in this volume, discuss the impact of famous and creative pathologists on invertebrate pathology, the reader should know that I am very conscious of the debt to noninvertebrate pathologists—whether they study the diseases of plants or human beings—that have gone before. Accordingly, I urge the reader interested in the subject of this treatise to become at least familiar with the works and contributions of such greats as Cornelius Celsus, Clarissimus, Galen, Antonio Benivieni, Giovanni Battista Morgagni, John Hunter, Matthew Baillie, Marie-Francois-Xavier Bichat, Carl Rokitansky, Rudolf Virchow, Julius Cohnheim, and a host of others. By all means read the well-known books *A History of Pathology* by Esmond R. Long (1965) and *An Introduction to the History of Medicine* by Fielding H. Garrison (1929). Even less comprehensive histories, such as *Pathology* by E. B. Krumbhaar (1962) are not without their value in helping one to appreciate more fully the morbid biology that occurs in all life forms. Especially not to be forgotten, in considering comparative pathology, are the men and works of veterinary pathology (see Schwabe, 1969, for a stimulating treatment of some major aspects of the subject) and of plant pathology (see Walker, 1969).

As one contemplates the relatively long existence of intelligent and civilized man on our planet, he cannot help but wonder —as with medicine and other areas of science—what really were the factors that delayed his progress in knowing more about the diseases of invertebrate animals, or was early knowledge, as in most branches of learning, simply lost in time. If we say, as some historians of science do, that his progress was slowed by the absence of adequate precision instrumentation, such as the compound microscope, then why was it so long before man invented the microscope and other instruments and methodology—the questioning can be pushed back further and further until we become lost in the unknowns of prehistory. Edelstein (1967), in interpreting this aspect of ancient science, conjectured that because of the intimate relation between science and technology, scientists would have been provided with all the instruments they needed, had they only asked for them. He concluded that "the view most widely held" is that science decayed, or in some fields, did not start, because of the existence of slavery or because of definite political measures imposed to abridge the freedom of science. Not only did the ancients seem to stagnate after the first early thrusts of what might be called science, but they loitered on the threshold of modern scientific methods for hundreds of years. According to Farrington (1947), when we look for the causes of this paralysis it is obvious that it is not due to any failure of an individual. The failure was a social one. "The ancients rigorously organized the logical aspects of science, lifted them out of the body of technical activity in which they had grown or in which they should have found their application, and set them apart from the world of practice and above it. This mischievous separation of logic from the practice of science was the result of the universal cleavage of society into freeman and slave." And there have been other scholars who have concluded that it is difficult for science to emerge or to flourish in the presence of slavery. Moreover, Farrington has argued that science can develop only if it is closely connected with the world of practice and techniques.

It is not my intention to get into a philosophical discussion of these points, except to remind the reader that most of our early knowledge of disease in invertebrates came to us by free men and because of the practical importance of the invertebrates involved—the honey bee and the silkworm.

2. OF BEES AND THEIR DISEASED BROOD

Actually as far as I have been able to ascertain, the honey bee, *Apis mellifera*, was the first invertebrate to have its maladies and death recorded—both as mythology and as fact. (I would agree with F. C. Pellett [1938], who, in his *History of American Beekeeping*, remarked that it is probable that bee diseases were discovered by men almost as soon as they learned to look to the honey bee as a source of sweets.) The Greek myth telling of Aristaeus, the son of Apollo and Cyrene, who as a keeper of bees, lost his hives through disease, may have originated as early as 700 B.C., although some authorities date the story as of 100 B.C. Edith Hamilton (1942), in her *Mythology*, recounted the story as follows:

> When his [Aristaeus's] bees all died from some unknown cause he went for help to his mother. She told him that Proteus, the wise old god of the sea, could show him how to prevent another such disaster, but that he would do so only if compelled. . . . Aristaeus followed directions. . . . he seized Proteus and did not let him go in spite of the terrible forms he assumed, until the god was discouraged and returned to his own shape. Then he told Aristaeus to sacrifice to the gods and leave the carcasses of the animals in the place of sacrifice. Nine days later he must go back and examine the bodies. Again Aristaeus did as he was bid, and on the ninth day he found a marvel, a great swarm of bees in one of the carcasses. He never again was troubled by any blight or disease among them.

Leaving mythology, Aristotle's (384–322 B.C.) generally cryptic descriptions of certain diseases of the honey bee and other insects in his *Historia animalium* may conveniently be taken as the beginning of the recorded history of insect pathology.[3] *Historia animalium* was probably written between 330 and 323 B.C., presumably with the help of a thousand or more colleagues and disciples. Although his mention of disease and the "suffering" of bees was brief and lacked the detail of his discussion of numerous other biological facts, it nonetheless was a perceptive account. The frequent references to bees in *Historia animalium* is understandable when we remember that, to the Greeks, honey was very precious when no other form of sugar was available. That he and his contemporaries should have occasion to note that the honey bee was subject to disease is, therefore, not surprising.

In his use of the word "disease" Aristotle did not always make clear what he meant. Broadly speaking, it definitely included invasion of the hive by what we now designate as the greater wax moth, *Galleria mellonella*, both as larva and as adult. He incriminated a "spinning worm" (presumably the larva of the wax moth) called "cleros" as causing "sickness in the hive"; he also referred to it as producing a web which entrapped some of the bees and allowed the covered combs to decay. Still another kind of "caterpillar" called "teredo" was found invading hives. He may also have been including the foulbroods in his generalization that bees "suffer most when flowers are covered with mildew [or, in the Creswell, 1878, translation, infected with rust], or in seasons of drought." The same may be said when he referred to bees suffering from lassitude and a "dispirited" condition, or when he wrote of a disease that caused a "strong smell in the hives" (Cresswell, 1878). Vertebrates (reptiles, birds, amphibians, and mammals) that preyed on bees were also designated as enemies of the honey bee by the Greek philosopher.

Aristotle also recorded the maladies or deaths of insects other than bees, as the following quotation from the D. W. Thompson (1910) translation indicate:

> Many insects are killed by the smell of brimstone; ants, if the apertures of their dwellings be smeared with powdered origanum and brimstone, quit their nests; and most insects may be banished with burnt hart's horn, or better still by burning of the gum of styrax.
>
> Their death is due to the shriveling of their organs, just as the larger animals die of old age. Winged insects die in autumn from the shrinking of their wings. The myops [a tabanid] dies from dropsy in the eyes.
>
> The grasshopper lays its eggs at the close of summer, and dies after laying them. The fact is that, at the time of laying the eggs, grubs [possibly of some ichneumon or other parasitic insect] are engendered in the region of the mother grasshopper's neck; and the male grasshoppers die about the same time . . . The attelabi or locusts lay their eggs and die in like manner after laying them.
>
> All insects, without exception, die if they be smeared over with oil; and they die all the more rapidly if you smear their head with the oil and lay them out in the sun.

Regarding invertebrates other than insects, Aristotle referred to unnatural death among mollusks; considered it unusual that the purple murex, a gastropod, would die within a day after being transferred from sea water to fresh water; attributed mass mortality among shellfish (clams) to desiccation, and noted that freezing temperatures would also kill these invertebrates;

and related the general health of oysters, scallops, octopuses, cuttlefishes, and lobsters to the condition of the spawning area or period.

In closing with these words of tribute to Aristotle for having astutely recognized and recorded affliction and death of insects, we should also mention that other early Greeks wrote on apiculture. For example, in the first part of the third century B.C., Nicandros of Colophon published a treatise (*Melissurgica*) on beekeeping, but I have been unable to ascertain to what extent he was concerned with the diseases of bees.

According to Sarton (1959), the first Latin treatise on beekeeping was chapter 16 of Marcus Terentius Varro's (127–16 B.C.) *De re rustica*, but the level of his knowledge of the subject did not extend beyond the Aristotelian. In spite of his lack of advanced information on beekeeping, he must have been a most remarkable man. He devoted most of his life to public affairs—politics and war. When Varro was sixty-eight years old, Caesar made him head of his Greco-Latin Library, and most of his tremendous literary activity was accomplished after this. Some have referred to him as the "most learned of the Romans."

The best known of the early classical authors is Virgil (70–19 B.C.), who, in his *Georgica* (37–29 B.C.)—a didactic poem on farming and farm life—described a disease of bees in the fourth book, a treatise on beekeeping. His description began, "Since life brings to the bees the same bad luck as to humans, they may suffer illness" (Day trans., 1941). He then went on to recommend a "tonic food" to aid the bees to regain their health. Virgil was born in Northern Italy in a village near the Po River. His father was a farmer who earned his living by beekeeping, which probably accounted for Virgil's knowledge of farming practices. Being a good student, he was able to attend excellent schools in both Milan and Rome. Due to civil wars and various vicissitudes, his father's farm was seized, but later Virgil was given a villa in Nola near the Bay of Naples and there he wrote his *Georgica* and his immortal *Aeneid*. He is considered the leading naturalist of that period and his *Georgica* the greatest didactic poem ever written (Sarton, 1959). It must be pointed out that from the standpoint of its introduction to apiculture and its practical value, the *Georgica* was one of the most important documents of its time and for seventeen or eighteen centuries thereafter. Some authorities consider it the only ancient document on beekeeping worthy of study. Another early Roman, Pliny (A.D. 23–79), referred in his writings on natural history (A.D. 77) to the maladies of the honey bee. He expressed in anthropomorphic terms the "grief" experienced by the bees when one of their queens (which he designated as "kings") was carred off by the pestilence.

At this point one cannot help wondering if the great Roman personage Galen, who lived in the second century A.D., had any influence on man's understanding of disease in the honey bee or silkworm. Even though he was the originator of experimental research methods in medicine (but who, through his dogmatism, brought the advancement of medical science to a virtual standstill for a thousand years or more), there appears to be no record that he or his contemporaries in any way influenced the understanding of disease in insects.

Before leaving the concern expressed by the ancients for the welfare of the honey bee, it seems appropriate, on behalf of insect pathology, to include among the mottoes we might adopt for the branch of science that shall concern us in this book, Aristotle's comment "Each of us may add a little, and from the whole there may arise a certain grandeur." In its modest way, it is my hope that this mottled historical account may be one of the little things that this Greek intellectual suggests each of us may add.

As time went on into the Middle Ages in Europe little was added to the observations made by the ancients with regard to the diseases of the honey bee. Of interest, however, is one of the first records of what might be considered an abnormality or teratology in insects. It is that of Albertus Magnus (1193–1280) in book 1 of his *De Animalibus* (1270), where he asserted "some animals, such as bees, produce poorly developed larvae." The same learned Dominican, who wrote at considerable length regarding bees and their activities, also attributed emotions to these insects. For example, when the queen (which he too designated as the king) died, Albertus Magnus discerned that "profound sadness" filled the hive, and several of the inhabitants "succumb to their own sorrow." In considering this observation, it would be unfair to assume that this zoologist had imbibed too freely of the "myod" or other intoxicating beverages made of honey. However, he may have only been elaborating on the writings of Pliny.

As already mentioned, references to the seriousness of honey-bee diseases had been recognized long before the causes were established. In 1586, the German apiculturist Nickel Jacob not only described certain bee diseases, which he believed had their origins in putrefactions, but also suggested methods of combating them. In the years to follow, other European beekeepers became interested and joined in the discussion of the various afflictions to which their bees were subject.

Schirack (1771) was among the first to use the name "foul brood" (modern usage, foulbrood) in reference to disease in the honey bee. Indeed, he designated five additional diseases of this insect, as follows: dysentery, disease of the antennae, queens laying drone eggs only, sterile queens, and queenless colonies. "Foul brood," which Schirack considered very dangerous and highly fatal, he attributed to two causes: improper food and a faulty queen which permitted the brood to be arranged in the cells so that the heads pointed inward. In 1860, Leuckart reported a fungus disease of the honey bee; at first he thought it might be responsible for what was being called "foul brood," but because he did not find the fungus always associated with the disease, he concluded that the "infectious foul brood" was not caused by a fungus and that probably various forms of disease were involved. Preuss (1868) and Schönfeld (1873, 1874) were inclined to ascribe the infectiousness of the virulent form of "foul brood" to a fungus which Preuss named *Cryptococcus alvearis.* Others (e.g., Molitor-Mühlfeld, 1868; Dzierzon, 1880; McLain, 1887; Lortet, 1890; Mackenzie, 1892; Howard, 1894, 1896) were among the nineteenth-century bee culturists who recognized that more than one infectious disease was involved in "foul brood," or who were concerned with bacteria they found associated with diseased brood. A summary of the conclusions reached by these men, as well as by others, during the nineteenth century and the first eleven years of the twentieth is presented in a bulletin by Phillips and White (1912) titled *Historical Notes on the Causes of Bee Diseases.*

One of the first bacteria to be described as an insect pathogen during the nineteenth century was *Bacillus alvei*, first mentioned by name by F. R. Cheshire during a talk at a conference of the British Bee-Keepers' Association on 25 July 1884, and then published in the 15 August 1884 issue of the *British Bee Journal.* Cheshire considered this sporeformer to be the cause of European foulbrood in the honey bee; however, the true cause of the malady is now generally considered to be *Streptococcus pluton* (see Bailey, 1957, 1963). *B. alvei* is presently considered by most authorities to be a saprophyte living on the dead remains of larvae dead of foulbrood. (*Bacillus larvae* White, the cause of American foulbrood, was not discovered until 1904.) *B. alvei* was described by Cheshire and Cheyne in 1885, at a time when the various brood diseases were not clearly differentiated, although at one time in 1884, as had few others before him, Cheshire recognized two kinds of foulbrood. However, later in the same year he reverted to the belief that only one form of foulbrood existed.

In a way, Cheshire had an opportunity to be the Pasteur of bee diseases, an appelation we might more aptly grant to G. F. White in the early part of the twentieth century. For a brief moment in history Cheshire was looked upon as the White Knight who had come to solve the foulbrood problem. And a serious problem it was. There are numerous stories in insect pathology for a Paul de Kruif—the story of foulbrood is one of them. In introducing Cheshire to speak before the British Bee-Keepers' Association in 1884, the chairman read, as typical, a plaintive letter from a cottager in Buckinghamshire:

> I cannot get rid of foul brood; nearly all my bees have got it, and I thought I had got rid of it. I don't think there are many free from it in this district. I shall have to clear right out I expect; only had two swarms. It makes me disheartened to keep trying. I hope Mr. Cheshire can cure it, it will be a

> blessing to bee-keepers. I have found it in Surrey, Hampshire, and Berkshire, and I have known thirteen places round me that have had it, some have lost all, others some. I should be glad to know how to stop it, for it has ruined me. . . . All I bought last year I shall have to destroy this year.

At this conference, the discussion following Cheshire's talk took almost as many pages to relate as did his talk itself—so great was the concern over, and interest in, the problem. Had only Cheshire been more attentive to Koch's postulates, or a little luckier, his vivid stage appearance might have been more lasting. Or, perhaps, had he allowed the micrococci found in diseased larvae by a German, Schönfeld, to whom he referred, to titillate his thinking, perhaps he would have come closer to the truth as to the cause of European foulbrood. Nevertheless, Cheshire brought the idea of specific bacterial etiology—then so earth-shaking in the case of certain human and domesticated animal diseases—to the biological level of the insect.

The honey bee apparently was introduced to North America from Europe, but, as chronicled by Thomas Jefferson, "The Indians concur with us in the tradition that it was brought from Europe, but when and by whom we know not." The insect was known as the "white man's fly" to the Indian to whom it signaled the approach of a settlement of whites. A number of early reports stated explicitly that the insect was brought to the New World (i. e., the eastern seaboard) by the English, although some writers believe it was previously brought to St. Augustine by the Spaniards. According to Nelson (1967), bees were in Virginia in 1622; Eckert and Shaw (1960) cite the date as 1621. In the case of some of the early reports there was obvious confusion between the honey bee and certain wild native stingless bees of North America and Brazil. In any event, one of the earliest records of the presence of the honey bee in North America is that by John Josselyn (1865), who lived in New England in 1638 and later wrote in his *Voyage in New England*, "The honeybees are carried over by the English and thrive there exceedingly." Early official documents of what is now the state of Massachusetts record the presence of bee colonies at least by 1660, and probably as early as 1640, perhaps even earlier (see Pellett, 1938). Nelson (1967) commented that beekeeping in the American Colonies declined rapidly after 1670 "probably because of the disease now known as American foulbrood."

It is a popular conception that Moses Quinby, who took up beekeeping in New York in 1828, and L. L. Langstroth, who began keeping bees in Andover, Massachusetts, in 1837, were perhaps the first American apiculturists seriously concerned with the diseases of *Apis mellifera.* At any rate, the publication in 1853 of two separate books, which went through many editions, by these men had a strong influence on apiculture in America. Moreover, they suggested steps for the control of bee diseases as logical as we could expect for that pre-germ-theory period.

Langstroth (1853) devoted most of his chapter on "Enemies of Bees" to the activities of the bee moth (now known as the greater wax moth, *Galleria mellonella*), the ravages of which were known to the ancients. He described the fatal effects of "dysentery," which he attributed to inadequate ventilation (and in so doing cast deserving barbs at inadequate ventilation in buildings occupied by human beings, especially female human beings), improper feeding (one should not feed bees late in the season on liquid honey), and inadequate protection of hives from moisture. As to foulbrood, he professed to know nothing by his own observation. Lastly, he referred to a "peculiar kind of dysentery," which attacked small numbers of bees, causing them to become irritable, then seemingly "stupid," and unable to fly. Their swollen abdomens were filled with an offensive-smelling yellow matter. He expressed no opinion as to the cause of this disease.

Moses Quinby (1853) appears to have been one of the first to recognize the infectious nature of brood diseases and to find a somewhat practical method of control, which, actually, was essentially the same as that used by the old German master of beekeeping, Nickel Jacob (see Pellett, 1938). In his book *Mysteries of Bee-Keeping Explained* Quinby began his chapter on bee diseases with the statement that the subject was new, inasmuch as he had never seen a chapter of any previous book devoted to this subject. From his account it would appear that he first observed diseased brood around 1833. Remarkably, Quinby experimented as to the infectiousness of the disease, as he observed it, by transferring honey from diseased to healthy colonies and in so doing recognized the source of the infection. Thus, in a general sense, Quinby demonstrated the infectious nature of one of the brood diseases (possibly American foulbrood) in a manner not unlike that by which Bassi showed white muscardine to be infectious for the silkworm, except that Quinby used infectious feedings whereas Bassi used direct inoculation. His notable experiment is described succinctly as follows:

> To test this principle still further, I drove all the bees from such diseased stocks, strained the honey, and fed it to several young healthy swarms soon after being hived. When examined a few weeks after, every one, without an exception, had caught the contagion.

Although "quaint," Quinby's chapter titled "Diseased Brood" is a delight to read. It is a refreshing example of the study of cause and effect and the value of experimentation. With obvious intellectual vigor he broke away from the prevalent thought of the day—that the destructive brood disease was caused by the chilling of the bees during cold weather (and other causes were cited; see Kohnke, 1882)—and likened the disease to the contagious diseases among humans. He admitted that mortality caused by the chilling may occur occasionally (one case in twenty), but that for the most part the disease was caused by a "contagion" and that "the virus was contained in the honey." He determined that the "seeds of destruction" were spread to healthy bee larvae by bees robbing the stores of the diseased hives. Accordingly, a major part of his program to control the disease was the destruction of the combs containing diseased breeding cells. Hives containing diseased larvae were detected by diminished numbers of bees and larvae "stretched out at full length, sealed over, dead, black, putrid, and emitting a disagreeable stench."

It is interesting that although disease of one sort or another has been known in the honey bee since the time of Aristotle, the very knowledgeable A. I. Root published a perspective in 1877 in which he assumed that disease did not appear to be a very serious matter in beekeeping. He stated in his *The ABC of Bee Culture*, "I am very glad indeed to be able to say, that bees are less liable to be affected with disease than perhaps any other class of animated creation." He did recognize "one" disease of the brood that was greatly to be feared, namely, foulbrood, a "dysentery" that he attributed to "bad food" and that he doubted should be considered a true disease, and a rather vague condition known as "spring dwindling" which he believed was also largely caused by "bad food." Toward the end of his section on diseases of bees he wrote, in his typical homespun manner, "when a bee is crippled or diseased from any cause, he crawls away from the cluster, out of the hive, and rids the community of his presence . . . if bees could reason, we would call this a lesson of heroic self-sacrifice for the good of community."

It would appear (Phillips, 1918) that in the United States the distinction between American and European foulbrood was first recognized in New York State in 1894. Prior to that time a sort of general recognition of two diseases, one called virulent and the other mild, was made, but their distinctness was not really always clear nor was their etiology recognized as different.

F. W. Alexander, a beekeeper at Delanson, New York, suffered from extensive outbreaks of what was apparently European foulbrood; he attempted to fight the disease by the use of various chemical disinfectants, but to no avail. Eventually he found that Italian bees appeared to have a greater resistance to the disease than did other strains. He observed that strong colonies of these bees performed such good housekeeping that they virtually eradicated the disease. Alexander published his method of disease control in the November 1905 issue of *Gleanings in Bee Culture*. Other articles, pamphlets, and books on the control of the diseases of bees were available by the turn of the century. The concepts of etiology they portray make fascinating historical reading.

Of the many books and booklets on beekeeping published in North America prior to 1900, most, but not all, referred to the problem of diseased brood. Both North American authors and those from abroad were content to record theories of disease dating from the first century B.C. (e.g., Columella's conclusion that "dysentery" of adult bees was caused by feeding on nectar from the blossoms of elms and spurge). One can also notice the plagiarism of the chapter "Diseases of Insects" by Kirby (Kirby and Spence, 1826) by many authors as they discuss "dysentery" and "vertigo" of adult bees as well as the "fouled brood" and "mildewed brood" of the combed larvae. Although one must be tolerant of some such writings that preceded scientific studies of bee diseases, it can nonetheless be said that it would have been well had more commenced their chapters or discussions on the diseases of bees as Munn (1870) did in his *Bevan on the Honey Bee*: "It is probable that much that has been written on this subject is fanciful."

According to Pellett (1938), the first record in America of an attempt to control the diseases of the honey bee by legislation was an act passed in 1877 by San Bernardino County, California. Later, in 1883, the state legislature passed a similar act applying to all of California. It provided that the Board of Supervisors of any county was authorized to appoint bee inspectors. Whenever the inspectors found diseased hives, the hives, combs, bees, and all were to be burned "the following night"; or, if burning was impractical, these materials could be well buried in the earth. Michigan in 1881 passed the first statewide bee-disease law, that required the burning of diseased hives, but this law did not, until years later, provide for inspectors with statewide authority. The province of Ontario, Canada, passed an act for the suppression of foulbrood, and for inspection, in 1891. Where legislation did not intervene, some beekeepers attempted methods of controlling the disease prior to destroying entire hives or colonies. A common rule of thumb was to attempt to save the colony if less than ten percent were diseased; if more than ten percent were diseased, the hives were completely destroyed.

L. O. Howard (1930) stated that the scientific study of the honey bee began in the Bureau of Entomology in 1885. These investigations, which included the entire spectrum of apiculture, were carried on largely by Nelson W. McLain, who was initially stationed at Aurora, Illinois. This work apparently ceased about 1888. It was initiated again in 1891 for the federal government by A. J. Cook at Lansing, Michigan, by J. H. Larabee in Vermont, and by Frank Benton in Washington, D. C. Benton's manual of bee culture was published in 1895 as Bulletin No. 1 of the New Series of the Bureau of Entomology. Benton was succeeded in office in 1907 by his assistant, E. F. Phillips, who had joined the federal service in 1905.

Although some sources credit W. B. Hayford with being the first—in 1856—to bring bees to the other side of the continent, that is, California, Pellet (1938) maintained that there were "well authenticated shipments prior to that date." Watkins (1967; 1968*a*, *b*, *c*,) credited C. A. Shelton with having imported the first bees in March 1853. Only one colony of the original twelve purchased survived the long trip via boat and the Panama railroad. It was brought to San Jose and threw off three swarms the first year. William Buck imported in 1855 eighteen colonies, becoming the first large importer of bees in the state. Harbison, who was actually the first real experienced beekeeper, made importations in 1857 and 1858 and did much, along with Appleton, to make beekeeping successful. According to Pellet (1938), Parks in 1917 stated that a monk named Cherepenin brought honey bees in straw keeps to Alaska from Kazan, Russia, in 1809 and about 1830 to Fort Ross in California. Both Essig (1931) and Watkins (1968a) doubted the importation of Russian bees into California. There is no record as far as I have been able to ascertain that John Sutter had bees at his New Helvetia fort.

The first record of disease in colonies of honey bees in western North America is clouded with uncertainty. During the 1850s, in California, high mortalities of some colonies were reported, but in most of these cases one has the impression that factors other than disease (e.g., unfavorable weather, long journeys) were responsible; still, in some cases the hives could well have been destroyed by brood diseases or by infestation with the greater wax moth, *Galleria mellonella*, which, in the United States, was first reported by the *Boston Patriot* in 1806.

Eckert and Shaw (1960) stated that 6,000 colonies of bees were brought to California during the winter of 1859–60, "one-half of which were said to have been infected with foulbrood. This is the first record of foulbrood in California." At any rate, diseases of honey bee became a serious problem in the West about the same time that they did in the East, as evidenced by the fact that the first record of an attempt to control diseases by legislation is the act passed in 1877 in San Bernardino County.

While the early literature pertaining to the diseases of the honey bee is plentiful and interesting, the investigations it reported do not appear to have assumed as pivotal a role in the development of insect pathology during the nineteenth century as the studies on the diseases of the silkworm. This, in spite of the fact that as of today, and as asserted by Bailey (1968), "more diseases of the honey bee (*Apis mellifera*) have been described than for any other insect species." During the twentieth century the study of bee diseases assumed a much more significant role. But the stimulus for the idea of using microorganisms to control harmful insects came largely from the silkworm-disease studies rather than from investigations of the diseases of the honey bee.

3. FROM "NINE DISEASES OF SERICULTURE" TO AGOSTINO BASSI AND LOUIS PASTEUR

> Having failed by so many different procedures to produce muscardine in silkworms . . . I resolved to search for this deadly being, to discover its nature and habits.
>
> *Agostino Bassi, 1835*

I

Inasmuch as sericulture was imported into the Western world from the Orient, it would seem logical to suppose that whatever knowledge possessed by the Chinese of diseases of the silkworm would be imported about the same time or shortly thereafter. Silk was mentioned in Western literature by both Herodotus

(c. 484–425 B.C.) and Xenophon (c. 430–355 B.C.) when they referred to the luxurious "medic" dress worn in Persia (Weibel, 1952). Silk was also referred to in Aristotle's *Historia animalium*; not only the kind of silkworm but also the spinning of silk by Pamphila, daughter of Plateus on the Island of Cos, was described. But neither he nor Pliny, who also described the silkworm, recounted anything about disease in the insect. It appears that the species of silkworm concerned in many early accounts was not *Bombyx mori*, the Chinese silkworm, but rather a "wild" or native species. The quality of the silk was quite distinctive, probably similar to the tussah silk of India. Fabrics made from it were very thin and almost transparent. A type of wild silk was apparently known in Syria-Palestine in the early sixth century B.C. since it was mentioned by the prophet Ezekiel (Ezek. 16:10, 13).

Let us here yield to the temptation to consider the beginnings of sericulture even though these beginnings tell us virtually nothing about the diseases of the silkworm. One of the best short historical accounts of the silkworm and silk production is that by Liu (1952). Liu begins one section of his work as follows:

> The origin of sericulture in China will always remain a mystery. The domestication of the insect was a slow process and so was the development of a new industry therefrom in those early days. Both took place long before there were any written records and consequently our knowledge today on their respective origins must always remain either the child of some later speculations or the outcome of hearsay. Concerning the who, when, where, and how of this problem, we can never expect to get any definite answer.

Liu goes on to relate some of the Chinese legends and folklore as to how the silkworm was first discovered and the use of silk first realized. He tells of how the utilization of silk was perhaps forced on the Chinese, how there was a shifting of the center of sericultural activity from the north to the south, and how it was that the industry developed into three regional types.

Although its origins are remote, most writers place the historical beginning of practical sericulture with Empress Seeling-see (Si-ling-chi), wife of the famous emperor Hoang-tee (Huang-ti), about 2,700 years before the time of Christ—more precisely 2640 B.C.[4] However, Liu concludes that the Empress "can hardly be credited as the discoverer of the art of raising silkworms." He found that according to some Chinese references, sericulture was in practice long before her time. For example, the *Levy Records of the Han Dynasty* stated that silk was already in use during the reign of Shen Nung (3218–3079 B.C.). According to Liu, we are told that Emperor Fu Shih, who might have lived hundreds of years before Shen Nung (the exact period of his sovereignty apparently is not known), was the first one to extract silk from the cocoon, and he might have invented the cloth.[5] At any rate, having developed in China over a period of centuries, the art of silkworm rearing spread south to Tibet and elsewhere. Knowledge of silk and the silkworm reached Japan through Korea about A.D. 300.

References in Sanskrit literature indicate that the silk industry existed in India about 1000 B.C. or, according to some, even possibly 4000 B.C. It is probable that, in the earliest periods in India, the silkworm concerned was a native species (producing the famous tasar or tussah silk) and not *Bombyx mori*. From the few Indian historical accounts we have seen, it seems clear that tasar silk developed most likely independent of China as a craft of "hill-folk and aboriginals" and down through the centuries became an important part of tribal culture (Jolly, 1968). While India presently ranks fourth among nations so far as the production of "mulberry silk" is concerned, she is second only to mainland China so far as the tasar industry is concerned. Tasar silkworms are usually reared on forest trees by seven or eight tribes in the central plateau from Dodavari to Ganges. Although a number of species of *Antheraea* moths may be found in India, *Antheraea mylitta* is the principal, and virtually the only, species that has been exploited commercially. As of today the principal diseases of *A. mylitta* are a nuclear polyhedrosis and certain imprecisely known bacterial diseases. It does not appear to suffer from microsporidian and fungal diseases to the extent that *Bombyx mori* does. However, insect parasites and predators do commonly attack it, presumably because rearing is generally an "outdoor practice."

In volume 16 of the *Chinese Repository* (1847, p. 224) we find the following statement:

> It has been generally supposed that the people known in ancient history as the Seres, were identical with the Chinese, both because of their eastern position, and that the principal silk manufacturers were believed to have been brought from thence, on which account the Romans named the country Sericum, or Serica, or Sereinda. This fact, however, is not at all certain; on the contrary, there is strong, and almost conclusive reason for allowing that the trifling quantity of silk imported into Rome came, not from China, or Sereinda, but from Persia. . . . It might be added . . . that the Chinese never traded, negociated [*sic*] nor were even known to the Romans, that the most learned ancient geographers conceived Serica to be identical with Tartary, not with China Proper; and, in their charts it adjoins Scythia.

The article goes on to say that if the Romans secured their silk from Persia then there is no proof that silk culture originated in China.

There are varying accounts of how *Bombyx mori* was introduced into Europe, and the question of how enough know-how concerning the technology of the reeling and spinning of silk was brought with the insect to establish the industry is not answered (Needham, 1961). According to Patterson (1956), the production of raw silk was first attempted under the Byzantine emperor Justinian. Most historians agree that the Western sericulture industry began with the bringing of the eggs of the silkworm about A.D. 530 (or some say A.D. 555) to Constantinople, probably from Khotan, where eggs, concealed in her coiffure, had been carried by a Chinese princess betrothed to a Turkistanian prince. Two Persian monks are credited with bringing the secrets of silkworm rearing to the West. They were probably Syrian Nestorians, who always were welcome in China and Middle Asia at this time. They are said to have concealed silkworm eggs in their hollow staffs and also to have taken the mulberry, which was so essential for the rearing, at the same time. The secrets of silk's manufacture and source are said to have been jealously guarded by the Chinese, although Liu questions that there was such a policy of secrecy. It was a mystery to easterners who sometimes described silk as fine down gathered from certain trees or flowers. Even those who eventually knew it to be derived from an insect had no clear idea as to its production. Although silk culture was well developed in Spain by the twelfth century, Italy became the chief center of the European silk industry by the thirteenth century; during the Middle Ages it began to develop and flourish in France and on a smaller scale in England.

Unfortunately as far as disease is concerned we know little of what must have been known to the early oriental sericulturists. Undoubtedly a more penetrating probe than I have been able to make into pre-Christian literature of China and India might throw more light on this. We know that at least some of the present known diseases of *Bombyx mori* occurred, but precisely when they were recognized as diseases is difficult to discern. The insect itself, especially the pupa, was used as food. The excrement was used as fertilizer, forage, and even as an insecticide and weedkiller (Liu, 1952, pp. 150–51). However, perhaps its greatest use, outside of silk production, was in medicine. Liu cites one source (*Pen Tsao Kang Mu*) as listing eighty prescriptions. Among the forms used was that of silkworms that had been killed and covered by what we now know as the entomogenous fungus *Beauveria bassiana*, the cause of muscardine disease. Of the eighty prescriptions, twenty-nine involved caterpillars killed by this fungus, and called *kiang-tsan* or "stiffened caterpillars." To quote again from Liu's interesting account,

> They [the muscardined silkworms] are generally used in a powder form, either in combination or independently, as a kind of antiseptic or antitoxin for such trouble as sore throat, tooth pain, wounds, abscesses, children's convulsions, and also as a stimulant for milk secretion of mammalian glands. It is even recommended as a cosmetic.
>
> For wounds and for sore throat, they are generally used independently and the effect is said to be almost instant. For removal of milk congestion and for children's convulsion, the effect is asserted to be also good, especially the first one. But here the caterpillars are used in combination with some other materials and hence

it is difficult to say whether the effect is produced by them alone. The same is true in case of the tooth pain in which the stiffened caterpillars have to be baked with fresh ginger and to be administered with the decoction made from the pods of *Gleditscha sinensis*. As a cosmetic, however, the author did not say whether it is also efficacious.

The preparation of these caterpillars for using in medicine is interesting. Opinions differ as to the kind of caterpillars to be selected. While some would accept any caterpillar that is straight and white, others prefer only those from the first crop for the reason that caterpillars of the first crop are usually larger and comparatively free from the infestation of muscid parasites.

Liu continued his account by describing the use of silkworm excrement and the use of the healthy moths themselves in the treatment of a variety of human ailments. The pupa and cocoon also had some medical uses, as did the chorions of the spent eggs that were used for contraceptive purposes.

Liu finished his account of *The Silkworm and Chinese Culture* with lists of domesticated silkworm species (five species of *Bombyx*) and wild silkworms (six species of *Antheraea*, three of *Attacus*, two of *Actia*, and eight miscellaneous, two of which belong to *Bombyx* but are considered to be wild species). Nothing concerning their possible diseases is mentioned.[6]

In volume 16 of the *Chinese Repository* (1847) we find expressed an appreciation of the harmful effects caused by factors we know to be noninfectious in nature and factors which, in one form or another, were for centuries considered to be detrimental to the rearing of the silkworm. In these descriptions, however, we detect what must have been the complications of infectious disease. For example:

Nothing is more necessary to be guarded against in the rearing of silk-worms than the effects of noise and cold; a sudden shout, the bark of a dog, even a loud burst of laughter, has been known to have destroyed whole trays of worms; and entire broods perish in thunderstorms. . . . It is this necessity for the formation of an artificial temperature that creates the great difficulty of rearing silk-worms (p. 231).

In volume 18 (1849), we find:

> It [the silkworm] belongs to fire and dislikes water, therefore it eats, but does not drink; it is injured by noise and by dampness. . . . The place where they repose must be dark and dry, but made light when they awake; when feeding, supply them constantly with fresh leaves, but do not expose them to the wind on awaking, nor give them too many leaves at first. Smoke, and the smell of spirits, vinegar, and all sour things; as also the odor of musk and oil, the stench of mouldy clothes, should all be removed; in feeding, the worms dislike leaves that are damp or hot; and harsh noises, such as pounding rice. . . . The proper heat should be applied gradually, otherwise the worms will become yellow and soft; when cooling them admit the wind slowly and gradually through the windows, otherwise they will blanch, or become rigid; . . . by eating damp leaves the worms are purged and turn white, while hot leaves produce costiveness, and swelled heads; in either case they produce no cocoons, nor come to maturity. . . . If from one third to a half of them are of a yellow color [nuclear polyhedrosis?], or have glistening skins, then diminish the leaves. . . . If there be a few light green or white worms which do not sleep, throw them away. . . . On awaking, do not jostle them, for their skins are easily injured (pp. 307–9).

Mitani (1929) stated that pébrine must have been prevalent in China for a long time and at the time of his writing there was still a great deal of loss due to the disease; also, the Chinese government did little to aid in the prevention and extermination of pébrine, but despite this lack of action on the government's part, the Chinese sericultural industry had not suffered losses as heavy as those in France and Italy.

Long before the germ theory of disease was conceived, the Chinese sericulturist believed that healthy offsprings could be reared only from healthy parents. On this basis they selected the healthiest appearing insects for their future stock. Upon emergence from the cocoon, moths showing any deviation from the normal were discarded. Following mating, the resulting eggs were also carefully scrutinized, color and shape being the primary criteria on which to base the selection.

I am told by my friend Tosihiko Hukuharo that in a Japanese code of laws consisting of fifty volumes, called *Engishiki*, and compiled between A.D. 905 and 927, there was a description of diseased silkworms called *hakkyo-byo*, which was obviously

NMU LIBRARY

white muscardine. This code of laws governed the control of these muscardined silkworms as a medicinal, ordering that each year a certain quantity of these diseased worms be delivered from the eleven countries or provinces to the Emperor Daigo as a medicine to be used in the treatment of palsy and paralysis. According to Sano (1898) only those muscardined silkworms that were white and straight were suitable for medicine, while those that were black were considered forgeries. In this context the Japanese did not look upon muscardined silkworms as a calamitous disease since the diseased and dead larvae were used to pay taxes and tribute to the emperor by the rulers of the several countries that comprised Japan at that time.

According to Ishikawa (1940), the first Japanese technical book of a comprehensive nature on sericulture may have been that by Dogen Nomoto, who wrote *Kaiko-shiyoho-ki* in 1702. Since that time numerous books and pamphlets have been written on silkworm rearing, virtually all of them containing chapters or sections on the diseases of the silkworm. Ishikawa mentioned that a brief explanation of *hoso-san* ("thin larvae") appeared in *Kaiko-yoshi-tekagami* [Review of silkworm breeding] by Shigehisa Bab (1712).

In 1756 a book with the intriguing title *Shinsen-yosan-hisho* [Newly selected secret book on silkworm rearing][7] was published by Yoemon Tsukada; in 1766 Tomonobu Sato published *Yosan-chawa-ki* [Essays on silkworm rearing]; and in 1803, Morikuni Uegaki published *Yosan-hiroku* [Secret notes on silkworm rearing]. Most of these early writings, as they related to the diseases of the silkworm, were concerned with empirical observable aspects of diseased larvae. However, in Tsukada's book, one can discern, from the descriptions given of the diseases, some of the afflictions of the silkworm as we know them today. Muscardine, already mentioned, was one of these, but Tsukada also wrote, "If silkworm larvae are warmed after the first feeding and molt on the fourth day, some larvae are found white and knotted. They excrete fluid [white in color] if you touch them." This disease was called *fushidaka.* The description strongly suggests the well-known polyhedrosis of the silkworm, for the Japanese strain of silkworm has transparent or pale greenish hemolymph as compared with the yellowish hemolymph of a European strain (Hukuhara, personal correspondence). There were other maladies (slim in head and sooty in body color, shrunken larvae and transluscent larvae) that Tsukada ascribed to excessive heat in the heated rearing rooms. Uegaki (1803) assumed the cause of fushidaka to be a cool and moist environment. The contagiousness of the disease was described in *Sanshi-kinu-burui* in 1812 by Narita. Failure of first instars to molt or of later instars to have an appetite was thought by Shimizu (1847, in *Yosan-kyoho-roku*) to be due to overcrowding, lack of adequate mulberry leaves, inadequate ventilation, or improper rearing temperatures. Parasitic flies (maggots?) also afflicted silkworms if adequate care against them was not taken.

In view of the substantial contributions to insect pathology made by Japanese scientists concerned with the diseases of the silkworm, a testimony of honor should be made to what may be the first Japanese publication devoted solely to the diseases of the silkworm. It consisted of a small, blue, cloth-covered folder in which was a small (about 12 x 23 centimeters) paper envelope containing a single, loose, folded sheet of mitsumata paper (made from *Edgeworthia chrysantha*) about 33 x 42 centimeters in size. The printing on the sheet was made from woodcuts. The author of this comprehensive one-page document was T. Fukushima, the date of its publication is 1828, the translated title reads *Propagation of Knowledge of Nine Diseases of Sericulture*, and the publisher was Tokuhoin, Oshu. Inasmuch as this publication preceded verification of the germ theory of disease, its contents are of particular interest. They fell under nine headings that, translated into rough English, may be rendered as follows: slim silkworm; not-molting silkworm; red ecdysis silkworm; going-bad silkworm; clear-head silkworm; falling-forward silkworm; fast-working silkworm (agitated silkworm). From these subject headings it is not difficult to speculate as to what some of today's equivalents might be.

Another of my favorite Japanese publications, one I consider to be a classic, is Omori's *Modern Treatise of Japanese Silkworm Disease*, first published in 1899, and reprinted in several (five,

I believe) subsequent revised editions. When this book appeared, bacteriology was a new science just coming into its own—this fact is amply evident in Omori's excellent treatment of his subject. It is unfortunate that this, and other, early Japanese works were not translated into Western languages so that more Western workers could appreciate the value of these pioneering treatises.

A famous Japanese scientist and educator, Professor Chujiro Sasaki, during a visit in 1910, recommended silkworm rearing as an appropriate industry for the southern areas of the United States. Of special interest is the fact that this gentleman's father, Chojun Sasaki (1830–1919), after studying sericultural science under Johan Bolle in Austria, returned to Japan in 1874, and, among other contributions, pioneered modern Japanese silkworm pathology. Chujiro Sasaki wrote an important treatise entitled (translated) *Study Book of Japanese Silkworm Pébrine*, which was published in 1900.

Sano (1898) mentioned the fact that pébrine was epizootic in China at that time and that the disease "was present in Japan from old times." In 1866, of the eggs being sent to France by the Shogunate government, only one-third were found to be pathogen free. However, apparently it was in 1875 that the spores of pébrine were first found in the larval stage of the silkworm in Japan. In 1886 the government instituted a regulation that the eggs of the insect be regularly and systematically examined. This governmental action is particularly significant because it necessitated the expansion of the silkworm-disease experiment station established in 1884 in Yamashita-machi, Tokyo. In order to train technicians to examine *Bombyx mori* eggs for the presence of pébrine spores, the station was enlarged and named *Sangyo-shiken-jo* (Sericultural Experiment Station), which was the precursor of the present Sericultural Experiment Station concerned with all aspects of sericulture. According to Mitani (1928) the station in 1886 published the results of the first research in Japan on the prevention of muscardine.

During the seventeenth, eighteenth, and nineteenth centuries, the matter of disease also came increasingly to be included in writings on the techniques of sericulture in Europe. In 1830, Count Dandolo declared, in a chapter on disease in a book on the rearing of the silkworm, that "hundreds of works" had been written on the diseases of this insect. Mention of the subject also occurred in general works that included discussions of the silkworm, such as a book written on butterflies in 1679 by Maria Sibylla Merian. Published works on sericulture became more and more scientific in character, and the need for a more factual comprehension of the diseases involved became evident.

It is noteworthy that as the concept of scientific method developed and biological science began to emerge into the realm of experimental science, among the first problems to gain the attention of biological investigators were those pertaining to the diseases of the silkworm. The significance of these studies from the standpoint of their influence on the understanding of human disease and such concepts as spontaneous generation is great. One of the earliest of what might be called a scientific consideration of the diseases of the silkworm appeared in 1808. This was a treatise, *Recherches sur les maladies des vers à soie*, by P. H. Nysten, who became an oft-quoted authority on the subject during the years to follow. Among other early European writers whose works included discussion of diseases of the silkworm were Boissier (1763), Aymard (1793), Montagne (1836), Audouin (1837 *a, b*), Robinet (1845)—who, incidentally, referred to such smaller works as those by Bérard, Carrier, Raynaud, Robert, Vincens de Saint Laurent, and others—Robin (1847, 1853), Guérin-Méneville (1847, 1849), Maestri (1856), Cornalia (1856), Lebert (1858), and de Quatrefages (1859). Masera (1956*a*) stated that the first to mention the disease we know as muscardine in the sericulture circles of Europe was Antonio Vallisnieri (or Valisneri, as he often signed his name), who, around 1708 and 1710, advanced certain ideas as to its cause. Noteworthy because it was among the first to approach the diseases of the silkworm from a pathophysiological (or physiopathological, if you prefer) viewpoint is a monograph by Auguste Chavannes (1862). He attempted to "elucidate the changes occurring in the urine and blood of insects in the course of the disease" (Martignoni, 1964). Martignoni related that Chavannes was a physician and a professor of zoology at the Académie (now the University of Lausanne) and that for his work on the diseases of the silkworm he received a gold medal from the Instituto Lombardo di Scienze e Lettere.

There is little doubt that it was the Spaniards who first attempted the rearing of silkworms in North America. According to Borah (1943), who wrote a thesis on silkworm rearing in colonial Mexico, the Spanish first tried to introduce silk culture into Hispaniola in the early years of the sixteenth century. In 1503, an ounce of silkworm eggs was shipped to Fray Nicolás de Ovando, the then governor of the area, with instructions to persuade the settlers to hatch the eggs and to raise the insects. The eggs of this first shipment failed to survive. A second shipment was made but, according to Borah (1943), "Whether or not they arrived in good condition is unknown; neither is there evidence to tell whether Ovando ever raised silk." In 1517, Bartolomé de las Casas persuaded King Charles to adopt a plan to produce silk in the West Indies; but in the course of making the necessary arrangements, he alienated influential nobles and royal councilors, and the plan was abandoned. A similar plan, in the 1520s, by Lùcas Vazquez de Ayllón was proposed for settlers on the southeastern coast of the present United States. He and his colony of two hundred men landed on the Carolina coast near Cape Fear, but they were unable to maintain themselves, and the expedition and silk-rearing plans failed.

In the meantime the Spaniards were spreading into southern and northern Mexico, and silk culture had begun, although who started it is not clear. Most authorities credit Hernán Cortés with this achievement, but as stated by Borah (1943), "Hernán Cortés contests the honor with his enemy, Diego Delgadillo, oidor of the first audiencia [governing official], and the shadowy figures of two lesser men, Hernando Marín Cortés and Juan Marín, emerge to challenge their claims." Hernán Cortés requested silkworm eggs from the Spanish government in 1522 (he received them in 1523) with which a small endeavor at silk culture was initiated a few miles south of Mexico City. In 1530 he had additional eggs brought over from Spain; in the meantime Delgadillo had successfully begun his silk-raising operation.

During the 1500s, contrary to the European method of rearing silkworms, which was usually a small-scale peasant-family operation, the Mexican rearings developed into large-scale enterprises. It became one of the country's most important ways of becoming wealthy, limited only by the number of Indians and the land available. The crown under Antonio de Mendoza's auspices encouraged the planting of 100,000 mulberry trees in the provinces of Atlixico and Huejotzingo to expedite the industry. Motolinia (Borah, 1943, pp. 15–16) in his travels through Mixteca stated: "There are many Spaniards who have seven and eight [silkworm] houses more than two hundred feet long and very wide and very high. These hold over ten or twelve thousand trays, and when the worm has spun, all the houses are full of silken cocoons from the floor to ceiling, like a forest full of roses." This may have been somewhat of an exaggeration, but it does indicate that silkworms were being reared on a very large scale. In fact, it was quite common to have two silk crops a year, whereas in Spain there was only one. In the years to follow, sericulture in colonial Mexico went through varying periods of rapid development, boom, stabilization, decline, revival, until sometime in the nineteenth century it "peacefully disappeared." But when was the presence of disease in the New World silkworms first recorded?

Apparently, infectious diseases of the silkworm caused so little difficulty in Spain that they were virtually unknown to Spanish sericulturists, while disease seems to have been a problem from almost the beginning in New Spain. Differences in rearing methods probably accounted for this. Mortalities from excessive heat or cold were noted, and considerable efforts were made to understand these factors in relation to the health of the silkworm and to acclimate the insect to the climate of New Spain. A leading sericulturist of the Mixteca area, Gonzalo de las Casas, along with a number of other quantity raisers, was very interested in developing better and more economical ways of rearing worms and producing silk—predating by two or more centuries European raisers' efforts to increase yield and productivity. Las Casas (1581) wrote a manual on silk raising titled *Libro intitulado arte para criar seda desde que se rebiue vna semilla hasta scar otra* [Art of growing silk from the time an egg hatches until another egg is laid] printed in Granada, Spain. He based his manual of silkworm rearing on his own observations and those of other writers and sericulturists of his time. He discussed the processes and the problems of mass-producing silk, including the silkworm diseases.

Relying on Borah's (1943, pp. 53–66) interpretation of las Casas's sixteenth-century description of signs and symptoms of the diseased silkworms, the following picture emerges: Inasmuch as las Casas did not describe the typically white-covered mummified larvae characteristic of the fungus-caused disease muscardine and did not refer to the dark "pepper" spots typical of the protozoan-caused infection known by the French name *pébrine*, one can assume that in all probability neither of these diseases was among the first maladies observed. If outbreaks of these diseases, particularly pébrine, had occurred, they probably could not have been controlled with methods known at that time and would have wiped out sericulture centers; las Casas made no mention of severe epizootics.

On the other hand las Casas did refer to the signs of yellowness and swelling with fluid "like rotten fruit." Such manifestations are those of the disease known by the French as *grasserie*, and in English as *jaundice* or *nuclear polyhedrosis*. The disease is caused by a rod-shaped virus typically occluded in polyhedral inclusion bodies, which may be discerned by ordinary light microscopy in certain of the disintegrating and liquefying internal tissues of the afflicted larva. Las Casas's reference to diseased silkworms with watery stools and a shinier or smoother than normal skin appearance would indicate a type of dysentery—possibly the notorious flacherie that Pasteur and others have considered to be caused by certain bacteria, some (such as Paillot) to be caused by a combination of a nonoccluded virus and bacteria, and others, especially in recent years, to be caused by a specific nonoccluded virus or by *Bacillus thuringiensis*. In the absence of the germ theory of disease, las Casas decided that the cause of the difficulty was the fact that Mixteca's raising season came too early for the sun to give the larvae enough heat for them to sweat out the moisture from the mulberry leaves that they ate.

It should be mentioned that the ingenious las Casas devised remedies for the maladies that, though not disastrous, did at times cause serious losses and were a nuisance—they annoyed rather than threatened the industry. He advised the silkraisers to build fires in their rearing rooms, thoroughly "baking" them, before moving the worms in—a procedure which indeed might have had a disinfecting effect. In addition, he recommended practices of segregating and removing ailing larvae, especially those with dysentery; keeping dust away from them; removing litter and maintaining cleanliness; carefully controlling the temperature and the quality of food—all procedures that would tend to reduce the spread of the infecting agents to healthy insects. In light of Pasteur's later similar recommendations, it is of interest that las Casas advised the selection of eggs from females spinning the largest and "best" looking cocoons and to make this selection from those woven early, thus giving some assurance that the hatched larvae would be healthy. According to Borah, Gonzalo de las Casas's advice was disregarded by the majority of Mexican silkworm rearers; as a result silk yields were generally far below the possible maximum because, in addition to disease, the careless rearer also did not cope adequately with bad weather, vermin, theft, and fire, which occurred frequently.

With regard to that part of North America now the United States, the development of sericulture has a fascinating, if abortive, history. However, inasmuch as we are concerned here with the diseases of the silkworm rather than with sericulture as an industry, we can but say that the reader will have little difficulty learning of the development of silk culture in reference works found in any major library. Not so with respect to the diseases of the silkworm, however. Not only were the diseases little understood in American sericulture, but the reports of their occurrence were few and imprecise.

I have already alluded to the unsuccessful plan by the Spaniard Lùcas Vazquez de Ayllón in the 1520s to introduce sericulture when he landed with a colony of men on the Carolina coast. The next attempt made in what is now the United States was in 1609 by the colonists and emissaries sent to the American continent by King James I of England. A shipwreck thwarted this effort. However, a subsequent try did establish a temporarily successful industry in Virginia in 1619. Unfortunately, it did not flourish as rapidly as England desired so in 1622 the government commanded the colony to accelerate its silk production, offering rewards for those who made the effort and punishment for those who failed to do so. James II did not approve of the use of tobacco and had hoped that silkworm culture would replace the growing of tobacco as a cash crop for the colonists. An interesting tract (Calvert, 1655) titled "The reformed Virginia silk-worm, or, a rare and new discovery of a speedy way, and easy means, found out by a young lady in *England*, she having made full proof thereof in *May*, *Anno* 1652" appeared extolling sericulture. During the remainder of the seventeenth century and throughout the eighteenth century, sericulture was taken up by other colonies; indeed, the records of the colonial period of the United States show that sericulture found favor virtually everywhere along the Atlantic seaboard and the Gulf of Mexico. Efforts in these areas, and eventually in California (in 1856 and again after World War I), gave initial hopes of success but all in time failed as industrial ventures largely for economic reasons.

A silk manual was authored in 1832—with subsequent editions—by J. H. Cobb, who, as a manufacturer in Dedham, Massachusetts, is frequently credited with really "establishing" the silk industry in the United States. His manual included a section on the diseases of the silkworm.

On the other side of the continent, in California, the established introduction of silkworms apparently did not occur by way of colonial Mexico. According to Hittell (1881), a Mr. H. Hentch of San Francisco was the originator of the silk culture in California. After several unsuccessful attempts (beginning in 1856) to import silkworm eggs from China, he was successful in obtaining viable eggs from France in 1858. It was also in 1856 that Louis Prevost at San Jose planted mulberry trees preparatory to his efforts to rear silkworms, apparently in 1860. From these beginnings California, along with Kansas and Mississippi, became the leading producers of silk in the United States, at least during the 1880s. However, the industry had its ups and downs. As related in a 1921 leaflet circulated by the Seriterre Company of Oroville, California, "Prevost's enthusiasm gained recruits and an ill wind in France blew success to the infant industry. Pebrine, a disease of silkworms, threatened to wipe out the European growers. The healthy eggs from California came as a godsend." Local growers realized as high as $1,200 an acre of mulberry trees selling eggs of the silkworm to unlucky Frenchmen and Italians. A boom was on and the state offered bonuses to mulberry tree growers.

Then another Frenchman, Louis Pasteur, as we shall see later, brought his test tubes and microscope to the succor of his countrymen and the ill wind, pébrine, suddenly stopped blowing. The European market for eggs was cut off, the state legislature refused to pay the bonus it had promised, Monsieur Prevost died, and the local growers, leaderless and without the modern facilities of getting together, broke ranks and the "cause" in California was essentially lost.

During the California boom in silk production, a number of groups or affiliations were formed for the purpose of promoting the silk industry. For example, one finds records of the California Silk Center Association of Los Angeles (founded in 1869), the Southern California Colony Association (1870), the California Silk Culture Association (1880), the State Board of Silk Culture of California (1883), and the Ladies Silk Culture Society of California (1885). In 1928, the American Silk Factories, Inc., bought out the San Diego Country Silk Company and began operations at San Marcos. Although Essig, in 1945, wrote that the latter company planned "very extensive expansions," these never seem to have been realized, and the industry gradually died out in California as well as in the remainder of the United States. The demise of the industry in North America seems to have been caused not by diseases of the silkworm but, rather, by the high cost of labor as compared to that in Europe and the Orient, by the relative scarcity of mulberry trees to support a great amount of silk production, by a basic lack of interest in the industry by those likely to be attracted to such activity, and by the fact that synthetic fibers came on the market with such high acceptability that silk production in the United States could not maintain itself as a competitive, viable industry.

As one reads the early literature and popular accounts of silk culture in the United States, he is struck by the fact that reports of outbreaks of disease are virtually nonexistent. One finds rather dogmatic statements relating to this point by so eminent an entomologist as C. V. Riley (1879), who stated, "Eggs raised in this country are free from disease." (Later, in 1885, he issued a warning against the use of diseased larvae in the eventual production of eggs for marketing.) Most discussions of disease were rehashes of information gained from European sericulturists and recommendations or admonitions relating to the methods for, and need of, keeping diseases out of American silkworm colonies. That the silkworm was subject to disease there was no doubt, but it would appear that, for the most part, American sericulturists considered their situation fortunate as compared to that in other parts of the world, and their main concern was to keep things that way.

Careful study of early sericulture manuals and reports indicates, however, that some diseases did exist if even to a limited extent. It seems clear that grasserie (nuclear polyhedrosis) and an undefinable flacherie were found in rearings from time to time, but these could usually be cleared up by the prompt removal of the diseased larvae and the maintenance of cleanliness in the rearing rooms. The protozoan-caused pébrine was also present but apparently not to the extent that it had occurred in France. Some diseased material may have been introduced from time to time, but the consequences did not appear to be serious. Thus, for example, Brocket (1876) mentioned that at Silkville, in Kansas, E. de Boissière obtained some eggs from France but they "developed insects suffering probably from *pébrine*, and they did not produce a supply of good cocoons."

In the case of California, where at one time it appeared as though sericulture might become an established industry, one finds the literature inconsistent with regard both to the kinds of diseases occurring in the silkworm and to the extent of their occurrence. Some silkworm-rearing enthusiasts (e.g., Hazarabedian, 1938) stated flatly, "I have never seen a sick larva in my experiments in the San Joaquin Valley of California. This is because the climate here is very adaptable to the culture of silkworms. . . . " Others did not consider the maladies they observed to be of the nature of infectious disease but rather the result of "extreme conditions" (heat, ventilation, light) of the rearing sheds. In fact, such an idea was early espoused by Prevost (1867), who, apparently, had all the qualifications for membership in today's Chamber of Commerce of California. Read his reassuring declarations:

> In all other works on silk raising, they have chapters on *the diseases of the worms.* We have no need in our California Silk Manual of such chapters, because, as long as we shall be able to give our worms fresh food from mulberry trees that are growing under the genial rays of our sun, no disease can be expected, as *the disease* is in the food. It is my positive opinion, that these watery leaves, taken from trees, growing most of the time *in the shade*, in a wet, damp atmosphere, are what create the disease. . . . Sudden changes from cold to heat are very injurious . . . in California, where they cannot be diseased. The dryness of our climate protects them from the disease, which proves that California is about the very best spot on this globe to raise silk.

However, that disease did in fact occur is revealed by Prevost himself by such statements as, "I have been obligated many times to feed with wet leaves, which was enough to make any worms sickly. . . . The leaves coming in bags were withered and nearly all the time heated—this last condition was sufficient to kill them all." Changing their food from mulberry to osage orange "started the disease, and they commenced dying by large numbers everyday."

In spite of claims obviously over enthusiastic, it seems likely that California sericulturists did enjoy a relatively low incidence of disease in their rearings. Nonetheless, infectious disease did occur, although apparently never in the sustained epizootic proportions as that which almost destroyed the silk industry in parts of Europe. Nevertheless, we have the cryptic statement in Essig's (1931) *A History of Entomology*, under the date of 1874, that "pebrine disease of silkworm [was] common in California."

II

It was during the nineteenth century that certain maladies of the silkworm were destined to play an important role in man's understanding of infectious disease. These same advances were momentous and highly significant in the development of insect pathology. As early as 1546, Fracastoro (and two hundred years later, Plenciz), without any experimental proof, had propounded theories that diseases were caused by seeds of infection. However, the idea of spontaneous generation—that is, of the generation of life out of dead and inert matter—remained a popular conception until finally overthrown in the latter part of the nineteenth century. Among those whose work not only aided in overthrowing the theory of spontaneous generation but helped to establish the germ theory of disease was Agostino Maria Bassi (1773–1856), one of a pair of twins, sometimes referred to as "Bassi di Lodi" after the town in northern Italy where he did his work and lived most of his life. The importance of Bassi's work has to a great extent been overshadowed by that of Pasteur's a few years later, but his genius deserves greater recognition than it has received. It was Bassi's discoveries that ushered in the dawn of the studies of infectious diseases. Certainly he stands as a giant among those early workers who may be said to have laid, albeit unknowingly, the first foundations for insect pathology.

Around 1800, in Italy as well as in France, a disease known as "mal del segno" or "calcino" (or, in France, "muscardine") began to reach serious proportions in silkworm *ménageries*. This was a time when the matter of contagion was being given considerable philosophical and medical attention. It is not surprising, therefore, that the contagious nature of this disease of silkworms was recognized even though the cause of the disease was unknown, and it was thought, by many, to arise de novo. In 1821, for example, Foscarini reported on a series of experiments by which he was able to show that silkworms dead of the disease, as well as tools used in handling them, were infectious and that the infectiousness could be destroyed by passing the tools through, or repeatedly over, a flame. Foscarini, however, did not recognize the true cause of the disease; he considered the cause to be a contagious miasma. It remained for Bassi and others to clarify this basic mystery.

The fact that the true causes of silkworm maladies were not known—although, to be sure, the proponents of the "bad-air" and other etiologies probably were convinced as to their validity—did not prevent the development of an ample number of cures, even before Bassi recommended his methods of disinfection. For example, in July of 1807 E. Cutbush sent to the editor of *The Philadelphia Medical and Physical Journal*, where it was published in 1808, an account on the use of "oxygenated muriatic gas for the purification of rooms set apart for Silk-Worms." The article was a translation of an Italian article that had been written by Sig. Paroletti in 1803 and published in *Biblioteca Italiana.*

It seems that muscardine and grasserie (jaundice), and possibly pébrine, were the diseases involved. According to Paroletti, the "vitiated state of the air in the rooms" in which the silkworms were reared "was the most common cause of their disease." He found fault with such corrective measures as lighting fires in the rooms, burning perfumes, and the use of various ventilating systems. However, because he had some success in preventing and curing disease by immersing the larvae in vinegar, he decided to try the fumes of mineral acids. Thus, he used "regulated doses" of oxygenated muriatic acid to fumigate, for a period of fifteen minutes, the rooms in which diseased larvae were found. "*In two days the disease disappeared,*" and the insects went to the bosco (spinning bundles) "happily!" Significantly, in a postscript to the translation, the American, Cutbush, alluded to three silkworm-rearing establishments (apparently in America) that he knew were abandoned "in consequence of the diseases, which those persons were found attacked with, who were engaged in rearing Silk-Worms." He suggested that perhaps the method used by Paroletti would be "the means of arresting some dangerous fevers to which those men are frequently victims, who are occupied in rearing Silk-Worms."

With remarkable ingeniuty Bassi showed that muscardine is caused by a vegetable parasite (the fungus we know today as *Beauveria bassiana*);[8] that this parasite grows and develops in the living silkworm; that it eventually causes the insect's death; that the diseased insect is, by virtue of the parasite's presence, rendered infectious; and that the infectious agent could be transmitted by inoculation, by contact, and by contaminated food. He showed that the characteristic white covering, or efflorescence, of the diseased larvae consisted of a mass of the causative fungus. He explained how the "seeds" conceivably produced by this mass were readily disseminated and how they were responsible for the disease in new individuals. He correctly ascertained that warm, humid conditions facilitated the growth and development of the fungus. He demonstrated that the pathogen can be destroyed by certain chemical and physical means. Indeed, his recommendations for controlling and preventing the disease (through disinfection with lye, wine and brandy, boiling water, burning, and exposure to sunlight) were of such an advanced caliber that he might also be considered as one of the founders of modern disinfection. Thus it was Bassi who, for the first time, showed experimentally that a microorganism (the fungus) was the cause of an infectious disease in an animal (the silkworm)! Although he submitted an application to do so in 1833, Bassi first presented his findings in 1834 in a communication delivered before a commission comprised of nine members of the Faculties of Medicine and Philosophy of the University of Pavia. He also successfully performed appropriate, and critical, experiments before the commission (see Yarrow's [1958] translation for details concerning the commission and its conclusions.) For this accomplishment he is regarded by many as the founder of the doctrine of pathogenic microbes or the germ theory of disease. It is true that some time earlier Nysten (1808), upon microscopic examination, saw the fungus associated with the dead silkworms, but he did not associate it with the cause of the disease, which he recognized as being of a contagious nature, but believed it to be the result of certain chemical changes in the afflicted insects. It is the glory of Bassi's work that his findings were based on sound experimentation.

Bassi was, in many ways, a remarkable and romantic man. He was born on 25 September 1773, in Mairago, near Lodi, and educated in law—but he also studied natural science in Pavia. Although from time to time he held various civil posts, and was invited to fill others, he gave up most of them because of failing eyesight and general ill health. In dire financial straits, he turned for a living to farming and the raising of sheep—an unsuccessful venture except that it induced him to write informative articles on potatoes, cheese, wine, and sheep raising. Fortunately, through the inheritance of a legacy from a relative, he was able to discharge his debts and to give more attention to scientific pursuits. The threat of calcino (muscardine) to the sericulture industry made him determined to find means of preventing the disease. His studies on the disease covered a period of about twenty-five years, and out of them came his great work *Del mal del segno calcinaccio o moscardino*, published in Lodi in 1835 and 1836. From this study, Bassi proceeded to extrapolate his findings and thinking into the field of human diseases. He wrote noteworthy treatises on contagion (1844), pellagra (1846), and cholera (1849). He became totally blind in 1856, the year in which he died. Before his death, extensive recognition and many honors came to him, especially from Italy and France. After his death, much of his work was generally forgotten, but in 1901 the city of Lodi transferred his ashes to a new cemetery, named a street in his honor, and designated his old home with a plaque. Italian scientific societies have since honored him in various ways, including the reprinting of many of his published works. The Sixth International Congress for Microbiology (Rome, 1953) was honored by an Italian stamp portraying Bassi with a silkworm. An occasional paper (e.g., Major, 1944; Ainsworth, 1956; Masera, 1956*a*, *b*; Steinhaus, 1956) properly attests to the significance of Bassi's work. Nevertheless, most writers of treatises and textbooks on infectious disease have yet to accord him the honor and credit he deserves. It is to the credit of the

American Phytopathological Society that, in 1958, on the occasion of its golden anniversary, they published an English translation of the first volume of Bassi's *Del mal del segno* (Yarrow translation, Ainsworth and Yarrow, 1958).

Although recent years have seen more appropriate credit being given to Bassi for his contributions to science, there has been one locality where he was not forgotten even though most of the world's scholars and writers of science were overlooking him. I speak of Lodi, Italy, his "home town." In 1954, while my family and I were driving from Milan to Padova, we decided to make a short side trip to the town of Lodi, somewhat off the beaten path as far as tourists were concerned. Perhaps, we thought, something might remain of Bassi's home, or we might find other evidences of the man's having once lived and worked here. We had not driven far into town before our eyes caught an inscription toward a corner of the top of a relatively new brick building: "Istituto Tecnico Agostino Bassi." As I entered the building it was clear that this was a technical school of some kind named in honor of the illustrious Bassi. Inasmuch as my "tourist Italian" was inadequate to explain the nature of my quest, the secretaries in the main office soon found a young lady in class who could serve as an interpreter. It was obvious that having an American on their hands who was interested in Agostino Bassi was something unique. Over my protests they sent someone to bring the director, Professor Giulio Lodi, from his home.

In their library, the director and his assistants showed me copies of the printed works of Bassi. They explained that not enough of Bassi's home site remained to show anyone; instead Professor Lodi insisted that I go with him to the town square, where the city library was located. There, in a special case, under lock and key, were original writings of Bassi's, as well as early editions of his published works. Although his major works have been published and are available, it is my impression that to this day, in the Community Library of Lodi, there remain other works and manuscripts of Bassi that have not been published or adequately circulated.[9] What an opportunity for the historian of science who is interested in placing this man's work in proper perspective with that of Pasteur and other contemporaries! In the meantime, the people of Lodi are justifiably proud of their illustrious former citizen. "He preceded Pasteur with his discoveries!" was one of Professor Lodi's last comments to me as reluctantly I left, regretting I did not have more time to browse among the Bassi memorabilia. This pride of the Italians in Bassi is amply reflected in the commentary by Alfieri and others (1925) when, in celebration of the centenary of Pasteur's birth (1822) and the one hundred and fiftieth anniversary of that of Bassi (1773), they initiated the reprinting and republication of certain of Bassi's writings. Thus, Alfieri declares, "It is the duty of every Italian when exalting the formidable work of Louis Pasteur, not to forget Agostino Bassi."

Bassi's discoveries lead to the first proof that a microorganism, *Trichophyton schönleinii*, could cause an infection—favus—in the human body. Schönlein (1839) found the fungus that causes favus, and his assistant, Remak, was able to reproduce the disease in himself in 1842. Other pathogenic skin fungi were found about the same time (between 1841 and 1846) by such men as Vogel, Bruby, and Eichstädt. These fungi-caused diseases are now catalogued under the designations aspergillosis and actinomycosis. As pointed out by Long (1965) the importance of Bassi's work was recognized in Germany by Jacob Henle almost immediately. His well-known essay *On Miasms and Contagion* (1840) made this clear and proposed postulates regarding the etiological relation of microbes to disease. These were later extended and became the famous Koch's postulates.

It must be admitted that at first Bassi's findings relative to the "mal de segno" were not universally accepted. They were especially attacked by the proponents of the theory of spontaneous generation and by others who acknowledged the presence of the fungus in the diseased larvae but who did not believe that it in itself could cause the disease. Eventually, however, the skepticism gradually was overcome after Audouin's (1837*a*, *b*) confirmation experiments of Bassi's results and especially after 1859 when Vittadini described the spore of the fungus, isolated and cultivated the fungus on nonliving media, and demonstrated conclusively that is was the specific and only cause of the disease. These developments and others confirmed Bassi's motto: "When fact speaks judgement is silent, because judgement is the daughter of fact, not fact the son of judgement." Thus Bassi joined the ranks of those like Redi and Spallanzani who set the stage so that the great Pasteur, of whom he was indeed a precursor, could finally, in the 1860s, deal the death blow to the ancient belief of spontaneous generation, and John Tyndall, a few years later, could lay its ghost. Bassi's achievements in this regard are marred only by the fact that in later years he had some misgivings about his stand against spontaneous generation and apparently believed it to be possible under certain conditions. Recent thinking in theoretical biology indicates that spontaneous generation of a sort may still be taking place, but of course in a different context than was being argued by nineteenth-century scientists.

III

Like Bassi, Louis Pasteur (1822–1895) approached and entered the field of microbial disease and pathology in animals by studying the diseases of the silkworm. The diseases studied by Pasteur, however, did not include the fungus infection elucidated by Bassi, but instead were concerned with maladies caused by protozoa and bacteria (and, unknown to him, probably a virus). Although he knew of Bassi's work, and of Audouin's (1839*a*, *b*) and Vittadini's (1859) confirmation of it, there is no indication that Pasteur's experimentation was significantly affected by that of the Italians. Nevertheless, Pasteur had the benefit of the observations by many sericulturists and biologists from which he profited and on which he could project his own discoveries.

The breeding of silkworms for the manufacture of silk began in France probably in the twelfth century. During the first half of the nineteenth century, sericulture flourished at an accelerated rate, annual production rising from six million kilograms in 1788 to twenty-six million kilograms in 1853—about one tenth of the world's production. Unfortunately for France, this progressive increase in production seemingly brought about the same kinds of problems the Mexican sericulturists had encounted in the 1500s when they introduced mass-production methods. About the middle of the nineteenth century there appeared throughout southern France a combination of devastating epizootics among silkworms that threatened the entire silk industry. The protozoan-caused disease pébrine was particularly serious, apparently making itself evident about 1840, reaching epizootic proportions by 1845. Production fell to about four million kilograms per year by 1865. Disease-free eggs ("seed") of the insect could no longer be produced in France but had to be imported. Then outbreaks spread to Italy, Spain, and Austria, then to Greece, Turkey, and the Caucasus.

In desperation, 3,574 owners of silkworm nurseries petitioned (in 1865) the government of France to cope with the disastrous pestilence. They further requested that measures be taken to reduce taxes, to supply silkworm breeders with reliable strains of eggs, and to provide for a study of "all questions related to this persistent epizootic, as much from the point of view of pathology as from that of hygiene." This petition reached the Senate, where the deliberations were led by Jean-Baptiste Dumas, friend and former teacher of Pasteur. While preparing his report for the Senate, Dumas persuasively asked Pasteur to undertake the investigation requested by the petitioners. After professing some apprehension over his ignorance of silkworms and sericulture, Pasteur accepted the challenge, probably largely out of respect for his master but also because he felt that the investigation might come within the range of his studies on fermentation and the "diseases" of wines.

Incidentally, not only had Pasteur never seen a silkworm prior to Dumas's request, but his naïveté regarding the metamorphosis of insects was considerable. The story is told by the famed naturalist Henri Fabre (e.g., see Cuny, 1967) about Pasteur's first encounter with a pupated (cocooned) silkworm. Pasteur held the object in his hand, placed it close to his ear, shook it, and exclaimed with surprise, "A kind of knocking noise . . . is there something inside?"

"Yes indeed!" replied Fabre.

"What can it be?"

"The chrysalis [pupa]."

"How do you mean, the chrysalis?"

"I mean the sort of mummy which the caterpillar changes into before it becomes a moth."

"And there's one of these things inside every cocoon?"

"But of course! It's to protect the chrysalis that the worm spins the cocoon."

"Ah!" was all Pasteur had to say.

Pasteur left Paris on 6 June 1865, going directly to the Department of Gard, the center of the area where the diseases reigned in greatest intensity. Here, in the town of Alès (Alais), he began his investigations, making numerous general observations of the nurseries and interviewing silkworm breeders in the surrounding territory. Many isolated facts were known and observations had been made, but there appeared to be no common thread to link them up or to unite them. In September he submitted a report of his observations to the Academy of Sciences. The situation was a discouraging one; truly a catastrophe had struck this leading agricultural pursuit of southern France. To one with lesser confidence in himself than that which Pasteur possessed, the matter might have appeared insoluble. Skeptical breeders were obviously disappointed that the government would send a "mere chemist" to investigate their trouble. The deaths of his father and his youngest daughter at this time added to Pasteur's problems. Nevertheless, in 1866 he returned to Alès with two assistants, Gernez and Maillot (and later Duclaux), and, in a lonely house a short distance out of town, established a laboratory at Pont Gisquet at the foot of the Mount of the Hermitage. Here, burdened with the death of another of his daughters, he began the intensive investigations that were not only to save the silk industry of France, but to add one of the most brilliant chapters to man's understanding of infectious processes and to the scientific development of insect pathology!

At the time Pasteur began his experiments, a clear distinction was not made between two of the diseases afflicting the silkworm. Pasteur himself did not recognize the differences until his studies had proceeded for about two years when he differentiated the disease known as "pébrine" from that called "flacherie." Although historical accounts vary somewhat as to the part of the world in which pébrine was first recognized (probably largely because its identity as a distinct disease was not appreciated until quite late), a version recounted by Rienzi (1885) represents the general conception. Apparently pébrine in Europe was first found to occur among silkworms in Provenza (Provence) in southeastern France bordering on the Mediterranean Sea in 1840; in 1845 it appeared at Covailinni; in the following year at Avignone, and at Nimes in 1849; in Buduza and the Valley of St. Martin in 1852, and in Spain during the same year; the next year it appeared at Benaco, then in Berne and in the Tyrol in 1855. In 1858, silkworm eggs in Prussia were infected at a time when a similar state of affairs was occuring in silkworm rearings in Tuscany, Italy. By 1862 it was found in Bucharest, and in 1864 at Capazzi. It had spread to Turkey and India by 1859. As we indicated earlier, the first appearance of pébrine in the Orient is not clear, but Sano did say that it was present in Japan from "old times," and, according to Essig, pébrine was "common in California" in 1874, but when it first occurred is uncertain.

Pébrine derived its name from the small spots, resembling grains of black pepper, on the integument of diseased caterpillars. It is characterized by the presence, in the tissues of the diseased insect, of numerous small oval spores, which were called "corpuscles" throughout the early literature (the disease was sometimes dignified by the Latin name *ovoidalis corpusculis*). Before Pasteur began his work, these spores, or corpuscles, had been observed by such men as Guérin-Méneville (1849), who called them "hematozoides"; de Filippe (1851, 1852); Carnalia (1856); Lebert and Frey (1856), who considered them to represent the vegetable parasite that Lebert (1858) named *Panhistophyton ovatum*; Naegeli (1857), who, believing them to be a yeastlike fungus, gave to them the present name of *Nosema bombycis*; Osimo (1859); and de Quatrefages (1859). Brouzet (1863) compared the corpuscles with the animalcules seen by Rayer and Davaine in the blood of sheep dead of anthrax, and predicted that pébrine would be successfully combated when a means of destroying the corpuscles was found. Béchamp[10] (1867) correctly believed the corpuscles to be the spores of a parasitic microorganism. Today, following the observations of Balbiani (1882) and Stempell (1909), the corpuscles are recognized as a stage (the spore) in the life cycle of a protozoan in the order Microsporidia[11] of the class Sporozoa.

In spite of the earlier suggestions as to the parasitic nature of the corpuscles, Pasteur, strangely enough, approached the problem unwilling to accept the idea. Whether or not his opinions were affected by his chemical background or by such men as Chavannes (1862), who thought pébrine to be caused by abnormal metabolic changes, is not known. In any case, for two years he believed that the disease was caused primarily by "physiological disturbances" (which of course may be predisposing or contributing causes) and that the corpuscles were merely products of tissue disintegration. Eventually, however, Pasteur's assistants became convinced that the corpuscles were the cause of the disease; and subsequently Pasteur himself reached the same conclusion, recognizing their relation to the so-called psorosperms of that day being described by Leydig (1863) and others in invertebrates other than silkworms.

Of great importance were Pasteur's observations, like those of Osimo (1859) and Vittadini (1859) before him, that the pathogen could be transmitted through the egg of the insect, as well as by contact with diseased silkworms, and through the ingestion of contaminated food. On the basis of this information he was able to select eggs that gave rise to healthy larvae. If, by microscopic examination, the moth that laid a given batch of eggs was found to harbor corpuscles of the disease, the eggs and moth were both destroyed by burning. On the other hand, eggs from moths showing no corpuscles in their tissues would yield silkworms free of pébrine. At first, skeptical sericulturists disdained the idea of the microscope being an effective tool in the control of the disease. To this Pasteur retorted, "There is in my laboratory a little girl [his daughter, Marie-Louise] eight years of age who has learned to use it without difficulty." (Marie-Louise might be considered the forerunner of our modern laboratory technician. Also, Pasteur's was apparently the first laboratory to employ a microscope for the diagnoses of infectious diseases.)

The early skepticism of Pasteur's method (with due credit to the proposals along this line, in 1859, by Osimo and Vittadini) gradually dissipated as small lots of selected eggs distributed among producers almost invariably gave rise to pébrine-free silkworms. Added to this convincing evidence was the energetic "campaign" by Pasteur, who through enormous amounts of correspondence, articles in trade journals, and scientific papers finally subdued virtually all criticism and won the approval of the government, including the applause of his friend Dumas.

As mentioned before, Pasteur, in his work at Alès, was concerned with at least two diseases of the silkworm. Pasteur realized this when he found larvae free of pébrine corpuscles but dying in a soft, flaccid condition and becoming black and decayed. This disease was called *morts-flats* or *morts-blancs* and later *flacherie.* At first its presence discouraged Pasteur because of the confusion it caused—"Nothing is accomplished; there are two diseases!" he complained to his assistants.

Gradually, differentiation of pébrine from other diseases became easier, especially when Pasteur was able to associate the presence of certain bacteria with the flaccid condition. The flaccid disease occurred when these bacteria multiplied in large numbers in the digestive tract of the silkworm. One of these bacteria was a coccus arranged more or less in chains, which Pasteur spoke of as "ferment en chapelets de grains," and which today is known as *Streptococcus bombycis.* The other was a spore-forming bacillus, Pasteur's "vibrion à noyau," now known as *Bacillus bombycis,* and, in some instances at least, perhaps identical to one of the varieties of *Bacillus thuringiensis* so well-known today. Even now, however, the etiology of true flacherie is not entirely clear. That the bacteria are secondary but active invaders to a rather benign virus has been postulated but not generally accepted. The Japanese (Yamazaki, 1960, and others) have isolated a nonoccluded virus from silkworms afflicted with flacherie. No one has disproved the claim that this virus is the basic cause of the disease with which Pasteur was concerned. In any case, the spore-forming bacillus observed by Pasteur is associated with what was called "true flacherie," and the streptococcus is associated with a similar disease known as "gattine."

It is important to note that in conducting his researches on these diseases, Pasteur was well aware of the fact, now generally appreciated by insect pathologists, that the susceptibility of insects such as the silkworm is influenced by the conditions under which they live. Pasteur considered such factors as excessive heat and humidity, inadequate aeration, stormy weather, and poor food to be inimical to the general physiological health of the insects and capable of decreasing the resistance to infection (reminding us of Gonzalo de las Casas's [1581] observations). These concepts and others are brought out in interesting detail by Jacques Nicolle's (1961) "story" of Pasteur's "major discoveries." If for no other reason, Nicolle's account should be read because of its unique or "different" approach to the history that surrounds Pasteur. At any rate, it is different from those of most of Pasteur's biographers—e.g., Vallery-Radot (1937) and Dubos (1950).

It should be noted that it was in the midst of his study of the diseases of the silkworm that Pasteur suffered a paralytic stroke that developed into hemiplegia. This event occurred on 19 October 1868, when the French scientist was in his forty-sixth year. He recovered rather quickly from his cerebral hemorrhage, but he never did regain the full use of his left hand. I would judge he also retained some difficulty in walking—at least walking down a stairway. I remember puzzling, when shown through Pasteur's living quarters in 1958, at the full-length panel of mirrors on the wall directly opposite the stairway. When I asked the curator about it she explained that the mirrors were placed on the wall so that Pasteur could watch, and thus be sure of, each step he took whenever he descended the stairs.

Before leaving Pasteur, special mention of his famous memoir, *Ètudes sur la maladie des vers à soie*, published in 1870 should be made. This is indeed a most stimulating and revealing document! Emile Roux, his well-known contemporary, hailed it as "a veritable guide for anyone wishing to study contagious diseases." As we read its pages and look at its excellent illustrations, we can almost experience with Pasteur his initiation into the problem of infectious diseases in animals. We can feel him becoming acquainted with the variability and unpredictability of animal life and behavior. The dawning of his realizations with respect to the laws of contamination, transmissibility, and epidemiology becomes apparent; and as we finish the treatise, we can understand his later enthusiasm for the principle of preventive medicine. This two-volume work, as well as some of Pasteur's other writings on the diseases of the silkworm, also reveals much of the human side of this man, the details of whose life are already so well known and chronicled. At times confident and boastful, at other times he was uncertain and humble. Even the preface of his memoir is interesting because of its concern with his personal feelings and attitudes. He begins by making a simulated apology for having undertaken the research for which he was so little prepared. (As we have indicated, when called to this task he had never so much as seen a silkworm.) It is obvious, however, that he wrote these words with pride and indulgence, knowing that he was presenting to the world not only a report of a successfully completed program of research but a solution to a problem of great practical importance to his countrymen. Nevertheless, he did not refrain from reminding the reader of his sacrifices, especially insofar as his work on the maladies of silkworms kept him from his pursuits in chemistry and fermentation, nor from lamenting that more fame would probably have come to him had he stayed in fields of pure science. We cannot sympathize with him too much, however, when we realize that his work with silkworms greatly enhanced his insight into the phenomena of infectious disease generally. Indeed, were it not for his work on the diseases of the silkworm, who knows but that this French scientist might never have made his monumental discoveries on anthrax, rabies, septicemia, and other infectious diseases! Certainly these endeavors in medical science would have been delayed but for the fact that the lowly silkworm suffered from diseases that commanded the attention of Pasteur, who, with Bassi, shares the credit of initiating the real scientific development of insect pathology.

I believe that many readers will find it appropriate to end our all-too-brief treatment of Pasteur's contributions to invertebrate pathology with the following lines written by Dubos (1950):

> The struggle against error, always imminent in these studies on silkworm diseases, is of peculiar interest because it provides a well-documented example of the workings of a scientific mind. As Pasteur himself said: "It is not without utility to show to the man of the world, and to the practical man, at what cost the scientist conquers principles, even the simplest and most modest in appearance." Usually the public sees only the finished result of the scientific effort, but remains unaware of the atmosphere of confusion, tentative gropings, frustration and heartbreaking discouragement in which the scientist often labors while trying to extract, from the entrails of nature, the products and laws which appear so simple and orderly when they finally reach textbooks and newspapers.

One should keep in perspective how Pasteur's studies of the silkworm fit into the other monumental discoveries that marked his career. Garrison (1929) put it succinctly thus:

> As set forth in the inscriptions on the arches over his tomb, Pasteur is memorable for his work on molecular dyssymmetry (1848), fermentation (1857), spontaneous generation (1862), diseases of wine (1863), diseases of silkworm (1865), microorganisms in beer (1871), virulent diseases (anthrax, chicken cholera) (1877), and preventive vaccinations (1880), particularly of hydrophobia (1885).

4. ENTOMOPHAGOUS INSECTS AND OTHER ANIMALCULES

The notations of early history of insect pathology cannot be made without mentioning the fact that undoubtedly the first to see a "microscopic organism" in an insect was the famed Dutchman Anton van Leeuwenhoek, who not only, like Robert Hooke and others, examined with his lenses the parts and structures of insects, but found at least one of these animals to harbor representatives of his "animalcules." His best-known biographer, Dobell (1932), stated that Leeuwenhoek recorded in 1680 (though did not publish until 1684, *Letter 33*) in a letter to Hooke the following:

> I have also seen, in the summer [presumably of 1680], in a big horse-fly (which was a female, out of which I pulled many eggs), a lot of small animalcules; though these were not a sixth of the length of the animalcules aforesaid [spermatozoa], but a good 10 times thicker. They lay mingled with the thin matter [watery material] that was in the fly's guts, and moved forwards very quick.

Although he acknowledges that the microorganisms the Dutch lens maker saw in this instance may have been bacteria, Dobell feels that it is more likely that it was a protozoan—*Crithidia* or *Leptomonas*—that was observed. A similar conclusion was made by Schierbeek (1959), who also wrote of the life and works of Leeuwenhoek. This observation by "the father of protozoology and bacteriology" is apparently the first ever made of an entomogenous microbe of microscopic size, although it should be remembered that Leeuwenhoek did not use a compound microscope but a single lens only. Apparently, it is also the first observation of a flagellate of the family Trypanosomatidae. At any rate, a protist of some kind was observed in the intestine of a horse fly as early as 1680.

In 1695 Leeuwenhoek also reported animalcules in "the ordinary water that is in the shells of oysters"; these were probably protozoa. Animalcules he briefly mentioned, in 1680, in the juice of oysters and mussels he later (1712) thought might have been inanimate particles and not living organisms. In 1696, in referring to unborn mussels in water, he expressed the idea of the mussels being "eaten up by the little animalcules," thus probably revealing that Leeuwenhoek had an inkling as to the role played by microorganisms in putrefaction.

Because, strictly speaking, entomophagous insects (parasites and predators) are pathogens of their hosts, and cause pathological changes in the bodies of their insect hosts, the discovery of these organisms, as well as true worms, is not outside any consideration of the historical beginnings of insect or invertebrate pathology. Leeuwenhoek was among the early observers to record insects parasitic in insects. For example, in writing of an aphid (probably *Myzus ribes*) in 1695, he commented, "They were not only dead, but the hinder parts of their bodies were perforated with a round hole, and their entrails gone, whence I gathered that provident Nature had assigned these creatures their enemies, to prevent their species increasing too fast, and also for the sustenance of other animals." His description of the development of the parasite *Aphidius* has been called "classical" (Schierbeek, 1959). He also discovered the ichneumonid hyperparasite of a gall-producing sawfly. However, it must be stated that, although he apparently did not recognize the phenomenon as parasitism, Aldrovandi in 1602, observed the pupae of the internal parasite *Apanteles glomeratus* on the integument of its host, *Pieris rapae* (Silvestri, 1909). This association was first correctly interpreted as insect parasitism in 1706 by Vallisnieri, but this was after Leeuwenhoek recognized the phenomenon in the aphid and sawfly. And, as I have already said in the case of microbial disease in insects, Doutt (1964) has said concerning predaceous insects:

> Obviously no one knows precisely when man first became aware of the entomophagous habit in insects. It is reasonable to surmise that he first observed predaceous insects and well understood the meaning of predation many centuries before any notion of parasitism was conceived.

Because there has to be a limit to what can be covered in this volume, and because our semihistorical-semibiographical approach will, as we go on, become a rather freewheeling one, I gladly leave the coverage of parasitism and predation of insects in other, more competent, hands. The reader interested in the historical aspects of the relations between agents causing disease in animals and plants and their insect vectors or invertebrate hosts may find such accounts in books on parasitology and medical entomology. Regarding historical aspects of the study and use of entomophagous insects, such books as those by Clausen (1940), Sweetman (1958), Franz (1961*a*), DeBach (1964), and others will serve as adequate starting points.

5. "VEGETABLE WASPS" AND "PLANT WORMS"

Although disease as an abnormal condition in insects was in all probability first critically observed in the honey bee and silkworm, the first microbial parasite was not actually seen as a pathogen in either of these insects. As might be expected, because of their macroscopic as well as microscopic appearance, the first of microbial life found associated with insects were fungi—or "vegetable growths" as some of the early writers referred to them. The first published record as well as the first illustration of a diseased insect seems to be that of de Réaumur in 1726, who relayed a report by a Jesuit priest named Parennin, who had sent specimens of the diseased insects from the Orient to France. (Again note that an historically important "first" in insect pathology came from the Orient.) This "worm" (apparently the larva of a species of noctuid, probably an *Agrotis*) consisted of a larva from which emerged a stemlike vegetable growth characteristic of the group of fungi now designated by the generic name *Cordyceps.* In his original account de Réaumur interpreted the fungus to be the root of a plant. The species concerned in this instance appears to have been *Cordyceps sinensis.*

René-Antoine Ferchault de Réaumur was born at La Rochelle in 1683 and educated in the Jesuit College of Poitiers. He apparently studied law. It is clear that he had an early interest in natural history and mathematics. In 1708 he published two papers on mollusks and three on mathematics. For nearly half a century he served as a distinguished member of the Paris Academy of Sciences.

The so-called Chinese plant worm, considered to be of great value and a rare drug in China, was used by the emperor's physicians. According to Hoffmann (1947), hepialid and other caterpillars infected with *Cordyceps* were considered by the Chinese to be a tonic food. They were made into a broth—both the infected caterpillars and broth being consumed. Szechwan Province apparently was the most productive area in China so far as *Cordyceps*-infected insects were concerned;

from here they were exported to other provinces in China and abroad as well. (I have found that it is still possible to purchase *Cordyceps*-infected larvae in herb stores in San Francisco's Chinatown.) A rather romantic history is associated with the fungi concerned, and for an early account of them the reader is referred to Cooke (1892, 1893). An extremely interesting scientist, Mordecai Cubitt Cooke was born in England in 1825 and educated in the United States with an M.A. from Yale and a LL.D. from New York University. When he returned to England he worked at the Kew Gardens (and received distinguished honors) until he retired. He died in England in 1913.

Another early, or perhaps even earlier, record of an entomogenous fungus was that made by Christian Paulinus in the beginning of the eighteenth century when he wrote that "certain trees in the island of Sombrero in the East Indies have large worms attached to them underground, in the place of roots" (Gray, 1858). Undoubtedly he was alluding to an instance of *Cordyceps* infection.

Insects parasitized by *Cordyceps* fungi were frequently known as "vegetable wasps" and "plant worms" (or "awetos" in New Zealand). Among the most celebrated of the "vegetable wasps" were those (*Vespa* and *Polybia*) infected with *Cordyceps sphecocephala.* In an account titled "Apparato para la historia natural de Espana," Torrubia (1754), a Franciscan friar, told of finding, in 1749, some dead wasps in a field near Havana, Cuba.

> One day, I found in the fields several dead wasps with their skeletons and wings intact, and out of their bellies were growing little trees which when fully grown attain a size of up to five hands. This plant is called *Gia* by the inhabitants of these regions, and it is covered with very sharp thorns. The natives attribute these thorns to the wasps' bellies which produced the plant: for this reason, they say, the plant is covered with stingers.
>
> The existence of this shrub was not commonly known until I made it public. After careful observations which I made with a microscope, I sent with a young man called Centellas a dead wasp, perfectly preserved, with a rather fair sized tree growing from it, to the Treasurer of my Order and my principal benefactor, Senor don Martin de Arostegui.

Torrubia gave a diagrammatic representation of two wasps lying on the ground with a "tree" growing out of the base of each abdomen, while three other wasps, each having a similar tree affixed to it, are drawn above similar plants. In all probability these "trees" represent species of *Cordyceps.* Torrubia concluded his report with an entertaining poem about his findings. He had written the poem to accompany the specimen that he sent to de Arostegui.

It cannot be assumed that the earliest observers of *Cordyceps* on insects realized that they were concerned with instances of infectious disease. Some of them probably accepted the concept of the Chinese philosophers that the infected specimens were herbs during the summer, changing into "worms" when winter appeared. Others considered them to be plants that simply looked like worms, and still others thought they represented worms that had merely attached themselves to the plants or their roots. Nevertheless, one cannot escape the feeling that by the end of the eighteenth century the parasitic essence of the relationship between plant and insect was in some vague way appreciated. As early as 1769, Fougeroux affirmed that the plant "presses and takes hold" of the insect's body, attaching itself to it. In any case, it is worth our remembering that the first fungi reported in association with insects were pathogenic forms. (Saprophytic species growing on the bodies of insects immersed in water were first reported in 1760 [Ledermüller, 1763].) Most of these and other early reports on entomogenous fungi have been cited by Robin (1847), Gray (1858), and Cooke (1892), and their writings should be consulted by the reader interested in the nature of these early but fascinating accounts. Of particular interest to the insect pathologist, for example, is the generalized description of the pathogenesis of fungus infections in insects as presented by Gray toward the end of his pamphlet on "fungoid parasites" of insects.

The first record of a *Cordyceps* fungus attacking insects in the United States may be that by J. Dorfeuille, who in 1822 wrote a note pertaining to what he called "an insect plant"— an insect with a fungus growing from the head of what Dorfeuille thought might be a cicada. He apparently first examined the specimen, collected at Nachitoches, Louisiana, in 1803. Similarly, Cist in 1824 and Samuel L. Mitchell in 1827 wrote of a *Melolontha* infected with a *Cordyceps.* Mitchell mentioned that he first saw fungi "sprouting from the bodies of dead insects" in 1808 in material brought to him by his friend W. A. Burwell of Virginia. A North American journal carried an article in 1824 titled "Remarks on Certain Entozoical Fungi" by J. P. Allaire, but the fungi referred to were from South American insects (*Vespa* and *Gryllus*) that had been presented to the Lyceum by a Dr. Madianna of Guadeloupe.

Some of the early leading naturalists recorded the occurence of disease in insects that they were observing for taxonomic purposes. For example, in 1776, Degeer[12] published probably the first description of what we now know to be an *Empusa* infection in flies; he thought that the flies might have eaten poisonous food. Degeer accompanied his description with a figure, in the legend of which he listed as victims of the disease the "Honigfliege (*M. mieleuse*)," and the "Haus-und Stubenfliegen." In the 1782 German edition of Degeer's work, the translator, Goeze, appended a footnote in which he referred to observations by Winterschmidts, in 1765, on a disease of mites ("Milben") that often killed thousands of the arthropods. Latreille (1805), another early naturalist, also mentioned a disease of domestic flies that in all probability was caused by an *Empusa* fungus. In no real sense, however, was the microbial nature of these diseases realized.

After Bassi and others showed the white fungus associated with muscardine of the silkworm to be pathogenic, other species of fungi were rapidly reported as parasitic on insects. Among those whose early work and writings have had an important influence upon our basic understanding of entomogenous fungi in general are Robin (1847; 1853), Fresenius (1856; 1858), Gray (1858), Tulasne and Tulasne (1863–65), Brefeld (1870; 1873; 1877; 1881), Cohn (1855; 1870), Zopf (1890), Cooke (1892), Massee (1895), Giard (1896), and Thaxter (1888; 1896–1931).

Alfred Giard was an outstanding French mycologist who published forty papers on entomogenous fungi between the years of 1879 and 1896. Of particular importance were his contributions on certain Entomophthorales and on *Beauveria tenella* (= *Beauveria densa* = *Isaria densa*). He studied fungus diseases of flies, locusts, various Lepidoptera, and the European May beetle, *Melolontha vulgaris*. Most of his *Beauveria* researches were in connection with the latter insect. He studied the pathogenesis of fungus infections in insects and did much to clarify our understanding of the basic nature of these infections. He also offered (Giard, 1890; 1892*a*, *b*) thoughtful advice concerning the use of entomogenous fungi in the control of harmful insects.

Roland Thaxter (1858–1932) was one of the leading American mycologists of his time and certainly one of the world's outstanding students of entomogenous fungi. His life was one of dedication to scientific ideals as they could be applied to the groups of fungi that interested him. His contributions were largely taxonomic in character and included presentation on development, host range, distribution, and other basic biological information. He is best known for his two monographs: *The Entomophthoreae of the United States* (1888) and *Contributions Toward a Monograph of the Laboulbeniaceae* (five vols., 1896; 1908; 1924; 1926; 1931). When Thaxter began his work on the group, only a handful of species was known. Rouget in 1850 had published the first record of the Laboulbeniaceae, and J. Peyritsch (1873) had been an early contributor to the knowledge of the group. When death disrupted Thaxter's work in 1932, he had thoroughly described and magnificently illustrated literally hundreds of species. Because of their potential economic importance as insect pathogens, Thaxter's basic contributions to our knowledge of the Entomophthoraceae are especially valuable to present-day insect pathologists. Although the first published account of what was undoubtedly an *Empusa* infection appeared in 1776 (Degeer, 1776), and substantial contributions to the systematic knowledge of the group were made by Nowakowski (1883) and others, a real understanding of the group as a whole had to await Thaxter's monograph on the subject. We shall consider Thaxter's life and work more fully in the next chapter.

6. PROTOZOAN, BACTERIAL, AND VIRAL DISEASES OF NOXIOUS INSECTS

Although biologists had been observing protozoa in insects prior to the time of Pasteur's studies on pébrine, it is generally acknowledged that greater attention was directed toward these entomophilic microorganisms as a result of the ravages of this disease among silkworms. Similar protozoa were revealed in other insect species, as well as in certain other animals. In 1882 Balbiani proposed the name Microsporidia (see footnote 11) as an order of the class Sporozoa in which to place the pébrine organism and related forms. A few years later Thélohan (1895) authored a notable monograph that included the Microsporidia, and Labbé (1899) presented a synopsis of genera and species of this group.

Other Sporozoa associated with insects were also being observed. Although some historians believe that Redi may have seen a gregarine in the seventeenth century, and Cavalini definitely described one from a crustacean in 1787, it was Dufour who, in 1826 and 1828, reported the presence of gregarines in insects and presented an authentic account of them as a group. Between 1851 and 1891 Leidy reported about twenty-five species of gregarines from arthropods (see also Crawley, 1903). Other earlier observers who contributed significant information on species from insects included von Siebold (1839*a*), Lankester (1863, 1880), Bütschli (1882), and Léger (1892).

Joseph Leidy (1823–91), mentioned above, is an excellent example of the numerous scientists who, though not primarily concerned with the diseases of invertebrates, contributed to the foundations of invertebrate pathology. Many of the more than 600 scientific papers that he authored during his lifetime are related to insects and invertebrates, even though he is best remembered as founder of vertebrate paleontology. Osborn (1913) stated in his memoir on Leidy that his "combination of endowments [as a scientist] will never reappear in a single individual." He had an encyclopedic knowledge and a broad grasp of the whole field of natural history. Osborn felt that Leidy was an evolutionist before Darwin, though he may have not formulated clearly the theory in his own mind. Certainly Leidy was one of the very first American scientists to accept Darwin's theory of evolution; immediately after the appearance of the first edition of the *The Origin of Species* in 1859, he proposed Darwin's name for membership in the Academy of Natural Science of Philadelphia, to which Darwin was duly elected in 1860.

Coccidia were observed infecting insects (*Gyrinus, Tipula,* and *Tineola* larvae) late in the nineteenth century (Schneider, 1885; Léger, 1897; and Pérez, 1899). Smith and Kilbourne demonstrated the transmission of *Babesia bigemina,* the cause of Texas cattle fever, by the tick *Boophilus annulatus* in 1893; and malaria parasites were seen in mosquitoes just before the turn of the century by Ross, in 1895 and 1897.

Flagellates were first seen in insects about the middle of the nineteenth century. In most of these cases, however, the protozoan caused no appreciable harm to its arthropod host. Some of them, e.g., the flagellates of termites, are distinctly mutualistic. The first amoeba observed in an insect apparently was *Endamoeba blattae,* a commensal in the colon of the oriental cockroach, reported by Leidy in 1879. Amoebae truly pathogenic for insects were not discovered until after the turn of the century. The same must be said for the parasitic ciliates.

It might be appropriate here to mention that nematode infections in insects have been known since the beginning of the nineteenth century. The scattered reports of entomophilic nematodes were brought together by von Siebold in a series of contributions between 1842 and 1858. Other prominent early contributors to our knowledge of insect nematodes, including North American species, were Camerano, Diesing, Leidy, Leuckart, Linstow, and Lubbock.

Knowledge of the bacterial disease of insects might be said to have begun with Pasteur's (1870) study of flacherie of the silkworm. As we have explained above, he observed two kinds of bacteria associated with the diseased silkworms. Today we know these as *Streptococcus bombycis* and *Bacillus bombycis,* but their exact relationships to the diseases gattine and flacherie are still not entirely clear. (Some modern workers believe that the classical flacherie of the silkworm is caused by a nonoccluded virus or by *Bacillus thuringiensis.*) During the remaining thirty years of the century, a considerable number of bacteria were reported from insects (see Steinhaus, 1946), but only rarely was their true pathogenic role demonstrated. It should be remembered that some of the so-called bacterial diseases of Lepidoptera, as described by certain of the early investigators, in actuality were virus diseases in which the bacteria isolated were secondary invaders or merely adventitious forms. This explains why so frequently the isolated bacterium never seemed to possess the virulence or capacity to cause epizootics that the natural disease exhibited. Nevertheless, some of the nineteenth-century studies of real and purported bacterial infections (e.g., those by Forbes, 1886*b*; Krassilstschik, 1896; and Duggar, 1896) provided valuable factual information as well as stimulated interest in the diseases of insects from the viewpoint of controlling noxious species. In addition, they helped direct attention to the broader aspects of insect pathology and to the fact that insect pathogens could be found among any of the major groups of microbial life.

That insects are susceptible to infectious agents known as viruses was not clearly demonstrated until the second decade of the twentieth century (von Prowazek, 1907, 1912; Escherich and Miyajima, 1911; Glaser and Chapman, 1913; Acqua, 1919). Although the most rapid and impressive advances in our knowledge of the virus diseases of insects have come during the past fifty years, significant observations were made before this. Most of the important nineteenth-century developments were made while studying the polyhedrosis of the silkworm—a disease that has been designated by such names as jaundice, grasserie, giallume, Gelbsucht, and others. The Italian poet Vida may have referred to this disease in his poem *De Bombicum*, written in 1527, and Maria Sibylla Merian mentioned what is probably this affliction in a book on butterflies written in 1679. One of the earliest published descriptions of the disease itself is that by Nysten in 1808.

Early sericulturists confused jaundice with other diseases of the silkworm. Toward the end of the nineteenth century, however, it was generally recognized as a distinct entity, but the causative agent remained in doubt. Some workers (e.g., Hofmann, 1891; von Tubeuf, 1892; Krassilstschik, 1896) believed the disease to be caused by bacteria, but others recognized that the characteristic crystal-like bodies (polyhedra) regularly found in the tissues and body fluids of the diseased silkworms were somehow related to the cause of the disease. Among the first to make such observations and to associate these bodies with the disease were Cornalia (1856) and Maestri (1856)—over a century ago.

Cornalia described some of the symptomatological and histopathological manifestations of silkworm jaundice and reported that the polyhedral bodies which he observed in the blood corpuscles originated from some kind of alteration of the blood. Maestri also observed the polyhedral bodies in the blood cells as well as in other tissues and called attention to their location in the nuclei of the cells. As far as the characteristic dissolution of the tissues was concerned, Maestri believed that the action of heat on the respiratory system of the silkworm brought about an alteration and a complete melting of the adipose tissue. According to Haberlandt (1871, 1872), who referred to the polyhedra as crystals, Verson was the first to recognize their crystalline nature, and Panebianco (1895) studied them from a crystallographic standpoint, likening them to rhombododecahedral crystals.

Bolle (1894, 1898), at first, also considered the polyhedra to be simply crystals; then he decided that they represented the sporulated form of a protozoan parasite. He believed it to be a sporozoan that multiplied in a manner similar to that of coccidia, and his drawings depicted a coccidianlike oöcyst filled with polyhedra, which he apparently believed were sporocysts. In any case, he correctly associated the polyhedra with the causative agent of the disease. He showed the polyhedron to be soluble in the alkaline juices of the silkworm gut, as well as in other alkaline and acid solutions, and observed that upon dissolution the polyhedron consisted of a central granulated mass and a peripheral layer surrounded by a thin membrane. These observations are particularly noteworthy since today it is known that the virus particles are contained within the proteinaceous polyhedron, which some believe to possess a membranelike covering. The virus particles may be released from the polyhedron when the latter is treated with a dilute alkali.

Aside from the polyhedrosis of the silkworm, virus diseases of few other insects were being observed, and, if so, were usually considered to be caused by bacteria. One of these, the polyhedrosis (*Wipfelkrankheit*) of the nun-moth caterpillar, first came to the serious attention of European entomologists in 1889 and 1892, and was destined to receive considerable attention in the years to follow.

One of the first true, ecological or epizootiological concepts of disease in populations of insects of economic importance was clearly enunciated by S. A. Forbes in 1885 (see Forbes, 1886*b*, *c*), when he described the effects of what he called a flacherie, but which most probably was a granulosis, in populations of the European (imported) cabbageworm, *Pieris rapae*, in Illinois. (He mentioned that the disease was probably first noticed in this country in 1879, in the vicinity of Washington, D.C., apparently by C. V. Riley.) Although Forbes observed bacteria in microscope preparations of the diseased insects, he also described "myriads of very minute spherules" or "granules" that, from other parts of the description he gave (although partly inconsistent), could easily have been in actuality polyhedra. They might also have been the capsules characteristic of the granulosis known to occur in this insect, except that the size he gave for them is at least twice the known size of the capsules. At any rate, it would appear that he observed extensive epizootics of a virus disease in the cabbageworm in the field. Forbes (1898*b*) also described a polyhedrosis of the cosmopolitan armyworm, *Pseudaletia unipuncta*. Although he could not ascertain the cause of the disease, he did describe the polyhedral bodies. Forbes observed that the disease constituted an important natural check on populations of the insect. However, there is a good possibility that the first record of what may have been a virus disease in North America is that made by Peck in 1795 and 1796, when he described in New

England in an insect he designated as "the cankerworm *Phalaena vernata*" (now known as the spring cankerworm, *Paleacrita vernata*) a disease he called "Deliquim." From Peck's description, the disease was undoubtedly a nuclear polyhedrosis, a viral disease. He considered the cause to be a "fermentation" of the insect's food and stated, "In this disease the whole internal structure is dissolved into a liquid, and nothing is entire but the exterior cuticle, which breaks on being touched." Peck considered the disease, along with a species of bird, to be a significant natural "check" on the cankerworm, which he suggested may "have been introduced into New-England by the importation of [apple] trees from the southern States." Thus, although the etiological agent itself could not be demonstrated, it is interesting that the first recorded disease of a wild insect in North American was apparently a virus disease.

7. KIRBY'S "DISEASES OF INSECTS"

Shortly after the turn of the nineteenth century there appeared the first distinctive writing on the diseases of insects that might be considered to be in any sense comprehensive. This rather remarkable presentation appeared in 1826 as a chapter in volume 4 of the notable treatise *An Introduction to Entomology*, by Kirby and Spence. This particular chapter, titled "Diseases of Insects," was written by Kirby, who, in his opening remarks, dutifully explained why he was assigning an entire and separate chapter to a consideration of the diseases of insects. A brief review of this chapter is eminently worthwhile here since it so well portrays the concepts of disease in insects as generally held by biologists of that time.

Kirby divided the diseases of insects into two large classes: those resulting from "some accidental *external* injury or *internal* derangement, and those produced by *parasitic* assailants." Under the first of these designations he discussed wounds, fractures, mutilations, tumors, and monstrosities. Diseases resulting from an internal cause included those described as a "kind of vertigo" (believed to result from a derangement of the nervous system), a "kind of convulsions," "the stone" or calculus, and what we now know to be some of the various infectious diseases. Had he realized the infectious or parasitic nature of the latter group, Kirby would have undoubtedly placed them with the second large class of maladies, i.e., those caused by "parasitic assailants." In this group Kirby included those diseases caused by some of the more obvious fungi, nematodes, and parasitic insects. Entomophagous insects are discussed in some detail, and the results of their activities are considered to represent, and properly so, a diseased state.

It is clear that by the time Kirby wrote his chapter on disease, insects mortally affected by entomogenous fungi were being observed frequently. However these fatal diseases were described as resulting from a variety of causes. Mention has already been made of the records of Degeer and of Latreille in this respect. Kirby himself clearly described an affliction of Diptera that undoubtedly was caused by an entomophthoraceous fungus. That a fungus was involved in these cases was not, however, suspected by Kirby, who believed the disease arose "from a superabundance of the nutritive fluid, or of the fat, so that it seems to be a kind of *plethora*." On the other hand, Kirby, as did Persoon and others before him, recognized the fact that true fungi did grow upon the larvae and pupae of some insects. He sagaciously differentiated between the saprophytic growth of fungi on dead insects (such as first observed by Ledermüller, 1760) and the possibility of fungi growing at the expense of living tissue. He pointed out that Persoon (1801) did not make it clear whether or not the insects on which he reported two species of *Isaria* were dead or alive. If alive, Kirby suggested that "perhaps in these cases these plants may constitute an insect disease." A remarkable deduction for that day—and possibly derived from the fact that as early as 1799 Kirby had interested himself in, and had published on, fungi parasitic on certain grain plants. (Tillet [1755] was the first to prove the infectiousness of a plant disease, but he did not realize the true nature of the disease.) Had Kirby, like Bassi—to follow—experimentally tested his beliefs and hypotheses, he would have merited an even more hallowed place than he does in the historical beginnings of "scientific insect pathology." Unfortunately, neither he nor the others, of that time and before, who observed fungus-infected or otherwise diseased insects had the benefit of knowing or conceiving the famous "Koch's postulates."

Kirby's discussion included a brief consideration of the diseases of the honey bee and the silkworm. The causes of these diseases he attributed to poisonous food or mephitic and other types of noxious air. He ended the chapter with a very interesting account of the infection of insects with worms, such as *Gordius*, and recounted some of the early observations by Degeer and others along this line. One is impressed with the surprisingly advanced state of knowledge existing at that time relative to menatode infections in insects.

William Kirby (1759–1850) was educated for the clergy, graduating from Caius College, Cambridge. Early in life he became interested in natural history, and he pursued his interests in botany and entomology with thoroughness and perseverance but never, according to Freeman (1852), neglected his duties to the ministry, which he served efficiently and well for sixty-eight years. Kirby was a careful and meticulous worker and has always been esteemed for the clarity and scientific accuracy (for his time) of his writings.

8. OF DISEASED EARTHWORMS, CRABS, AND OYSTERS

The historical development of the study of disease in invertebrates could probably best be presented according to each phylum. However, to do this is not my intention here since the emphasis throughout this account is on the diseases of insects. Nevertheless, I do not wish to give the reader the impression that, of the invertebrates, only insects were being studied with respect to their diseases and abnormalities. Accordingly, from time to time I shall at least allude to some of the advancements and personalities associated with the understanding of disease in noninsect invertebrates.

Worldwide, it would appear that the first recorded parasite of an invertebrate other than an insect was an undescribed species of gregarine in an unknown species of crustacean by

Cavolini (1787). According to Rudolphi (1819), Cavolini reported a gregarine parasite in the "appendicibus ventriculi" of a crab, *Cancer depressus*. He indicated the parasite was possibly a *Distomata* sp. In 1851, Diesing named the parasite *Gregarina conformis*, but later it was renamed by Léger and Duboscq (1911) as *Cephaloidophora conformis* (see Kamm, 1922). So it would appear that this was the first recorded non-insect invertebrate parasite. However, whether or not "infection" by this rather harmless parasite constitutes a true disease state will depend on one's interpretation of the type of biological relations known to exist between this and other gregarines and their hosts. A gregarine said first to have been casually observed by Meckel was further described by Lieberkühn in 1853, and still further described and named by Bosanquet (1894) as *Monocystis herculea* and the invertebrate host was the earthworm *Lumbricus herculeus*. (As time went by, the earthworm was found to be the host of several species of gregarines.) Bosanquet observed that adult gregarines and the cysts lay in the coelom of the worm, but at times he found the cysts embedded in masses of tissue that seemed to consist of altered and degenerated nephridia. Another sporozoan was reported a few years later by Lenssen (1897) infecting the epithelial cells of the gut of the rotifer *Hydatina senta*.

An interesting early observation of a parasite parasitizing a parasite is that discussed by Metchnikoff (1892) of *Ascaris mystax* by *Mucor helmenthophorus*, which invades the intestine and genital organs. According to Metchnikoff, free Nematoda are also frequently subject to other members of the higher fungi as well as by the Chytridiaceae.

The well-known French biologist A. Giard (1897) made significant observations on the parasitic castration of crabs and an inhibition of their molting by rhizocephalans, which are largely parasitic on decapod crustaceans. Within the host the growth of the parasite occurs through the ramification of nutrient rhizoids. The female is eventually converted into a hermaphrodite in some cases. The development of the gonads is either retarded, or the gonads atrophy; varying degrees of change in secondary sexual characters occur. That Giard was thinking in a comparative-pathology context is indicated by his comparison of parasitic castration in crabs to the effect of castrating male chickens (capons).

Apparently one of the earliest reports of a rather great mortality among invertebrates was Giard's (1894) in which he described a disease ("maladie du pied") of the European oyster *Ostrea edulis.* The disease was caused by a fungus *Myotomus ostrearum* that attacked the shells and tissue. Infected oysters developed rubbery green spots on their shells and loose adductor muscle attachments. The disease is fairly well known as the "shell disease" throughout Europe (Korringa, 1948).

As noted by Steinhaus (1969) from the standpoint of parasitology, the first disease reported in an invertebrate (oyster *Crassostrea virginica*) in North America seems to be that caused by the trematode *Bucephalus cuculus* described by J. McCrady in 1873. A similar finding of *Bucephalus* in an oyster, *Ostrea edulis*—and in the cockle *Cardium tuberculatum*—had been reported previously from Europe by de Lacaze-Duthiers (1854).

Deserving of our attention is J. A. Ryder, who, in 1887, first described a "tumor" in an oyster. The apparent neoplastic growth was approximately one-third of the length of the oyster's body. Histologically, the cells in and near the tumor differed in basic appearance and structure from normal tissue cells. A few years later, Williams (1890) described a benign polypoid growth on a fresh-water mussel, *Anodonta cygnea.* Another early record of what appeared to be neoplasms was made by Collinge (1891), who found two abnormal growths in the same species of mussel.

In 1895 Cooke told of abnormal growths and monstrosities of the shells of several species of mollusks. Some of the teratologies are minor and almost superficial whereas others are quite pronounced and serious. (He also told of the presence of parasitic worms and mites infesting Mollusca, and referred to their host specificity and to their molluscan hosts as being intermediate hosts, the worms attaining their sexual stage upon reaching a vertebrate host.) In 1898 Stubbs described abnormal shells of *Planorbis spirorbis* and other freshwater mollusks.

Perhaps the earliest report of the bactericidal effect of an extract of an invertebrate is that of Picou and Ramond (1899), who made an aqueous extract of the tapeworm *Taenia inerme* and found it to be toxic to different intestinal microbes, both saprophytic and pathogenic, of man.

9. ELIE METCHNIKOFF: A FASCINATING PIONEER AND A CHAMPION OF COMPARATIVE PATHOLOGY

Elie Metchnikoff (1845–1916), who may be considered one of the "giants" in invertebrate pathology, was a Russian zoologist destined to become famous for his studies on phagocytic immunity and to receive the Nobel Prize for this work. However, every invertebrate pathologist, or general comparative pathologist, should recognize the fact that Metchnikoff's interest in, and study of, disease in insects and other invertebrates not only initiated his interest in disease generally but broadened his scientific interests to include his monumental works on inflammation, phagocytosis, and immunity. To understand this more adequately, I recommend the reading of at least two of Metchnikoff's major publications: *Lectures on the Comparative Pathology of Inflammation* (1892) and *Immunity in Infective Diseases* (1905). These are available in English, as well as the original French, editions.

The first of these works is a testimony to the declaration Metchnikoff made in the original French edition, "The principal aim of my work is to establish a lasting connection between pathology and biology in general." (This statement has also been translated: "My principal object in writing this book being to show the intimate connection that exists between pathology and biology properly so called.") A goal, incidentally, still being sought today (e.g., see Steinhaus, 1964) with even more justification than that available to Metchnikoff when he presented his lectures (which served as the basis of his book) at the Pasteur Institute in Paris, in 1891. As stated by Silverstein (1968) in his introduction to the Dover edition (Dover Publications, Inc.) of Metchnikoff's book:

> Modern pathologists and biologists in general have still to formulate some of these connections; but it is clear that Metchnikoff's development of the phagocytic theory, leading to a general biological concept of inflammation, succeeded in forging one of the very

> strong connections that he sought. . . . It was Metchnikoff's work on the evolution of the inflammatory response, culminating in the publication in 1892 of the original French edition of the present work, that helped set the stage for our modern understanding of the nature and significance of the inflammatory response.

Metchnikoff approached the study of inflammation by observing the response of the more simple unicellular organisms to injury and to foreign bodies, and then proceeded up the evolutionary scale to the study of vertebrate responses. (The same procedure can be used with other phenomena involving disease processes; Steinhaus, 1965, 1969.) Whereas such great pathologists as Virchow and Cohnheim considered inflammation to be a harmful process or reaction of no benefit to the host, Metchnikoff saw the process as a protective one and of benefit to the host. One could cite numerous statements from his book on inflammation that showed his broad concept of pathology, but the following should suffice to establish the point.

> As the comparative anatomy of former times treated only of man and the higher animals, so medicine has hitherto excluded all the pathological phenomena which occur in the lower animals. And yet the study of these animals, affording as they do infinitely simpler and more primitive conditions than those in man and vertebrata, really furnishes the key to the comprehension of the complex pathological phenomena which are of special interest in medical science.
>
> In deciding to give a few lectures on a subject belonging to the domains of pathology, I have resolved to do so solely in my capacity of zoologist. The complexity of the most important pathological processes, as studied according to the universal custom on vertebrates, is so great, even in so low a member as the frog, that it becomes impossible to analyze them or to attain any adequate conception of their real significance.
>
> In all organisms, starting from the simplest forms of life, we find infectious disease produced by different classes of parasites. It is therefore only natural to suppose that this parasitism gives rise to a definite series of disturbances in the infected organism, and likewise provokes phenomena of reaction in the latter.
>
> A branch of zoology—the comparative pathology of animals—may thus be formed, which would differ from the present comparative pathology in many ways. Whereas the latter, which has been

> mainly founded on veterinary science, is only concerned with the higher animals and chiefly with mammals, genuine comparative pathology should include the whole animal [and we might add, plant] world, and be treated from the widest biological standpoint.
> The groundwork of such a comparative pathology was laid about five-and-thirty years ago. About this time, in 1857 and 1858, the theory of natural selection was built up on scientific foundation by Darwin and Wallace, the biological theory of fermentation by Pasteur, and the theory of cellular pathology by Virchow.
>
> If, however, medical science may learn much from biology, of which it forms but a part, it may at the same time give something in return. General biology may extend its knowledge by including the study of the morbid phenomena of which pathology takes cognisance. . . . General pathology should go hand in hand with zoology or rather with biology, and form one branch of it, that of *comparative pathology*. This science is only in its infancy, and yet it is already in a position to render good service to medicine. By facilitating the analysis of the reactive phenomena, it indicates the elements which should be especially protected and reinforced in the conflict of the organism against its enemies, and thus contributes to the solution of one of the great problems of humanity.

These excerpts from Metchnikoff's lectures are still valid today even though since his time much has been accomplished in invertebrate (especially insect) pathology. Exhortations by this author (e.g., Steinhaus, 1967*a*) and by others—e.g., Schwabe's (1969) assertion that invertebrate pathology, as well as other aspects of the comparative study of disease in animals, should be included in the domain of veterinary medicine—do not seem to have really brought about and fulfilled Metchnikoff's concept of comparative pathology. Although, as these pages are written, some progress along these lines seems to be emerging. Nevertheless, because comparative pathology and veterinary medicine generally have not embraced the work being done by invertebrate pathologists, it has been necessary to have special departments and institutes of invertebrate (or insect) pathology formed, special journals published, and even the formation of a separate scientific society. Hopefully, recent efforts to bring all pathological (disease) phenomena under a single unifying umbrella—invertebrate, vertebrate, and plant pathology—by the formation of centers of pathobiology (Steinhaus, 1968) may help eventually to bring about the broad-spectrum comparative pathology that Metchnikoff championed.

Whether we read the somewhat flamboyant account of Metchnikoff by Paul de Kruif (1926), who saw him not so much as a sober scientific investigator but "more like some hysterical character out of one of Dostoevski's novels" who (understandably) insisted that he could experiment best when pretty girls were close by, or the more decorous account by his wife, Olga, whose *Life of Elie Metchnikoff* was published in 1921, or even the matter-of-fact biographical statements by medical historians —whichever type of biographical statement we read, the unusual qualities of this man emerge. In any case, invertebrate pathologists can take pride in the fact that this man not only saw the value in studying disease and immunity in invertebrates but used these animals to prove certain of his contentions regarding cellular immunity at a time when the merits of humoral immunity were being championed by others. One should not forget that he first observed phagocytosis in starfish (1882, 1883) and demonstrated the phenomenon in the crustacean *Daphnia magna* which he infected with the yeast *Monospora bicuspidata*. This, in addition to his work with the fungus *Metarrhizium anisopliae* and its possible use in the control of harmful insects, should indeed be adequate to grant him secular sainthood in the realm of invertebrate pathology.

The phenomenon of phagocytosis was actually first reported by Haekel in 1862. He injected granules of indigo into a mollusk —a species of *Thetis*—and observed the animal's leucocytes ingesting the granules. Later he similarly injected granules of indigo into snails (*Helix*) and crayfishes (*Astacus*). Although Haekel concluded that the leucocytes would engulf virtually any solid particle they encountered, he apparently did not consider this action to be a defense mechanism.

Unfortunately, except for the study of phagocytes, the record does not show that Rudolf Virchow (1821–1902), whom Long (1965) called "the greatest figure in the history of pathology," had much direct effect during his lifetime on the study of disease in invertebrate animals. Nevertheless, almost subconsciously, insect pathologists adopted many of the principles of cellular pathology that Virchow established. Certainly by the early part of the twentieth century his emphasis on disease as being the physical, chemical, or anatomical disturbance of cellular units was generally accepted by those few students who, although briefly in most cases, concerned themselves with some aspect of disease in invertebrates.

Although Metchnikoff had been what we could call an "experimental pathologist," he was not the only one who used the experimental method in late nineteenth-century invertebrate pathology. For example, in 1891 Herrmann and Canu (1891) studied the phagocytic response of the "blood" cells in beach fleas (*Talitrus*) after a fungus (*Oidium*) had been introduced into the animal's body cavity. Not only did phagocytosis occur but, in the process, the engulfed fungus usually was lysed. Most of the beach fleas inoculated with the fungus died in spite of the action of the cellular defense mechanism. In this respect their findings differed somewhat from those of Metchnikoff (1884*b*) who found that many of the *Daphnia magna* inoculated with the yeastlike *Monospora bicuspidata* were so well protected by the phenomenon of phagocytosis that they survived. In speaking of phagocytosis, mention should be made of the work of de Bruyne (1893, 1895) on the phagocytosis of particles in bivalves. His descriptions of phagocytosis were excellent, and, unlike others working in this field, he elucidated some of the discrepancies found in published reports on phagocytosis.

10. FROM LECONTE AND FORBES: THE BEGINNINGS OF MICROBIAL CONTROL

The serious observation and study of diseased insects had not progressed far before the idea of using disease to destroy noxious insects was conceived. Just exactly when, where, and by whom the idea originated are points that may never be absolutely certain, but perhaps the story can be constructed from the available literature. In any event, the information so gleaned is exceedingly interesting and revealing. As is the case with many ideas, this one probably occurred to several men, at various times, and in different parts of the world; and in all probability most of these conceptions went unrecorded. Nevertheless, it is fascinating to attempt to trace the development of the idea.

From the historical account that has already been presented, first came the observation that insects, such as the honey bee and the silkworm, are subject to disease; then developed the realization that these diseases are of a contagious nature, and, later, that outbreaks of disease occurred in insects in nature. Proceeding with the story it is well to keep this last point in mind, since the awareness of it constituted a significant background in the more direct contributions to microbial control arising from the work on silkworm diseases. Early mycologists, for example, were aware that insects serve as natural hosts for numerous species of fungi (Kirby, 1826; Gray, 1858; Cooke, 1875). Most of these early observers, however, were primarily interested in the nature and taxonomy of the fungus itself and paid little attention to the fungus-insect relationship. Nevertheless, many insect hosts were recorded, and the idea that certain fungi could be considered as natural enemies of insects gradually became established. Furthermore, some of these biologists reported the occurrence of "pestilential epizootics" among insects in nature. Audouin (1838*b*), for example, in 1838 observed the "disappearance" of *Galeruca calmariensis*, a coleopterous pest of elms, as the result of an epizootic caused by the muscardine fungus (*Beauveria*). Hagen (1879*b*) told of an epizootic of "the common dung-fly" that occurred in 1867.

> Not only those, but many other insects died in the same locality and in the same manner; also other species of flies and gnats, the caterpillars of moths and of Phalaenids, and the common hairy caterpillar of a moth which is very nearly related to the famous hairy caterpillar of the Boston Common. Of some species the destruction was so complete that the next year they were very rare. . . . Similar observations have been made in other places in Europe and here. . . . In Entomological journals are reported fatal epizootics of leaf lice, of grasshoppers, of the cabbage butterfly and of the currant worm, both imported here only a few years ago, and both very obnoxious.

Other early reports of natural epizootics include those by Frauenfeld (1849), Köppen (1865), Bail (1867*a*), Shimer (1867), Ratzeburg (1869), Cohn (1870), Brefeld (1877), and de Bary (1878).

The concept of using pathogenic microorganisms to combat insects appears to have grown out of the observation that the maladies were infectious and contagious[13] and that they could be transmitted from diseased to healthy individuals both in nature and experimentally. As soon as the transmissibility of the silkworm diseases was established, the phenomenon as it applied to silkworms was carried over to the diseases that occurred in other insects, as, indeed, Bassi (1835) himself accomplished experimentally in his efforts to prove the infectious nature of the white muscardine fungus, *Beauveria bassiana*. This pioneer not only artificially, by needle, infected numerous different species of insects (unnamed, but sometimes referred to as worms or caterpillars) but passed the fungus through long series of different insect species. Moreover, he declared that such transmissions could be accomplished whenever desired! Herein lies the germ of the thought that man can communicate agents of disease to susceptible insect pests at will.

Even more remarkable, however, is another contribution of Bassi's fertile and imaginative mind. In a footnote to a report of his on *negrone* (flacherie), Bassi (1836*a*) told of finding that although the hemolymph extracted from normal larvae caused no harm when inoculated into another silkworm, if the fluid was first allowed to putrefy, the inoculated insect inevitably died "in a state of *negrone*." The same phenomenon occurred when putrefying substances such as milk and urine were used as inocula. Then he wrote these important lines (freely translated):

> This fact considered, instead of using useless fumigations or medicated baths to kill the worms that destroy plants useful to us, one could try to spray their leaves with this water. The same worms nourishing themselves at any of the points touched by the poisoned liquid, even in the slightest portion, would unfailingly and quickly die. This bath or spray, far from being harmful to the tree, rather aids it, increasing its nourishment. Thus, by rotting a raw chicken egg, and after breaking it, throwing it

> into the water, one could prepare in this manner, if one wishes, several brentas [*brente*] of this exterminating liquid, with little expense.[14]

Thus, it would appear, on the basis of information available to the writer, that the credit and honor of having first suggested the possibility of employing the activities of microbial life to destroy insects harmful to man's interests belong to Agostino Bassi. To be sure, Bassi did not specify the use of microorganisms as such, but it is clear from the text accompanying the footnote referred to that he conceived of the putrefying fluids as being of an infectious nature similar to the muscardine fungus on which he made his monumental studies.

Following the infectivity experiments of Bassi, others reported successful attempts to transmit artificially the infectious agent of muscardine to insects other than the silkworm. In 1836, Turpin was able to infect noctuids and other lepidopterous species with the fungus. A year later Audouin (1837*b*, *c*) expressed the belief that muscardine was not peculiar to the silkworm but appeared among insects in general, and perhaps only among them. Later (1839*b*), he successfully transmitted the fungus by injection into other harmful insects including the gypsy moth. In the same year, Bonafous (1839) similarly transmitted the muscardine fungus to the larvae of several species of insects. Audouin (1839*a*) told of a sericulturist who emptied contaminated silkworm-rearing trays out of a window onto trees whose leaves were being attacked by unidentified defoliating larvae. All of these insects were attacked by the fungus and died of muscardine four days later. Apparently, this fortuitous happenstance is the first reported instance of harmful insects in nature being destroyed by the artificial dissemination of a microbial pathogen. Robin (1853) and Metchnikoff (1879), apparently referring to this same instance, ascribed the reporting of it to Bonafous. The original brief note was designated as a communication to Audouin from Bonafous. Careful reading of the note, however, reveals that the reference to the experience of the unidentified sericulturist was made directly by Audouin as additional proof of his belief that muscardine could be spread by contact. This is also made clear in a subsequent note in which Audouin

(1839*a*, p. 200) disclaimed having recommended the procedure or having said it had been successfully employed. Perhaps he was attempting to protect himself from any evil consequences of the unnamed sericulturist's careless sanitation. In any case, he did not deny that the incident took place. From the standpoint of the development of microbial control, the kind of transmission referred to here was more significant than the type, for example, described by Lebert (1858), in which he was able to transmit the muscardine fungus from diseased to healthy insects by placing them together in the same enclosure.[15]

Unfortunately, Bassi's visionary suggestion, as well as the Audouin report, apparently fell on barren soil, for it was not until more than three decades later that the suggestion was made again. This time it was made by the American entomologist J. L. LeConte (1874) before the twenty-second meeting of the American Association for the Advancement of Science, held in Portland, Maine, in August 1873. The paper he read was titled "Hints for the Promotion of Economic Entomology," and in it he suggested a "new system of checks" to be employed against harmful insects. One of these checks was "the production of diseases." He proposed the communication of the muscardine fungus affecting silkworms to other lepidopterous larvae, saying, "I am extremely hopeful of the result of using this method. I have learned of an instance in which from the communication of the disease by some silkworms, the whole of the caterpillars in a nine-acre piece of woods were destroyed."[16] Toward the end of his paper he presented a number of recommendations, among which were "careful study of epidemic diseases of insects, especially those of a fungoid nature: and experiments on the most effective means of introducing and communicating such diseases at pleasure." To the best of my knowledge, LeConte's recommendation represented the first clear-cut suggestion advocating the "artificial" use of disease agents as a means of insect control to appear in the English language. To be sure, he undoubtedly derived the idea from the observations of others on the diseases of the silkworm, but LeConte's proposal was definite yet broad in its concept and clearly envisioned the practical possibilities involved (see also Lesley, 1880). Moreover, one should not forget that the value of the occurrence of disease in nature, as distinguished from dissemination of infectious materials by man, had not been going unrecognized. For example, mycologists such as de Bary and Tulasne, around the middle of the nineteenth century, were calling the attention of agriculturists and orchardists to the importance of *Beauveria tenella* (=*Isaria densa*) as a natural check on destructive insects (Speare and Colley, 1912).

It is highly interesting and significant that another early, definite suggestion that microorganisms might be used to combat insects came from Louis Pasteur—undoubtedly as a result of, or an afterthought from, his classical work on the diseases of the silkworm. He published this suggestion in 1874 at a time when the grape phylloxera was threatening grape production in France. He had suggested the use of "les corpuscules de la pébrine" against this pest, apparently assuming that it would be susceptible to the protozoan. Then he recommended that a search be made for a fungus that was capable of destroying the insect and that such a fungus be introduced into vineyard populations of phylloxera. Although he never tried to establish the practical usefulness of fungi against phylloxera or other insects,[17] Pasteur apparently thought the idea had possibilities, for (as told by Dubos, 1950) in 1882, almost ten years later, he dictated the following laboratory note to his assistant Adrien Loir:

> To find a substance which could destroy phylloxera either at the egg, worm or insect stage appears to me extremely difficult if not impossible to achieve. One should look in the following direction.
>
> The insect which causes phylloxera must have some contagious disease of its own and it would not be impossible to isolate the causative microorganism of this disease. One should next study the techniques of cultivation of this microorganism, to produce artificial foci of infection in countries affected by phylloxera.

In 1884, however, Balbiani, also expressing the idea of using the pébrine organism to control the phylloxera, pointed out some of the problems that would be involved, such as how to infect the sucking insect—if, indeed, this were possible and which he, correctly, doubted—and how to distribute the pathogen in the soil of the vineyard. He then stated that the idea appeared to have been abandoned, and indicated that because of the nature of the particular pathogen involved, it deserved to be.

Scientific observations and reports are frequently important not because of the factual information that they may contribute but rather because of the influence that they may have on the thinking of others. Such a characterization may be applied to certain transmission experiments and reports, factually erroneous, conducted by Carl Adolph Emmo Theodor Bail of Danzig. His experiments were significant primarily because of the impression they made on H. A. Hagen, who later, in the United States, advocated the use of fungi to destroy insects. At European meetings of an association of naturalists, in 1861, Bail delivered lectures during which he exhibited a mold grown on a mash that had been sown with the "fungus of the house-fly," as well as a keg of beer brewed from such mash and a cake baked with what Bail considered to be the yeast form of this fungus. This Prussian worker maintained that the fungus was capable of killing such insects as flies, mosquitoes, and caterpillars brought in contact with the inoculated mash. There is no record, however, of his having advocated the practical use of fungi to control harmful insects in the field. (See also Bail, 1867*a*, *b*; 1868; 1869*a*, *b*; 1870; 1904.)

On the basis of present-day knowledge, it is obvious that, from a systematic viewpoint, Bail's mycology was considerably in error, as was pointed out by some of the botanists of that day. He believed that four different species were but different forms of the same fungus. Thus the house-fly fungus (*Empusa*) appeared as a "common mold" on vegetable matter; as a yeast under conditions conducive to fermentation, such as on a mash; and as a water mold (*Saprolegnia*) when grown in water. Presumably, either the form occurring on house flies or the yeast form was capable of killing insects. In light of our understanding of these forms today, four distinct species are represented, and of them only the house-fly fungus (*Empusa*) is pathogenic to insects; the yeast (*Saccharomyces*) referred to almost certainly was, in itself, quite harmless to insects.

Of interest is the fact that the so-called house-fly fungus was not always considered a boon as the result of its activities. Hagen (1879*a*, *b*) referred to it as "the vexation of every housekeeper. The dead flies stick in the fall firmly to the windows, or anywhere else, and are covered by a white mould not easy to be removed."

Impressed by the work and writings of Bail, Hagen (1879*a*, *b*) published a paper[18] in which he expressed "the conviction that a remedy for insect pests, offering several prominent advantages, could be found in the easy application of the yeast fungus." Also, "I believe I should be justified in proposing to make a trial of it against insect calamities." He then proceeded to make rather specific suggestions, as follows:

> Beer mash or diluted yeast should be applied either with a syringe or with a sprinkler; and the fact that infested insects poison others with which they come in contact will be a great help. Of course it will be impossible to destroy all insects, but a certain limit of calamities could be attained, and I think that is all that could reasonably be expected. In greenhouses the result would probably justify very well a trial, and on current worms and potato bugs the experiment would not be a difficult one, as the larvae of both insects live upon the leaves, which can easily be sprinkled. But it seems to me more important to make the trial with the Colorado grasshopper. I should recommend to infest the newly-hatched brood, which live always together in great numbers, and I should recommend also to bring the poison, if possible, in contact with the eggs in the egg-holes, to arrive at the same results, which were so fatal to Mr. Trouvelot's silk-raising. After all, the remedy proposed is very cheap, is everywhere to be had or easily to be prepared, has the great advantage of not being obnoxious to man or domestic animals, and if successful would be really a benefit to mankind. Nevertheless, I should not be astonished at all if the first trial with this remedy would not be very successful, even a failure. The quantity to be applied and the manner of the application can only be known by experiment, but I am sure that it will not be difficult to find out the right method.

Thus Hagen joined Bassi, LeConte, and Pasteur in being among the first to make original and concrete proposals to attempt the use of microorganisms to destroy and control noxious insects. Except for Bassi's (1836) suggestion, these proposals were made within a six-year period, from 1873 to 1879, during which none of these men actually conducted experiments to test their ideas. But experimentation, with an eye to the practical applications of the method, was not long to follow. In fact, cursory and inquiring experiments were conducted during the same year, 1879, in which Hagen made his proposals.

Although he made no specific scientific suggestions regarding the use of microorganisms in the control of insects, the rather quaint manner in which M. C. Cooke (1893, pp. 245–46) did make a suggestion along this line should be recounted:

> We presume that, having demonstrated that so obscure and apparently simple a form of fungus life as the fly-moulds are in possession of a process of sexual reproduction by conjugation, we have justified a reference to them here. How much more justified do we feel in having made known the various methods by which these simple organisms reproduce and extend themselves amongst a very destructive class of noxious insects; and, on the romantic side, to suggest, if we can do no more, the inoculation of plant-lice with a potent epidemic with a view to their destruction, and once more corroborate the axiom that "Knowledge is power."

Acting upon the suggestions made in Hagen's paper, J. H. Comstock, C. V. Riley, and J. H. Burns conducted separate experiments, in breeding cages, using suspensions of commercial yeast in efforts to destroy "caterpillars," "cotton worms," and "potato bugs" (see Hagen, 1879*b*; also Comstock, 1879). Comstock and Riley reported negative results, as did Prentiss (1880), Willet and Cook, and others in separate series of experiments. Burns wrote Hagen that insects sprinkled with the yeast solution died from eight to eleven days following treatment, whereas insects not so sprinkled showed little mortality. In the wing blood of some of the supposedly diseased specimens sent to him, Hagen found "spores of the yeast fungus in quantity." On the basis of these experimental results, Hagen came to the conclusion that "the application of yeast on insects produces in them a fungus which becomes fatal to the insects." Later, Hagen (1880; 1882*a*, *b*) reported additional successful results after the yeast had been used against aphids and "currant worms." However, since, in general, varied results were being obtained by those who tried the yeast, Hagen felt obliged to postulate that "a *certain stage* of the yeast solution is needed to make it effective." From 1882 on, very little was written concerning the use of commercial yeast to kill insects. This is understandable since certainly it cannot be considered a pathogen of insects. Unless the solutions tested were contaminated with pathogenic forms, it may be assumed that the mortality attributed to the yeast was coincidental or, at least, unrelated to this microorganism. It matters little that the results of the experiments with yeast were misjudged by Hagen, and—as referred to by de Bary—amounted to an "item in the history of error." The significant thing is that he, along with LeConte and Pasteur, saw and advocated the potentialities of microbial control within the limits of microbiological knowledge of that time.

About the same time that the American workers were concerned with the use of yeast as an insecticide, a notable development in the idea of applied insect pathology was also taking place in eastern Europe. Metchnikoff found himself concerned over the great amount of destruction that was caused to cereal crops in Russia by the "grain beetle," or wheat cockchafer, *Anisoplia austriaca.* He was impressed by the rise and fall of the populations of this pest in different years and believed that such oscillations could be caused by outbreaks of disease among the insects.

Beginning in the autumn of 1878, in the region of Odessa, Metchnikoff found three distinct diseases in *A. austriaca*; one of the maladies was caused by one or several kinds of bacteria or "vibrions," another by a nematode, and the third by a fungus that he named *Entomophthora anisopliae* (later [1880] he re-identified the organism as *Isaria destructor* on the advice of the mycologist Cienkowsky) and that is now known as *Metarrhizium anisopliae.* This fungus (since found infecting numerous insect species) causes a disease that Metchnikoff called "green muscardine," and that is characterized by the dark green color of its conidia, or spores. He studied the fungus from a mycological viewpoint as well as from a pathological one.

Especially noteworthy are the facts that Metchnikoff (1879)

appreciated the significance of natural epizootics in reducing insect populations, that he envisioned the practical use by man of disease agents, especially fungi, in the control of insects, and that he tested this possibility experimentally. According to his wife, he originally conceived his ideas from having previously observed a dead fly enveloped with a fungus that had evidently caused its death. He suggested scattering about the infested fields the bodies of larvae dead of green muscardine, the soil in which diseased larvae had been found, or the free spores of the fungus itself. Moreover, he recommended that for greater success in the conduct of the operation, nurseries should be established at various locations for the purpose of producing the pathogenic fungus. Metchnikoff believed that natural epizootics, in themselves, could not be depended upon to control a destructive insect but that with man's participation effective suppression of the pest might be attained; man might not only find means of effectively disseminating the pathogens but might also find means of intensifying their virulence and activity.[19]

From the standpoint of chronology, I wish to point out that although Metchnikoff's advocacy of microbial control methods followed by several years similar suggestions by Le Conte and Pasteur, and by students of forest insects in Germany, his initial experimental work apparently preceded by several months that of the American workers who tested Hagen's proposals. It is worth noting, however, that Metchnikoff (1880) himself referred to "similar results" obtained by de Bary in experiments with the fungus *Isaria farinosa.*

In his 1879 report, Metchnikoff told of experiments in which healthy *Anisoplia* larvae when placed in earth containing diseased larvae acquired the fatal disease. Furthermore, and obviously with an eye to the method's practical application, he was able to bring about the disease by mixing the fungus spores themselves with the soil into which the larvae were to be placed.

During the year 1879, Metchnikoff (1880) found infected *Anisoplia* larvae in other regions of southern Russia; he also found the disease affecting the sugar-beet curculio, *Cleonus punctiventris.* In the case of the latter insect he estimated that forty percent of the natural population was destroyed by the fungus. Experimental tests confirmed the susceptibility of the weevil. Upon Metchnikoff's suggestion to use the fungus to control the insects, entomological commissions in Kharkov and Odessa were assigned to investigate the matter further. In the meantime, Metchnikoff, after seeking methods to propagate the fungus artificially, discovered (upon the suggestion of the chemist A. Werigo) that spores could be produced on sterilized beer mash. (He appears to have been the first to realize the importance of the mass production of entomogenous fungi or their spores by "artificial" means.)

At this point, apparently, Metchnikoff's attentions were directed toward other problems. Just what took place is not clear. However, Madame Olga Metchnikoff in her *Life of Elie Metchnikoff* (1921) recorded the following cryptic paragraph:

> At first he confined himself to laboratory experiments; then a great landowner, Count Bobrinsky, placed experimental fields at his disposal. As the acquired results were very encouraging, Metchnikoff, forced to leave the neighborhood, left a young entomologist [Krassilstschik?] in charge of the application of his method. So far as he himself was concerned, this study proved the starting-point of his researches on infectious diseases.

I know of no scientific record of the field experiments to which Madame Metchnikoff alluded. From what is known, it appears clear that Metchnikoff's work and ideas (in all probability influenced by those of contemporaries such as Pasteur and de Bary) gave genuine impetus to the idea of microbial control and accomplished and inspired the first significantly practical results to be attained by such means. This was evident a few years later in the work of I. Krassilstschik; in the meantime, Metchnikoff's observations, as had those of Hagen, caught the speculative attention of a few entomologists and biologists (e.g., see Lankester, 1880).

Isaak Krassilstschik, following the lead of Metchnikoff, in 1884 organized a small production plant in Smela. After about four months of operation, fifty-five kilograms of *Metarrhizium* spores were produced (Krassilstschik, 1888). These spores, mixed with find sand, were scattered about in certain field plots in the vicinity of Kiev. After ten to fifteen days, from fifty-five to eighty percent of the *Cleonus* larvae in the plots were dead of green muscardine. In spite of this encouraging beginning, however, the work was not continued, apparently (according to a letter from Krassilstschik to Giard and quoted by Paillot, 1933) because of a rather sudden cessation in the production of sugar beets that made it unnecessary to control *Cleonus*. However, some Russian authors (Rubtzov, 1948; Pavlovsky, 1952) ascribed the abatement of the work to the failure of the method to give consistently successful results, which in turn was caused by a lack of understanding of variations in virulence and the basic epizootiological factors involved.

As a result of the stimulus of Metchnikoff's observations and recommendations, the next two decades were to see the gradual acceleration of interest in the possible use of fungi to control insect pests. In Europe, Brongniart (1888) advocated the scattering of pulverized fungus-infected insects as well as spores of entomophthoraceous fungi among the larvae of flies and other agriculturally important insects as a means of inexpensive control. Similar recommendations were made by Künckel de Herculais and Langlois (1891) with regard to certain grasshoppers. In 1892 the physiologist Franz Tangl (1893) attempted to use the white-muscardine fungus, *Beauveria bassiana*, against caterpillars of the nun moth, *Lymantria monacha*. His laboratory experiments were successful, but in nature the trees sprayed with spore suspensions were not protected against the insect, possibly because conditions of adequate moisture did not prevail at the time. Similar negative results were obtained by von Tubeuf, about this same time, using *Cordyceps militaris*. Among other nineteenth-century European workers who experimented with and wrote on microbial control methods for controlling insects were Giard (1890; 1892*a*, *b*; 1893), Dufour (1892), Prillieux and Delacroix (1891), Danysz (1893), Sauvageau and Perraud (1893), and Trabut (1898*a*, *b*; 1899). Giard's attempts to use *Beauveria tenella* (=*B. densa*) against *Melolontha* were particularly interesting.

Several methods of producing the spores in quantities for field distribution were tried by Giard and others (see also Kellog,

1894). The results of the field trials varied—sometimes the results were excellent and at other times discouraging. The outcome appeared to depend largely on environmental conditions, of which an adequacy of moisture began to emerge as of primary importance. Giard himself concluded that the use of the fungus gave favorable and encouraging results but only under the appropriate conditions. This worker also offered sage advice against unwarranted generalizations on the use of fungi against noxious insects. Especially did he protest dangerous popularizations of microbial methods; he believed strongly that the approach should be one that is serious, careful, and completely scientific—a viewpoint that can still be heartily endorsed. He was not without hope of eventual practical achievements, however, if man but diligently and carefully worked to reveal the secrets of nature involved. Also worth repeating is the assertion by Dufour, who pointed out that the difficulty in using fungi to destroy insects is not in finding the necessary entomogenous fungi (of which hundreds of species abound and are relatively well known) but in knowing how properly to use these fungi to cause epizootics at will.

Meanwhile, in the United States, there was beginning to unfold what was to become one of the best-publicized, and in some respects least-understood, chapters in applied insect pathology. Indeed, in all probability the final pages of this chapter are still to be written. And although man has not succeeded in mastering the use of the fungus concerned, so many basic lessons were learned as a result of the project that the time, money, and effort involved were eminently worthwhile. I am referring to the attempts made in midwestern United States to control the chinch bug, *Blissus leucopterus*, a serious pest of cereal crops, by means of *Beauveria bassiana* (=*Beauveria globulifera*=*Sporotrichum globuliferm*). Most of the salient features of this story have been told elsewhere, so it is my intention here merely to refer briefly to these earlier accounts.

Although a white fungus on chinch bugs was observed in 1865 by Shimer (1867) in Illinois, the first certain record of *Beauveria bassiana* on *Blissus leucopterus* was that by Forbes (1890) in 1887, in Clinton County, Illinois. Shortly thereafter it was reported from Minnesota, Iowa, Ohio, and Kansas. (The chinch bug was first noticed in the United States in North Carolina, in 1783—more than a hundred years earlier.) The first attempt to bring about an outbreak of the disease by artificial dissemination was that of Lugger, in 1888, who scattered diseased bugs about the fields in several localities in Minnesota. Although the experiment appeared successful, Lugger suspected that, since the disease spread so rapidly, the spores of the fungus were already present in the test fields and that he had only reintroduced them.

In 1888 F. H. Snow (1890, 1891, 1894, 1895, 1896), in Kansas, began his work on the chinch bug fungus. The Kansas state legislature established an "experimental station" at the University of Kansas (Snow later became president of the University) to propagate the fungus and to distribute it free of charge. It was placed under Snow's direction. Almost 50,000 packages of the fungus were distributed by this station, but the true value of the program was never determined with certainty. The reports of observers in 1891 and 1892 were very favorable, whereas those made during succeeding years were less favorable. Distribution programs were carried out in states other than Kansas, but in each case the work was eventually dropped. Some states, e.g., Nebraska (see Bruner and Barber, 1894), developed a plan whereby the interested farmer sent in collections of live chinch bugs for which, in return, he received a like number of infected bugs that were relied upon to spread the fungus to others in the fields in which they were distributed. Lugger in Minnesota tried the method again in 1895 but gave it up by 1902. It was similarly abandoned in Illinois, Nebraska, Missouri, Ohio, and Oklahoma.

Although natural outbreaks of the disease were frequently very effective in reducing large populations of the insect, the artificial distribution of the fungus did not appear to affect materially the incidence of the disease or the effectiveness of the control. Possible reasons for this were analyzed in a report by Billings and Glenn (1911), who made a comprehensive study of the disease and the distribution programs, and concluded that, because of its widespread natural presence, the artificial distribution of the fungus was of little or no value. This conclusion was soon echoed in Experiment Station bulletins and continued to be for some years thereafter (e.g., Swenk, 1925). In spite of the abandonment of the program, it is to the credit of F. H. Snow (1840–1908) that he did much to awaken entomologists, as well as farmers, to the potentialities of this type of biological control. (I have letters in my files from people who still remember Snow's work and who still testify that they saved their crops by "artificially" distributing the fungus or fungus-diseased insects provided them by Snow.) It is well to remember that Snow (1891) recommended that the state of Kansas should have "a bacteriological laboratory" in which the relation of bacteria to insects, as well as to human beings and domestic animals, might be studied.

Similar credit must be extended to S. A. Forbes (1844–1930), who for a time nurtured a deep interest in insect pathology. His concern with the diseases of insects antedated his studies on the chinch bug fungus and included bacterial diseases (and unknowingly some virus infections), as well as those caused by fungi (Forbes, 1882; 1883; 1886*a*, *b*, *c*; 1888; 1895*a*, *b*, *c*; and 1898*a*, *b*).[20] His efforts were "directed especially to the point of artificially propagating [the diseases] for the destruction of injurious insect species." His observations on the chinch bug fungus were, in general, careful and discerning, and at several points he anticipated the findings and conclusions of Billings and Glenn. Like Snow, he ably administered the distribution of fungi among growers in a manner designed to obtain the type of effective cooperation required in such endeavors.

Forbes, who organized the Illinois Natural History Survey, is described by Howard (1930) as a born naturalist and as the then "dean of American economic entomologists." "His writings," continues Howard, "have been broad and sound and far-sighted. He is an all-round naturalist and a deep thinker. He was probably the first entomologist in the United States to adopt the word *ecology*. . . . There is probably no American writer on entomological topics who is held in more respect and whose writings are as sound and as broad." (Words similar to these have been used to describe Howard's own writings.) However, Howard, apparently never too impressed with the possibilities of microbial control,[21] very probably did not have in mind Forbes's work with entomogenous fungi and bacteria, when, after Forbes's death, he wrote: "The value of Professor Forbes' work could hardly be overestimated. He was a sound worker and an advanced thinker throughout his whole career, and was a leader among all the American entomologists. . . . His career . . . has dignified the applied science and has helped no end to bring about its present important standing."

Forbes's scientific work was highly diverse in character. Ward (1930) states, "Men speak of him as the first economic entomologist in America, as the leader in the study of aquatic biology and as the founder of the science of ecology." Among the unique facets of Forbes's career is the fact that although he never received a bachelor's degree (he received a Ph.D. from Indiana upon examination and the presentation of a thesis), he was a thoroughly scholarly man. He was well versed in several languages (while he was a prisoner of the Rebels during the Civil War, he studied Greek and Spanish), wrote more than 500 publications in clear and handsome prose, spoke easily and well, was an officer in the Civil War at age twenty, almost completed the course in medicine at Rush Medical College (which fact no doubt enhanced his interest in the diseases of insects), and—last, but not least—was president of the first golf club organized at the University of Illinois. And it should be recorded that this first American general insect pathologist was arrested for speeding while driving his automobile on his eightieth birthday. Incidentally, Forbes was the Founder of Honor at the 1962 meeting of the Entomological Society of America.

There is an account (see Riley, 1883) of a paper presented by Forbes before an entomological club in which he suggested the possibility of using contagious diseases of caterpillars for economic purposes. In this paper he referred to Pasteur's 1869 experiments on the contagious nature of flacherie of the silkworm, indicating that his own ideas for microbial control were stimulated by this earlier work on the diseases of a beneficial insect.

Although the promises hoped for in attempting to control the chinch bug with fungi waned as the twentieth century began, the opposite appeared to be the case with regard to entomogenous fungi found on scale insects and whiteflies, not only in the United States but in other parts of the world. Petch (1921) stated that the earliest record of a fungus parasitic on a scale insect was made in 1848 by Desmazières, who collected his specimens from willow and ash at Caen, France. Nineteenth-century mycologists (e.g., Berkeley, the Tulasnes, Saccardo) were very much interested in the taxonomy of the fungi on these insects. Pettit (1895) issued a bulletin that included a long bibliography on entomogenous fungi including those on scale insects. In the United States, most of the interest centered in Florida, where climatic conditions favored the growth and development of these fungi especially on those insects attacking citrus. H. G. Hubbard (1885) was one of the first to take serious note of entomogenous fungi on scale insects in Florida. Separate observations by H. J. Webber, who, in 1895, discovered the first fungus parasitic on the citrus whitefly (see also Webber, 1897) and P. H. Rolfs, who, in 1897, presented a bulletin describing a fungus disease of the San Jose scale and its use in combating this scale insect in Florida, gave impetus to the flurry of interest and activity in the subject as the century turned.

The nineteenth century closed with the use of microorganisms just emerging as a potential method of controlling insects. The role of microbial pathogens in the natural control of many species of insects was recognized; the secrets as to how nature accomplished this, however, were inadequately known—as, indeed, they are today although to a much lesser extent. Virtually all of the thinking with regard to microbial control methods was concentrated on the possible use of fungi. Although understandable, this was to some extent unfortunate inasmuch as the successes and failures of entomogenous fungi colored the approach that was to be made with other infectious agents.

So it is clear that by the end of the century invertebrate pathology, and certainly insect pathology, were being pursued on both basic and applied levels, though mainly applied. True, except for governmental programs concerned with the diseases of the honey bee and the silkworm, most of the observations of note were isolated instances of observing disease or abnormality in a small group or a single species of invertebrates.

It was not until after World War II that we see insect pathology (and invertebrate pathology) come into its own as a unified discipline. At this point, I hope to have convinced the reader of primarily one point: that insect (or invertebrate) pathology does have historical roots. And, although the science with which we are concerned was not to flourish until the twentieth century, its precursors are to be found from centuries before Christ and up through the enlightenment that began to emerge in the nineteenth century. Now, let us see what the beginning of the twentieth century had to offer, if for no other reason, hopeful that our rendition of what takes place in the continued development of the pathology of animals lessens the "deceit" in Guy de Maupassant's aphorism "History—that excited and deceitful old woman!"

1. This chapter is a revision, with many significant additions, of an article written by the author in 1956 and published in *Hilgardia*, 26, 107–57 under the title of "Microbial Control—The Emergence of an Idea: A Brief History of Insect Pathology through the Nineteenth Century."

2. It is appropriate here to mention that when the principality of Monaco sought to pay homage to Schweitzer in 1961 through the means of a postage stamp, it portrayed over the phrase "Respect de la Vie," a simple outstretched hand holding a common form of insect life—an ant.

3. It is well, at the outset, to have an understanding of the definition and manner in which I intend to use the words *pathology* and *disease* throughout this book. Pathology is a biological science that deals with all aspects of disease; it involves the study of the cause, nature, processes, and the effects of disease. The word is derived from the Greek *pathos*, meaning disease, and *logos*, meaning discourse. (In a more limited sense, usually in medical literature, *pathology* is used to refer to only the structural and functional changes from normal found in tissues and organs. Although somewhat of a redundancy, the word *pathobiology* may conveniently be used to emphasize the broad biological nature of the science. Nevertheless, "pathology" is the "biology of the abnormal.") Then there are a variety of "classical" definitions such as that used by Long (1965): "Pathology has always been only the attempt to understand the nature of these deviations from normal health which we call disease." Moreover, textbook and dictionary definitions abound. The broadest definitions essentially describe pathology to include anything that goes wrong with life and life systems; the narrowest definitions limit the use of the word to the examination and description of diseased cells, tissues, and organs. *Insect pathology* is that branch of invertebrate pathology or of entomology that embraces the general principles of pathology as they may be applied to insects; *invertebrate pathology* has a similar relationship to invertebrates in general.

The word *disease* ("lack of ease") denotes any departure from the state of health or normality. It is a condition or process that represents the response of an organism's (e.g., an insect's) body to injury or insult; a disturbance of function or structure of a tissue or organ, or of the body in general. A disease may be infectious (caused by a living microorganism) or noninfectious (not caused by a microorganism). A healthy insect is one so well adjusted to its internal environment and to its external environment that it is capable of carrying on all the functions ultimately necessary for its maintenance, growth, and multiplication with the least expenditure of energy.

For definitions of other terms used in these pages, the reader is referred to *An Abridged Glossary of Terms Used in Invertebrate Pathology* compiled by E. A. Steinhaus and M. E. Martignoni.

4. On page 485 of volume 2 of the *Chinese Repository* (1834) we find this account expressed as follows: "Near the northeast corner of the western gardens [in Peking] is a temple consecrated to 'the discoverer of the silk-worm.' This discovery is attributed to *Yuenfe*, the wife of the emperor Hwangte, who began to reign, according to Chinese history, in the year B.C. 2636. . . . The empress dowager and other great ladies of the court assist in tending the worms, in order to encourage a branch of industry which is indispensable to the clothing of the inhabitants of China; and the empress herself comes in person to attend the annual sacrifice here 'presented to the genius that protects silk-worms.'"

5. Apropos the origins of sericulture, Hazarabedian (1938) made an interesting comment:

> . . . it was Emperor Fohi, the first Chinese ruler, who some authors contend was Noah himself, [who] discovered this most beneficial insect, the silkworm. However, it was only noted by Fohi, who admired [it] as a wonderful act of God, the sight of a tiny worm appearing on the mulberry bushes, growing wild at that time, and making a ball in which to hide before coming out as a moth.
>
> [Much later] Si-Ling-Chi, the Empress, however, was the first person to take interest in the silken ball and to find a way to unwind it and make silk threads.
>
> Emperor Hoang-ti, the Empress' husband, however, seeing this in 2602 B.C., ordered his wife to try to domesticate the silkworm by rearing it on the ground or inside the buildings. This Si-Ling-Chi, the Empress, did, and so started the industry. She then, under orders from the Emperor, devised means of not only reeling the fibre off the cocoon, but also a way to weave it into garments.
>
> Soon the faithful subjects of Emperor Hoang-ti admired the Empress and started to worship her as a goddess, accepting her as the Silk Goddess.

6. Liu (1952) listed the following pre-Christian era literature references to be "among the classical references": the *Book of Odes*, the *Book of Rites* (governmental policies on sericulture), and the *Chou Li*. There are two *Tsan Ching* [Silkworm classics], an early one by Liu An (22 B.C.), although this authorship is questioned by some scholars, and a later one by Huang Shen-tseng of the Ming Dynasty (1368–1644). Other (A.D.) principal publications include the *Chi Ming Yao Shu* by Chia Sheh, the *Tsan Sa* by Ch'in Kuan (1049–1100), the *Keng Chi Tu* by Lou Shou, the *Nung Sang Chi Yao* published in 1285 by the Ministry of Agriculture of the Yuan Dynasty, and the *Tsan Sang Hoh Pien* by Wan Chi in 1844, translated into English and published by the *Chinese Responsitory*.

7. These early Japanese references were not published in the modern sense, but were handwritten. The date of completion is hard to determine and the number of pages varies depending on the size of the letters. Thus these references will not be given in the main bibliography since all the information is in the text.

8. Bassi, handicapped by failing vision and lack of training in cryptogamic botany, had the assistance of Giuseppe Balsamo Crivelli in determining the nature of the fungus. Balsamo (1835) placed the organism in the genus *Botrytis* and first named it *Botrytis paradoxa*. Later he changed the name to *Botrytis bassiana* in honor of its discoverer. A still later (1912) revision by Vuillemin made the fungus the type species of the genus *Beauveria*.

It might be pointed out that according to Robin (1853), a paper by Lomeni appeared in 1835 claiming that before the reports of Bassi and Balsamo, muscardine was known to be caused by a fungus, and that certain other assertions by Bassi lacked proof.

Examination of records and published reports prior to those of Bassi, however, do not justify these retrospective claims by Lomeni.

9. The manuscripts I was shown were probably those Alfieri (1925) described as having been given to the library by Carlo Besana, who, in turn, had received them from a "modest electrician" by the name of Tomaso Cornelli. Besana had the manuscripts bound in several volumes, attempting to classify them according to subject matter. Not all of the manuscripts were autographic; because of his impaired and diminishing eyesight, Bassi was forced to dictate much of what he wished to record. Even so, according to Alfieri these manuscripts (primarily dealing with muscardine, mulberry trees, and contagion) were inscribed with corrections and marginal notes in Bassi's handwriting. Included with the manuscripts were notes on separate sheets of paper; most of these notes were in Bassi's handwriting, which is very difficult to decipher. In addition, one finds manuscripts by other authors on which have been transcribed observations by Bassi, personal letters both to and from Bassi, and diplomas and certificates of membership in various societies and academies to which Bassi belonged. It is generally acknowledged that many of Bassi's writings have been lost. Incidentally, also in the town hall of Lodi, one can find preserved two small biconvex lenses and a small telescope once owned by Bassi. His compound microscope is probably one of the two kept in the department of physics at the Lodi Lyceum; a large telescope Bassi had constructed in his home was sold to the Barnabite Fathers of Moncalieri in 1860, and later demolished "for reasons of security."

10. Antoine Béchamp was an implacable opponent of Pasteur, with whom he differed on a number of issues, including the germ theory and the nature of the silkworm diseases. He formulated a doctrine of microzymas, microscopic granules he believed to be the basis of life and the essential composition of protoplasm (see Bulloch, 1938). In the case of pébrine of the silkworm, however, it appears that Béchamp appreciated the parasitic nature of the disease before Pasteur did, even though they began their researches on the malady at about the same time. Despite rather fantastic claims made for Béchamp's work by one writer (Hume, 1923), however, Béchamp never really mastered the problem in all its facets. Certain details of his observations and conclusions were in error, so that in the end Pasteur's understanding of the disease and its control rested on a more firm and lasting scientific, as well as practical, foundation. In the case of the malady known as *flacherie*, Béchamp resorted to his microzyma theory, considering the disease to be caused by an abnormal development of the microzymas in the body cells of the silkworm.

11. [Recent taxonomic revisions based on spore structure separate the class Microsporida from the subphylum Sporozoa and place it in the subphylum Cnidospora. MEM.]

12. Concerning this famous entomologist's name it is useful to repeat a footnote concerning its spelling that Wheeler (1930) included in one of his books: "In the zoological literature the name appears in various forms—De Geer, Degeer, de Geer, von Geer, and Geer. Dr. Bequaert, while aiding me in correcting the proof of this book, calls my attention to a paper (Zool. Anzeig. 35, 1910 p. 521) in which E. Clément shows that the entomologist expressly wished his name to be written 'Degeer' because the 'De' is not the French 'de' or German 'von,' but a part of the surname."

13. The earliest observers of muscardine in France, e.g., Boissier (1763) and Pomier (1763), did not consider muscardine to be contagious among silkworms. However, Nysten, in 1808, did consider it contagious under certain conditions. According to Robin (1853), Bonafous, in 1829, observed that muscardine is contagious. Not only silkworms but other larvae (*Phaloena verbasci* Linn.) placed in contact with silkworms dead of muscardine became infected in two or three days.

14. Although I had long been familiar with Bassi's writings and contributions to insect pathology, this particular passage (in Italian) from Bassi's 1836 work was kindly called to my attention by Dr. Enrico Masera of the Stazione Bacologica Sperimentale, Padova.

15. Some authors (e.g., Cooke, 1892) incorrectly stated that Lebert's transmission experiments were done in trees, and implied that the work was done in 1826. Careful translation of the original account (1858, p. 178), however, reveals that the transmission experiments were done not in trees but "in einem Raume." The 1826 date referred not to the time of the transmission experiments but rather to the time when he first observed larvae of *Arctia villica* (=*Euprepia villica*) to suffer from muscardine.
An interesting type of transmission was postulated by Robinet (1845), who, as had others, observed small red "insects," called *Lentes* ("lice"), probably mites or other parasites, running about over the bodies of the silkworms, "stinging" them. He supposed that the spores of the fungus were carried by these arthropods and that infection occurred through the puncture wound.

16. LeConte may have referred here to an observation by Trouvelot (see Hagen, 1879*a*, *b*) in Massachusetts. Shortly after returning from Europe, in 1867, with silk-producing moths that apparently were infected with muscardine fungus, he observed the disease in a population of *Polyphemus* moths (and other species) he was rearing for silk on twelve acres of shrub land.

17. Pasteur was later, in 1887 and 1888, to advocate and test the use of a pathogenic bacterium (the fowl-cholera bacillus, *Pasteurella multocida*) to destroy rabbits. On the Pommery estate near Rheims, where rabbits had become a nuisance in a wine cellar, his assistant, Loir, conducted an antirabbit campaign Pasteur had outlined. Within three or four days after placing the bacteria on cut alfalfa around the burrow openings, thirty-five rabbit cadavers were found, and no living rabbits were in evidence. Later additional dead rabbits were found in the burrows. These promising results induced Pasteur to send Loir to Australia to organize an antirabbit campaign there in response to that government's plea for an effective method of exterminating the animals. This project was never carried out, however, because the necessary authorization was never granted by the Australian Department of Agriculture. In recent years in Australia, and in Europe, the mosquito-transmitted virus of infectious myxomatosis has been used to bring about epizootics in rabbits. In some areas marked declines in the rabbit populations were effected (see Fenner and Day, 1953).
Other efforts have been made to use disease organisms to destroy animal and plant pests. Best known perhaps are those attempts to control rats and mice with *Salmonella typhimurium* (e.g., see Danysz, 1893, 1900; Rosenau, 1901). The idea of destroying noxious weeds by means of pathogens originated at about the same time as did similar ideas of destroying insects, as evidenced by Peck's suggestions along this line in 1876.

18. This paper was first published in *The Boston Evening Transcript* on 11 April 1879, and then appeared in the May 1879 number of *Le Naturaliste Canadien*, vol. 11, pp. 150–55, and then in the June 1879 issue of *The Canadian Entomologist*, vol. 11, pp. 110–14. Following this, Hagen had the paper reprinted, with some revisions and additions, as a bulletin by the Cambridge University Press, December 1879.

19. Like Pasteur, Metchnikoff (1879) advocated the use of microorganisms to control the grape phylloxera. He believed that such control was possible because of Leydig's (1854, 1863) report of "pébrine corpuscules" in *Coccus hesperidum*, which he considered to be a "close relative" of phylloxera. It now appears that what Leydig and others had mistaken for microsporidian spores were in fact the yeastlike symbiotes characteristically present in coccids.

20. For an interesting and valuable list of annotated references to the American literature of insect pathology between the years of 1824 and 1894, the reader is referred to one of Forbes's (1895*c*) excellent reports as state entomologist of Illinois. From this list it is apparent that the American contribution to the beginnings of insect pathology was substantial. A similar, although less complete, list of references also appeared a few years earlier in *Psyche* (Forbes, 1888).

21. Doutt (1964) has recounted the antagonism that Howard held against biological control, particularly that done in California. In 1931 Howard apologetically retracted some of his criticism.

And, step by step, since time began
I see the steady gain of man.

John Greenleaf Whittier

2

Early Twentieth-Century Probes

As I have already indicated, the principal purpose of this account is to present a semihistorical record of the development of insect and invertebrate pathologies during the twentieth century in North America. In the previous chapter a brief and admittedly inadequate summary was given of early emerging ideas and discoveries that hopefully will serve as a latticed foundation for what I shall have to say regarding the early twentieth-century insect pathology. The well-known historian's joke that we know too much about the nineteenth century to write its history is even more applicable when we come to the twentieth century. At least we can try to chronicle some of the events even though we tend to lose objectivity as we approach our own times.

The more one studies the emergence of our understanding of the diseases, the more one is inclined to see the picture of the twentieth century as consisting of two large periods: the one from 1900 through World War II, the other from 1945 to the present. However, there is much interdigitation and intertwining between these two periods, and since I am not a professional historian and thus have difficulty in dividing history according to chronological developments, I shall not claim to proceed in a precise sequential manner. Moreover, this attitude will be reflected in the titles and arrangement of this and the chapters to follow.

Interest in the diseases of insects began centuries ago because of the ravages they caused among beneficial insects, especially the honey bee and the silkworm. To this day the diseases are studied because of their practical importance, as well as the guinea-pig role these domesticated insects so naturally assume. The side effects of these studies—such as Pasteur's work on the diseases of the silkworm—were also pragmatic in that they "led to discoveries which have opened out new departments of science, initiated a new era in medicine, and [given] a new view of the world of life" (Singer, 1959).

Twentieth-century invertebrate pathology began with a strong interest in its applications. There was concern not only about how to suppress disease in beneficial insects but about how to use microbial pathogens to control insects harmful or destructive to man's interests. In the United States, the momentum of the field observations and experimentation of such men as Forbes and Snow at the close of the nineteenth century carried across the century mark the hope that populations of destructive insects could be destroyed by initiating in them certain epizootics. The year 1900 probably represents the approximate date of the high-level mark of the aspirations of these early workers. By the 1930s, interest in the study of insect diseases generally decreased and, except for the work of a lonely few, remained at a low ebb until just after World War II. But if we carefully examine the period between 1900 and 1945, we shall discover that the slow ferment of man's curiosity not only kept alive the flickering flame of knowledge concerning the afflictions of insects—beneficial and harmful alike—but gave rise to several remarkable individual advances. Thus during the first half of the twentieth century a rather firm foundation was laid for the renaissance to come.

At the beginning of the century, one of America's greatest entomologists, L. O. Howard, sounded both the skepticism and the remaining hope that came to be characteristic of the early 1900s so far as microbial control was concerned. Howard (1901) pointed to the fact that the spread of the well-known fungal disease (caused by *Beauveria globulifera*) of the chinch bug (*Blissus leucopterus*) was so contingent upon favorable weather conditions that artificial dissemination of the pathogen was useless. Moreover, if the conditions were favorable the disease would break out naturally because the spores of the fungus were so widespread in nature. This conclusion was subsequently confirmed on a broader base and in greater detail by Kelly and Parks (1911) and by Billings and Glenn (1911). However, Howard went on to say that some work had been done which appeared to indicate that "there may be a practical side" to the artificial propagation and spread of the agents causing disease in grasshoppers. He encouraged research on the practical use of disease agents (especially certain fungi) but felt that the results obtained up to that time did "not justify very sanguine hopes."

1. ROLAND THAXTER'S CONTRIBUTIONS TO INSECT PATHOLOGY

In his 1901 paper, Howard referred to Roland Thaxter and his famous 1888 treatise (his Ph.D. thesis) "The Entomophthoreae of the United States." Most of Thaxter's contributions are of a systematic or taxonomic nature, but he also provided rich amounts of information concerning the general biology of these fungi and, in a detached manner, was well aware of the possible use of certain fungi in the control of insects. Although called the "greatest mycologist of his time" (Weston, 1933), Thaxter was well-versed in entomology, which undoubtedly was a factor in his decision to concentrate on the fungi associated with insects. Of course, Howard's reference to Thaxter's work on Entomophthoraceae was motivated by the fact that some species, such as the type species of *Entomophthora muscae* fatally infected the house fly, and *Entomophthora grylli* destroyed grasshoppers; but Thaxter's treatise was most valuable for its systematics, for he did little work on this important—and now quite large—group after 1888. (I shall defer comments relating to the historical aspects of Entomophthoraceae until a subsequent section.) Altogether he published eighty-eight papers and monographs, all with an excellence and precision rarely equalled. After his Entomophthoreae monograph, he began an investigation of *Cordyceps*, *Isaria*, *Aschersonia*, and several Fungi Imperfecti. But his greatest work was still to come, and it took him the remainder of his long life.

Confronted with Thaxter's magnum opus, one is overwhelmed by its magnitude, its precise and detailed descriptions of species, and its fantastically detailed and clear drawings. The work was modestly titled *Contribution Towards a Monograph of the Laboulbeniaceae*, but actually it consisted of what might be considered a series of five monumental monographs published in 1896, 1908, 1924, 1926, and 1931. Unfortunately, death prevented him from completing the last of his planned "contribution towards a monograph," which undoubtedly would have been an invaluable masterpiece. Ill health and the development of cataracts certainly made the work of his last days more burdensome and slower than otherwise would have been the case. It is sad, therefore, to read in the introductory note of his fifth monograph of the series that the subject matter planned for the fifth contribution "has become so unwieldy that . . . it has been necessary to shorten the text, excluding the largest genus, *Laboulbenia*. . . . Camera outlines for the genus *Laboulbenia* are already finished, and it may be possible at some later time to issue a final Part illustrating this genus as well as other addenda not here included, in connection with a general review, classification and host-index." Although, since 1931 others have reported on species and other groups of Laboulbeniales, I doubt that we have yet recovered the treasure that the Harvard mycologist had planned to present.

In the British journal *Nature* (1932, *130*, 84–85) there appeared a brief obituary statement which alluded to Thaxter's finished and unfinished work as follows: "Thaxter's method was to publish preliminary diagnoses of new genera and species and gather these together in monographs bearing the title *Contributions Towards a Monograph of the Laboulbeniaceae*, which were illustrated with numerous plates of admirable drawings by the author. Part V of the *Contributions* appeared this year, and he was busy preparing a sixth when he died. Some idea of the work entailed in the descriptions and figures may be gained from the fact that the five parts in all run to 1185 pages and 166 plates."

In a letter I received from Professor W. H. Weston in late 1967, I learned that through the services of Dr. I. M. Lamb and others at the Farlow Herbarium, much of what Thaxter was in the midst of preparing for part 6 still remains at Harvard. This material consists of notes, descriptive information, and camera-lucida outlines of new or not previously illustrated species of *Laboulbenia.* Since it appeared that this material still has value, Professor Weston suggested that someone like R. K. Benjamin, already known for his current work on Laboulbeniaceae, "work up" as much of Thaxter's remaining materials as might be possible. However, Benjamin, in 1951 (1967 personal correspondence) had examined some of Thaxter's unfinished work and found it in a state that would make it difficult for another mycologist to complete. The Harvard mycologist had left no manuscripts, and no record of host determinations could be found. According to Benjamin, Thaxter had a very peculiar and drawn-out method of preparing illustrations, hardly usable by anyone other tham himself. But slides and duplicate specimens remained that Benjamin was able to use.

Parenthetically, it should be mentioned that R. K. Benjamin has assembled what is probably the second largest (Thaxter's being the largest) collection of Laboulbeniales in the world. His slide collection numbers well over 6,000, and he has enough unmounted material to more than double this number. Benjamin's interest in Laboulbeniales began while he was a graduate student at the University of Illinois, where he presented his Ph.D. dissertation on them in 1951. As of this writing, Benjamin, at the Rancho Santa Ana Botanic Garden in Claremont, California, intends to continue working primarily on the comparative morphology of the Laboulbeniales, and hopefully to complete an illustrated atlas on the genera and a revised classification of the order.

A generally overlooked value of Thaxter's monographs on Laboulbeniaceae is the fact that, with certain exceptions they constitute monumental works on fungi of no apparent economic importance to man. In general, the Laboulbeniales are commensals, growing and multiplying almost entirely on the external surfaces of insects. It is known that, in some species at least

(Richards, 1953), the integument of the insect is penetrated by the fungus, which may then fill the interior of the insect with extensive rhizoidal processes. On the insect these small, frequently minute ascomycetes appear as scattered or as densely crowded bristles or bushy hairs. They may form a furry or velvety patch and be likened to a skin disease. But overt pathogens they are not. Yet they constitute an intriguing entomophilic relationship that seems automatically to interest any insect pathologist or insect microbiologist who has an opportunity to observe them. That Thaxter could devote so much of his life and effort to a group of parasites of insects without having to worry about their immediate economic value is a cogent testimony to science for science's sake, to the institution that permitted him to pursue this work, and to the man's own attitude toward the quest for knowledge.

On the other hand, Thaxter was more familiar with economically important forms of entomogenous fungi than his publications would indicate. He is known to have speculated and conversed on the possible use of certain entomogenous fungi, such as the Entomophthoraceae, to control insects. I have already pointed out how such economic entomologists as L. O. Howard kept watch on Thaxter's work, and the mycologist very probably inspired the work of Speare and Rorer in this applied field. Moreover, a number of Thaxter's smaller papers dealt with fungi that are true pathogens of insects. Thus he deserves a significant place in any historical account of insect pathology and insect microbiology.

The life and works of Thaxter have been well chronicled by William H. Weston (1933), G. P. Clinton (1937), and others. Thaxter was born August 28, 1858, in Newtonville, Massachusetts, to scholarly and cultured parents. He graduated from Harvard in 1878 as an outstanding member of his class. He received the degrees of M.A. and Ph.D. under W. G. Farlow in cryptogamic botany. His first employment (1888–91) was in plant pathology and applied mycology at the Connecticut Agricultural Experiment Station. In 1891 he returned to Harvard to an assistant professorship, becoming a full professor in 1901. In 1919, after Farlow's death, and at the age of sixty-one, he voluntarily retired, becoming Professor Emeritus and Honorary Curator of the Farlow Library and Herbarium in order to pursue his own research uninterrupted. Thaxter died April 22, 1932, at seventy-three years of age. Aside from being a man utterly absorbed with his work, this complete research orientation might have been somewhat encouraged by the subjective knowledge that he was not the best of teachers in the classical sense, not being a fluent and engaging lecturer. He was at his best, as a teacher, in the laboratory, where he developed in his students qualities of intelligent observation and independent investigation. His grasp of cryptograms as a whole was so great that his associates bemoaned the fact that he would not take the time to author a textbook that would reflect his extensive knowledge. Thaxter traveled widely, especially for collecting purposes, was well known and awarded numerous honors. His biographer, Weston, thought that perhaps his outstanding characteristic was his stern, undeviating loyalty to his work and to its ideals. Although an austere disciplinarian and generally reserved, he had a dry sense of humor that from time to time revealed itself. He was an accomplished musician and a discriminating lover of art and literature. He was strongly opposed to smoking, primarily, perhaps, because he believed it to impair the delicate control of the hand necessary to execute the type of exquisite and detailed drawings characteristic of his own work.

The greater part of Thaxter's collected and described materials remain housed in the Farlow Herbarium at Harvard; most of it is well preserved and catalogued. One day in 1963, having a free hour or two, I dropped in at the herbarium. The time was just after lunch and none of the staff had returned. A young assistant, probably a graduate student, kindly agreed to show me what he could of the Thaxter collection. I was impressed by the fact that the materials and records made by Thaxter reflected the same meticulous care one sees in his publications. While waiting for the assistant to bring me additional material, I noticed a small wooden box in a corner of the room. Out of curiosity I gently lifted the cover and found it filled with pinned specimens of insects from which were growing *Cordyceps* and other entomogenous fungi. They were labeled, apparently by Thaxter, who was designated as the collector, and named. Without thinking, I expressed my horror upon seeing that the material had deteriorated badly, mostly from the invasion of dermestids. My embarrassed host quickly and apologetically took the long-forgotten material from me, but my eyes caught the designation of one or two paratypes and, possibly, even a type specimen. This is an example of how often valuable material, even that of a man as great in science as Thaxter, is left unattended and forgotten after a scientist's death. I had a similar experience with scientific material left by the eminent French insect pathologist Paillot, which I shall recount later.

2. THE FLORIDA STORY OF THE USE OF THE "FRIENDLY FUNGI"

While Thaxter was producing one of America's greatest series of monographs describing a fascinating group of entomogenous fungi, the wave of interest in the practical use of fungi against insect pests (generated by the nineteenth-century work on the control of the chinch bug) maintained itself in another part of the country. In Florida, considerable attention was being given the apparent widespread destruction of scale insects (coccids) and whiteflies (aleyrodids) by fungi. Similar observations were being made in other parts of the world (see Parkin, 1906), but the situation in Florida was, and is, in many ways unique. It was the only general geographical area in the United States that provided the ideal warm, humid conditions for the growth and development of entomogenous fungi attacking the principal insect enemies of citrus plants. (Attempts to use fungi in a similar manner in California citrus orchards generally failed because the humidity is low and there is virtually no rainfall during the summer months when the temperatures would be conducive to the growth of fungi, and during the rainy winter months the temperatures are too low.) The possible promise of these fungi (*Aschersonia*, *Aegerita*, *Verticillium*, *Sphaerostilbe*, *Podonectria*, *Myriangium*, and others) loomed gradually greater in the minds of Florida growers and scientists during the early decades of the century in spite of the discouraging conclusions relating to the artificial distribution of other fungi to control the chinch bug in central United States.

As indicated in the last chapter, attempts to control citrus pests by means of fungi in Florida followed on the late nineteenth-century observations of such men as Hubbard, Webber, and Rolfs. The Florida entomologist E. W. Berger and his associates issued a series of reports beginning in 1906 (see also 1910, 1919, 1921, 1932, 1938, and 1942; Watson and Berger, 1937) encouraging the use of the so-called friendly fungi and reporting the successful control of scale insects. Similar encouraging results were reported from the West Indies by South in 1910. In 1908, H. S. Fawcett impressively described fungi parasitic upon the citrus whitefly, *Dialeurodes citri*, and implied the practicality of their use in the control of the insect. Indeed, prior to 1908, Fawcett had pioneered in growing the fungi in artificial culture in order that they might be distributed in a fashion more dependable than that provided by nature. About this same time Berger was perfecting a method of spraying fungal spores into whitefly infested trees. Inasmuch as in these early days of the work the Florida State Plant Board provided no funds to cover the expense of growing the fungi, they were dispensed at a nominal charge of seventy-five cents per culture. Complete instructions as to how to introduce the fungi were provided by Rolfs and Fawcett (1913) and Berger (1919, 1923).

However, not all entomologists and growers were as convinced or as enthusiastic about the efficacy of the friendly fungi in controlling citrus pests as Berger and his associates. The "inefficiency" of *Sphaerostilbe* (red-headed fungi) was reported as early as 1899 by F. S. Earle; and Forbes (1899), after visiting Florida, found, on the basis of experiments in Illinois, the fungi to be erratic in their coverage, and suggested that possibly only those scales comparatively deficient in vitality were actually parasitized and destroyed. A. W. Morrill, special field agent of the federal Bureau of Entomology, was a strong advocate of the use of chemical fumigation and achieved considerable success with his "tent" method of fumigation in Florida during the years 1906 to 1908. He and Back in 1912 published a bulletin on the natural control of whiteflies in Florida. They contended that the fungi could not be depended on to give satisfactory results and that chemical remedies should be relied on instead. They did, however, mention that, under certain circumstances, such as in citrus groves located in low-lying hammocks where the use of insecticides would be impractical, fungus parasites might be used to an advantage.

Years later, in evaluating the differences in the conclusions reached by Berger and by Morrill and Back, Fawcett (1944) suggested that the latter workers experimented at a period of the approximate nonsaturation point of spores for infection of whiteflies, whereas Berger probably experimented at periods of saturation in possible infection for prevailing conditions. Following this line of reasoning, a similar statement may apply to the contradictory results obtained in the case of scale insects and the attempts to enhance their control through the agency of fungi. Nevertheless, the work of Morrill and Back stands as an excellent attempt to examine critically a baffling and contradictory situation. They undoubtedly were representatives of chemical control advocates that to this day find zealots of biological control somewhat difficult to tolerate. Nonetheless, they ended their important bulletin with what in subsequent years has become a classic disclaimer: with more advanced knowledge effective use may be made of natural control methods. In addition, they left hanging a fascinating concept regarding mortality of whiteflies that could not be explained. Although they hinted that pathogenic bacteria might be involved, they designated this phenomenon by the obvious term "unexplained mortality." Fawcett (1944, 1948) has suggested that this unexplained mortality could be due to the fact that fungi frequently have the capacity to invade an insect and develop within it to a point where the insect is killed; but because of dry atmospheric conditions, the stroma of the fungus does not burst through the integument in a visible manner as is the case when the air contains ample moisture.

It is worth digressing to reflect a moment on the scientific interests of Howard S. Fawcett because much of insect pathology, both in this country and abroad, was actually performed by plant pathologists and mycologists—even when it came to the use of pathogenic microorganisms (mostly fungi) in the control of insect pests, although most of them collaborated with entomologists. Fawcett concerned himself with entomogenous fungi primarily during the years 1907 through 1910, and then in 1944 and 1948 wrote two lengthy reviews on fungal and bacterial diseases of insects as factors in biological control. During the intervening years he was involved with the more orthodox activities of a plant pathologist, leaving Florida in 1911 to become the California Comissioner of Horticulture from 1911 to 1913 and then a professor of plant pathology at the University of California from 1913 until his death in 1948.

I was fortunate to become acquainted with this very interesting man during his last years on the Riverside campus of the University of California, where he was energetically at work at the Citrus Experiment Station. This covered only the short period from 1945 to 1947 when I made occasional visits to Riverside from the Berkeley campus. Although a generally kind and busy man, he seemed to be especially gracious to me and generous with his time because of my expressed interest in the diseases of insects. His qualities as a Quaker were evident—he had gone to Russia in 1922 to aid in the relief work there by the Society of Friends—and the fact that he had passed the official retirement age of sixty-seven did not seem to slacken his activities or scientific interests. While he conceded that the climatic conditions of southern California would make the use of entomogenous fungi difficult in the control of citrus pests, he was convinced to the end that in Florida the friendly fungi not only served as effective natural enemies of whiteflies and scale insects but had great potential as control agents if properly disseminated by artificial means. The fact that the fungi lived on insects rather than parasitizing the citrus plants themselves seemed to pose no conflict of interests for this well-known plant pathologist. Fortunately, the same may be said for other plant pathologists and mycologists who "held the fort" until the advent of the insect pathologist in his own right.

In Florida during the early years of the twentieth century and in some quarters up to the present, despite the differences of opinion that prevailed, the interest of growers and entomologists alike was maintained in the practical use of fungi. Efforts to distribute the whitefly fungi (*Aschersonia aleurodis*, *A. flavocitrina*, and *Aegerita webberi*) artificially throughout the citrus-producing areas of Florida were made by the Florida Experiment Station (see Rolfs and Fawcett, 1913) and by private agencies that offered the fungi for sale. Growers usually obtained scale fungi, which were not produced commercially, from neighboring groves. Orchardists were convinced that it was to their advantage to have the fungi, in adequate numbers, in their groves whether introduced naturally or artificially. The conviction was supported by the advice of the entomologist (Watson, 1923) of the Florida Experiment Station, who asserted that it was very important that the grower have a good supply of these fungi in his grove and that if they were not already present in abundance, "it will pay him to make a particular effort to introduce them." Citrus growers in California also became interested in the role of fungi in the control of citrus pests. As related earlier, climatic conditions unfavorable for the growth of these entomogenous fungi kept the interest somewhat subdued. Later in this volume, we shall take note of interesting attempts by private individuals to promote the use of the fungi in spite of their unlikelihood of success in California.

Suspicions questioning the efficacy of these fungi and the early claims made for them are seemingly sustained by the work of Holloway and Young (1943) on the purple scale, which, on a somewhat different basis, supported the conclusions of Morrill and Back (1912) with respect to whiteflies. An interesting relation between the efficacy of the friendly fungi and the spraying of bordeaux mixture (copper sulfate, lime, and water) had been noticed by a number of observers (e.g., Fawcett, 1912; Winston et al., 1923; Hill et al., 1934; Osborn and Spencer, 1938). It appeared that the use of the fungicide to treat certain diseases of citrus trees themselves also killed the entomogenous fungi. Thus the use of bordeaux mixture caused a marked increase in the number of scale insects on the trees. Other workers found somewhat similar results to occur after the use of certain other chemicals, such as copper sprays and sulfur.

On the other hand, W. L. Thompson (1939, 1940, 1942) on the basis of a spraying program using copper and zinc compounds concluded that the red-headed fungi had little or no effect in controlling purple and Florida red scales. He also concluded that general tree vigor was a contributing factor. Incidentally, although I find no published report to substantiate it, there seems to be little doubt, on the basis of what I have been told, that it was Thompson who, in 1931, did the first work on the residue theory. He apparently also was of some help to Ziegler (1949), who was interested in comparing the effect of a so-called nonresidue-forming copper fungicide mixture called Coposil with those of bordeaux mixture in 1933.

While studying the influence of fungicidal sprays on entomogenous fungi on the purple scale in Florida, Holloway and Young (1943) obtained data indicating that the scarcity or abundance of fungi did not influence the rate of total mortality of scale insects. Furthermore, despite the claims of some earlier workers, no abnormal increase in scale insects appeared to be associated with the fungicidal properties of the sprays, but increases were instead associated with the residues from applied materials. These residues apparently either interfered with the activity of natural enemies or aided the young crawlers to attach securely to the leaves rather than, as is frequently the case, fall to the ground where they perish. Although he expressed certain qualifications to the idea, Fawcett (1944) acknowledged that there was some merit to the claims of a number of workers in Florida that road dust, lime, and other residual materials not considered as fungicides can cause large increases in purple scale insects and spider mites, and that "not all the increase of scale insects following bordeaux spraying is due, as previously thought, to a killing of the [entomogenous] fungi." Others (e.g., Griffiths and Fisher, 1949, 1950; Griffiths and Thompson, 1950) continued the interest in the role of residues and of orchard-growing conditions on the insect and mite problems of citrus in Florida, mostly reflecting diminished importance of the friendly fungi—especially the red-headed fungus, *Sphaerostilbe auranticola*, and the "pink fungus" (actually usually yellowish white to pinkish red, with the perfect form orange red or bright red), *Nectria diploa*. These Florida workers reached somewhat different conclusions than did Holloway and Young; they found that any of the materials used resulted in increases in purple scales that were greater than would have occurred on unsprayed trees and that where amounts of residues were similar, the chemical nature of the residue is the determining factor in the extent of the duration of the abnormally high populations. At the 1949 meetings of the Entomological Society of America, the Florida workers were critical of some of the methods used by Holloway and Young. However, I was the beneficiary of verbal rejoinders Holloway made of some of the latter work done in Florida. Also, attached to a reprint I have of a paper by Fisher and Griffiths (1950; see also Fisher, 1950*a*) on the fungicidal effect of sulphur on entomogenous fungi found on purple scale are notes by Holloway questioning some of the conclusions of the Floridians and alluding to what he considered contradictory statements. (For example, the authors found sulfur paste to be very fungicidal, and trees treated with it showed severe scale infestations; whereas Holloway, considering their data on different sulfur compounds, including lime-sulfur, thought that the residue effect of the compounds was more important than the fungicidal effects against an endoparasitic fungus, *Myiophagus*.) By this time, Holloway was engaged in his classical work on the control of weeds with certain insects and did not feel inclined to express his reservations in print; moreover, at the time he and Young did their work on residues, the existence of the fungus *Myiophagus* in the purple scale was not known. Interestingly enough, however, in spite of disagreements with regard to some of the details involved in the role of fungicides, residues, and *Myiophagus* in determining scale populations, Holloway and the Florida investigators were not far apart in their general conclusions and in their evaluation of the role of the so-called friendly fungi. One can only regret that Holloway and the Floridians were not able to get together to resolve their conceptual differences (never expressed in print) by collaborative experimental work.

Examination of the data obtained by Holloway and Young also reveals a strong indication that mortality of the scale insects during Florida's rainy season, the time when the entomogenous fungi flourish, was associated with the wet weather rather than due directly to the fungi. There was an implication in their data to the effect that certain entomogenous fungi did not invade the living healthy insect but attacked only those that have been weakened or made unhealthy by other influences associated with wet weather or with certain unknown extrinsic factors. Such a conception is reminiscent of views held more than 100 years ago when certain observers (e.g., Gray, 1858, p. 19) expressed the belief that the development of entomogenous fungi "does not depend altogether on being nourished by warmth and moisture . . . but rather on the insect becoming sickly and feeble by the effect of heavy rains that fall at stated periods in the intertropical regions, or from the extremely humid seasons which prevail occasionally during certain months of the year in most extratropical countries." A somewhat similar view was expressed by Forbes in 1899, as well as by others.

Here, again Fawcett (1944, 1948) attempted to reconcile the several opinions as to the nature of the mortality of scale insects. He suggested that there "is a complex of possible fluctuating factors that need to be unscrambled by experiments with controlled conditions for the insects, for the fungi themselves, and for the complex fungus-insect relationship, before it can be decided what part is played by the deposits or residues from applied materials, by nutrition of the tree and thereby nutrition of the insects, and by parasitic organisms."

Then, in 1947, there appeared in the *Annual Report of the Florida Agricultural Experiment Station* a brief, but fascinating, account titled "Insect Disease Studies" by a young assistant plant pathologist at the Citrus Experiment Station at Lake Alfred. Miss Fran (Francinia) Fisher, to whom we have already referred with regard to the matter of residues, began her report with the declarative statement "No evidence has been found that any of the so-called 'friendly fungi' actually parasitize any of the scale insects." She went on to say, "Numerous recordings have been made of the occurrence of *Microcera* sp. ('Pink and Red-Headed' Fungi) on the dead bodies of red scales, purple scales, chaff scales, and long scales. *Microcera* sp. and *Podonectria coccicola* have also been recorded growing entirely on armors of purple scales which had been killed by the endoparasitic chytrid, *Myrophagus* sp." *Myiophagus*, meaning "devourer of flies," was originally misspelled by Sparrow, hence the incorrect form *Myrophagus* in some of the early literature.) In a subsequent publication Fisher (1950*b*) made it clear that the fungus was first found parasitizing purple scale insects on citrus by W. L. Thompson in 1934 at Babson Park, Florida. Moreover, in addition to the purple scale, she also reported it in Florida red scale, *Chrysomphalus aonidum*, the chaff scale, *Parlatoria pergandii*, and the Glover scale, *Lepidosaphes gloverii*. Fisher's 1947 statement not only lent support to those who questioned the value of the friendly fungi as control agents, but introduced an entirely new explanation for some of the natural mortality found in scale insects—especially in the purple scale, *Lepidosaphes beckii*.

About the same time, Waterston (1947) discovered a chytridiosis in females of the purple scale, as well as of *Lepidosaphes newsteadi*, on citrus and cedar trees in Bermuda; according to Karling (1948), Waterston also found the disease affecting the oystershell scale, *Lepidosaphes ulmi*, in Ontario, Canada. Karling considered the species of chytrid concerned in both the Florida and Bermuda findings to be *Myiophagus ucrainicus*, first observed by Wize (1904) in Coleoptera (*Cleonus* and *Anisoplia*) larvae and pupae collected in the Ukraine. Sparrow (1939) found apparently the same fungus in dipterous pupae collected in 1902 by Thaxter (who had made notes on, and sketches of, the fungus) in Maine in the United States. Similarly, Petch (1939, 1940*a*) in 1934 found in Ingeleborough, Yorkshire, England, the same organism parasitizing dipterous pupae. (In our own laboratory after studying specimens from Bermuda and Florida, we found fungi indistinguishable from the latter in purple scale sent to us for disease diagnosis from Hawaii by P. W. Webber, and in the lepidopteran *Chilo suppressalis* sent to us from Japan by I. Tateishi [Steinhaus, 1951; Steinhaus and Marsh, 1962]. We were also able to infect experimentally nearly mature California red scale, *Aonidiella aurantii*, with the material obtained from Hawaii.) The chytrids belong to the order Chytridiales, which includes some of the more primitive of the phycomycetous fungi.

As indicated, Miss Fisher was not the first to observe friendly fungi growing on the dead bodies of scale insects in Florida. W. L. Thompson (unpublished manuscript referred to by Fisher, 1950*b*) presented evidence to show that the red-headed fungi were not parasites but were only saprophytes living on the armors and dead bodies of the insects. He noticed that large numbers of purple scales had been found with the red-headed fungi growing on the armors but not attacking the living bodies of the insect. Another individual who found the red-headed fungus "growing on the empty pupal cases of male purple scale . . . [and] in a few cases on the scale covering live healthy purple scale" was L. W. Ziegler (1949), who concluded that "this organism is not an active pathogen. . . . The theory of fungus control of purple scale appears to have no scientific background." The history of the publication of Ziegler's paper is a fascinating one and probably reflects some of the strong differences of opinion and power existing in Florida at the time. The paper, titled "The Possible Truer Status of the Red-Headed Scale-Fungus," was published in *The Florida Entomologist* in 1949, with its first page carrying a footnote, "Manuscript received August, 1935." Probably a record for the length of time between receipt and publication by any scientific journal! At the end of the article is an interesting editor's note indicating that the publication of a paper on the fungus diseases of scale insects by Fisher, Thompson, and Griffiths (1949) appearing earlier in the same year in the same journal reminded "some" that Ziegler's manuscript had been sent to *The Florida Entomologist* for publication. A search of old files turned up the manuscript of the paper that had been read before the 1935 meeting of the Florida Entomological Society. Since the paper "foretold some facts which have since been substantiated by the work of Fisher, Thompson, and Griffiths," the editor felt it only fair to publish the paper even though fourteen years had elapsed. One cannot help but wonder just how or why a paper that would, in 1935, be quite controversial (perhaps Watson and Berger would have found it so?) could become "lost" or why, on the other hand, the author did not in the meantime demand its return for publication elsewhere.

Back to Fisher's observations on the endoparasitic *Myio-*

phagus and the implication that, in some cases at least, the well-known friendly fungi probably merely overgrew scale insects that already had been infected and killed by the endoparasite. In the 1948 *Annual Report of the Florida Agricultural Experiment Station*, she related her findings that *Myiophagus* infection occurred in samplings from throughout virtually the entire citrus belt of Florida and that the incidence of the disease and the ensuing mortality were greater in the purple scale than in the Florida red scale. The disease was present in groves that were sprayed with fungicides as well as in unsprayed groves; however, during the summer concerned, the "effectiveness" of the fungus in sprayed groves was delayed until almost all of the fungicide had been washed off the trees. Further reports (Fisher, Thompson, and Griffiths, 1949; and Fisher, 1950*b*) had the ring of conclusiveness as far as the authors were concerned: *Myiophagus* sp. (Fisher hesitated to accept Karling's assumption that the fungus she was working with would prove to be *M. ucrainicus*, although she had no concrete reason to doubt it.) was a true parasite and was acting as a factor in the biological control of purple and Florida red scales on citrus in Florida. However, dependence on *Myiophagus* as a natural control agent was not advocated; its effectiveness depended too greatly upon the density and distribution of the scale population and upon the presence of moisture. Inasmuch as mass-production methods of growing the fungus had not been developed, no conclusion could be made as to its potential when artificially applied. After two years of testing, no evidence was obtained to show that the red-headed fungi *Sphaerostilbe auranticola*, *S. flammea*, and *Nectria diploa* were parasitic on red and purple scales. In the course of this work, Fisher (1950*c*) discovered two fungi, *Hirsutella besseyi* and *H. thompsonii*, that appeared to be parasitic on purple scale nymphs and rust mites respectively.

By 1952, Fisher determined that *H. besseyi* was a pathogen of the purple scale; indeed, she stated that after six years of investigation, while she could find no pathogenic properties associated with the famous red-headed fungi, she did find three organisms infecting scale insects and causing their death. Two are fungi, *Myiophagus* and *Hirsutella*, and "one appears to be a bacterium." The bacterial disease appeared to be widespread in Florida and up to that time had been found attacking only Florida red scale. She concluded that fungal diseases (apparently caused by the two fungi just mentioned) "materially aid growers by reducing scale populations in both sprayed and unsprayed groves. For this reason they play an important part in the control of scale insects from a practical standpoint. Low populations are far easier to clean up with a scalicide than high populations. Because of uncertain weather conditions growers should not rely entirely on diseases, predators, or insect parasites for scale control. Until more is known about these natural enemies, growers should continue to use insecticides to prevent damage by heavy scale infestations." In 1961, Muma and Clancy reported that the seasonal incidence of disease caused by *Myiophagus* was high during one summer and fall when they studied it, but this peak did not occur during the succeeding two years. They considered the disease of crawlers caused by *H. besseyi* to be "of minor importance" as a natural control factor. This last conclusion conformed somewhat with a statement made previously by Muma (1955) that although *H. besseyi* is common during peaks of crawler activity "and undoubtedly plays a part in the natural control of the purple scale, it has not been proved to be directly associated with reductions in scale infestations." In the same paper Muma found chytridiosis to be prevalent in the relatively dry summer of 1945, and scale mortality was high.

As of this day in 1968 as I write these words, questions put to entomologists and others in Florida bring forth no certain knowledge of the overall role of all species of friendly fungi on all species of scale insects or whiteflies. There is no doubt, however, that the general consensus is that the friendly fungi of pre–World War II days are indeed soprophytes, that they grow primarily on the dead remains of the insects concerned, and that their presence is merely an indication of the degree of mortality, from other causes, in populations of scale insects. Any conditions or treatments that increase the number of dead scales also increase the visible "infection" by the fungi. Entomologists at Lake Alfred, though finding *Myiophagus* infection a factor in making grove evaluations, have been unable to produce epizootics with this fungus either in the laboratory or in the field. On the other hand, without specifying the fungi, there are entomologists in Florida that tell me that "fungi are of great importance in keeping whiteflies, scales, mealybugs, aphids, and mites at levels with which we can live. Except for the insect parasites *Aphytis lepidosaphes* and *A. holoxanthus* on purple scale and Florida red scale respectively, the fungi are probably more effective biological control factors than the arthropods. Whitefly populations seem to be held to rather constant low to moderate level throughout the year by fungi" (personal correspondence). Currently, entomologists at the Lake Alfred Citrus Experiment Station recommend against "the excessive use" of sulphur, copper, zinc, and other inhibitors of entomogenous fungi (Simanton, 1967, personal correspondence).

If Miss Fisher and her contemporary workers at the Florida Citrus Experiment Station are correct in their conclusions regarding the inefficiency of the friendly fungi, the efficiency of *Myiophagus* (which completes its life cycle within the insect's body), and the role of fungicidal sprays and spray residues—has the house of Rolfs, Berger, Fawcett, Watson, and others largely collapsed? What a change from 1908, when Rolfs and Fawcett could confidently say, "There can no longer be any question" as to the efficiency of *Sphaerostilbe* and certain other fungi in controlling scale insects in Florida! What of the positive declarations of hundreds of orchardists and growers who had confidence in, and treated their groves with, fungi to control both scale insects and whiteflies? If the friendly fungi, and their many relatives, do not in fact parasitize the insects upon which they are found, what causes them to be entomophilic? What is the basis of the entomic association of all the allied species of fungi described by Petch and others? Where does this leave us with the fascinating story of Florida citrus and its entomogenous fungi—pathogens and nonpathogens? I sincerely hope that my readers are less perplexed than am I. There appears to be little reason to doubt the essential thrust and principal conclusions of Fisher and her associates; but we must remember that they did not study the pathogenicity of *all* species of friendly fungi on *all* species of citrus insects, under *all* environmental conditions. So, there is indeed a great deal more research required before we can know the full and true role of the entomogenous fungi associated with these insects. Until this research has been accomplished, the Florida story remains incomplete, controversial, and irresistibly tantalizing!

Interestingly enough, the fungus described by Fisher (1950) as *Hirsutella thompsoni*, and early reported as a pathogen of the citrus rust mite, *Phyllocoptruta oleivora*, by Speare and Yothers (1924), Yothers and Mason (1930), and Fisher, Griffiths, and Thompson (1949) apparently exerts effective control of the arthropod when conditions are favorable (Muma, 1955, 1956). However, I have had, in 1968, reports from entomologists who feel that whereas the citrus rust mite is at times greatly reduced in numbers by *H. thompsoni*, it is seldom held below economic level and apparently chemical control for this arachnid is necessary much of the year.

Speaking of mites, the Texas citrus mite, *Eutetranychus banksi*, in Florida has been found infected with a fungus described by Weiser and Muma (1966) as *Entomophthora floridana*. Apparently this is the same species originally recorded by Fisher (1954) on this mite but is not the same as the unidentified species (1951) she reported as attacking the citrus red mite, *Panonychus citri*, although the latter arthropod, as well as the six-spotted mite, *Eotetranychus sexmaculatus*, are susceptible to *E. floridana* in the laboratory (Selhime and Muma, 1966). Another mite found infected with a fungus is the red-legged earth mite, *Halotydeus destructor*, in western Australia; Petch (1940*b*) observed the infected mites and later named the fungus *Entomophthora acaricida*. However, few such observations have been made in North America.

One cannot leave an account of the friendly fungi of the citrus orchards of Florida without acknowledging the important role of the systematic mycologist in attempting to properly classify these organisms. Some of this work was accomplished by those (e.g., Fawcett, Fisher) we have already mentioned;

much of it was done by nineteenth-century taxonomists (e.g., Berkeley, the Tulasnes, and Saccardo). But several important corrections and a great amount of taxonomic labor on the fungi found on scale insects generally were rendered by the British mycologist Tom Petch (1921), who in 1920 chose as the subject of his presidential address before the British Mycologial Society "Fungi Parasitic on Scale Insects." Petch is worthy of mention in a book on the history of North American insect pathology not because on rare occasions he worked with material from this continent but because the effects of his work were felt by all those concerned with entomogenous fungi and especially fungi associated with scale insects. He was born in 1869 (Gadd, 1949) or 1870 (Brooks, 1949), lived a fruitful and productive life of seventy-eight or seventy-nine years, and died on Christmas Eve in 1948 at his home in Wooton, England. A constant stream of papers emanated from his pen, published chiefly in the *Transactions of the British Mycological Society*. Short biographical accounts of Petch have been published in *Nature* by Brooks and in *Tea Quarterly* by Gadd.

I can vouch for the accuracy of Gadd's statement (1949) that whatever Petch wrote, "whether in official correspondence or for publication, was precise and clear, occasionally trenchant." My contacts with Petch were by correspondence, which began in a manner he must have considered brash. In 1944, while preparing a textbook, I wrote him requesting a set of reprints of as many of his papers as might be available. Four months later I received a brief handwritten note saying, "I regret that I am unable to comply with your request. My stock of earlier separates is exhausted, and my later ones are reserved for dispatch when conditions permit. . . . As my aim has been to straighten out the systematic side of entomogenous mycology, you would probably not find much suitable for your purpose in my papers." His point was well taken except that I was interested in his work; for in the process of attempting to organize our Laboratory of Insect Pathology, I was hopeful of building as complete a library as possible. A year later I had occasion to write Petch again; this time I offered to purchase any extra reprints he might have of his work. No reply was forthcoming.

Then, quite unexpectedly, one day in February, 1947, I received from him virtually a complete set of reprints. I was surprised, delighted, and grateful. Late, in October of the same year, he patiently wrote a detailed letter providing me with his records of the number of species of *Hypocrella* with their corresponding *Aschersonia* (*Aschersonia* is the conidial form of *Hypocrella*). The letter poignantly began, "Since 1940, owing to the infirmities of old age and travel difficulties due to the war, I have been unable to make periodical visits to London [from King's Lynn] to keep in touch with mycological literature." However, the remainder of his long letter gave little evidence of his then being seventy-seven years of age. Taxonomic descriptions and revisions that occupied most of Tom Petch's life may not sound like exciting stuff to the sophisticate, but his works stand as a testimony of a man who not only must have loved his work but who worked to serve others in a manner and in a field that attracted few and has benefited many. He was not particularly interested in disease itself as caused by entomogenous fungi and saw little value in their use as agents to control insect pests. His quiet, retiring disposition, his distaste for publicity, and his preference for working quietly and unobtrusively in the background undoubtedly were to some extent responsible for the fact that his work was little known outside of a small group of taxonomic specialists.

3. *Beauveria bassiana* AND ITS MANY ROLES

While so great an amount of attention was being focused on the friendly fungi in Florida by mycologists, entomologists, and plant pathologists, what had become of the work on white muscardine[1] caused by *Beauveria bassiana* (=*Beauveria globulifera*) that Snow, Forbes, and others—just before and at the turn of the century—had hoped to use to control the chinch bug in the Midwest and other regions? Did the 1911 reports (published one week apart) by Kelly and Parks and by Billings and Glenn (later confirmed by a similar study by Headlee and McColloch, 1913; Packard and Benton, 1937) to the effect that the fungus was already so widely distributed in nature that artificial distribution of the fungus was of little or no value and that epizootics would break out naturally if the conditions of humidity and temperature were right put an end to the hopes of the practical use of the fungus by man? Apparently they did in the control of the chinch bug. In fact, so far as microbial control (if not insect pathology itself) is concerned, except for a few individuals, the period between 1910 and 1944 might be considered the period of "unenthusiasm." In spite of the great names in the United States (e.g., J. L. LeConte, Hagen, Howard, Snow, and Forbes) that at one time lent their prestige to the idea of using microorganisms to control insects, most entomologists during this period became discouraged with this particular application of insect pathology or they considered it a fanciful, fringe hope; consequently, there was not as much concerted effort made to study the diseases of insects as might otherwise have been the case. A story told by F. F. Dicke, who was with the USDA Bureau of Entomology, illustrates the attitude of these times. According to Dicke, in the early 1930s, a congressman became very enthusiastic about the prospects of microbial control of insects and was putting pressure on government officials to begin work on this aspect of control. The officials were not only uninterested but also against microbial control; but a crusading congressman is not to be ignored, so Dicke was placed in the library and told to compile a bibliography of grasshopper diseases. This was done so that thereafter when the congressman confronted the USDA officials they could reply, "Oh yes, we have a project going on that."

Of course, as it was with the diseases of the silkworm abroad, so in North America, considerable effort was made in investigating the diseases of the honey bee. Several outstanding pieces of research in insect pathology were accomplished by individuals or as small projects by scientists in different parts of North America interested in the same goals. However, just as the monks of old kept the thread of literature and knowledge intact through the Dark Ages, so too were there a few in North America and abroad who continued the "silent vigil" of studying the diseases of insects primarily because of their deep basic interest in the subject.

Despite the discouraging reports by Billings and Glenn and others, *Beauveria* was not entirely forgotten. In Europe and the Orient it still was a mortality factor in the rearing of silkworms, and in the United States—even in the case of the chinch bug—it was recognized as one of the most destructive enemies of certain insect pests whenever appropriate conditions of moisture and temperature prevailed. Moreover, there were those (Speare, 1912; Speare and Colley, 1912; Glaser, 1914) who quickly pointed out that merely because a particular fungus had failed as a man-manipulated control agent against the chinch bug, there was no reason to assume that other fungi pathogenic for other insects would likewise fail.

By this time considerable uncertainty had arisen as to the correct classification and nomenclature of the fungi that caused white muscardine in insects. Were all cases of this muscardine caused by the same species of fungus, by different species, by different varieties, or by combinations of these alternatives? In all likelihood we still do not have the final answers, nor do we know precisely the number of species belonging to the genus *Beauveria*. Fortunately, a comprehensive study by the Canadian Donald M. MacLeod (1954) has simplified the nomenclature for the time being. After examining numerous isolated strains of fungi that had been placed in the genus *Beauveria*, or its close relatives, he concluded that fourteen species possessed the characteristics that placed them in this genus. These fourteen species he placed in two species, *Beauveria bassiana* (Balsamo) Vuillemin and *Beauveria tenella* (Delacroix) Siemaszko. Thus synonymous with *B. bassiana* are *B. stephanoderis*, *B. laxa*, *B. globulifera*, *B. effusa*, *B. doryphorae*, *B. delacroixii*, *Botrytis acridiorum*, and *Isaria vexans*. The following he considered to be strains of *B. tenella*: *B. densa*, *B. brongniartii*, *B. shiotae*, and *Botrytis melolonthae*.

Although this listing of species may appear a bit pedantic for our book, it will help to clarify some terminology that we will have to deal with later. At the moment, however, it is worth the reader's attention to remember that Balsamo first studied and described the fungus that caused white muscardine of the silkworm, and by placing it in the closely related genus

Botrytis, and with the specific epithet *bassiana* he honored the pioneering work of Bassi (chapter 1). In 1911, Beauverie (1914) studied the fungus of muscardine and decided that it was similar to another strain, *Botrytis effusa*, found on silkworms, but distinct enough from the type species of *Botrytis* (*B. cinerea*) to be placed, along with *B. effusa*, in a separate genus. In 1912, Vuillemin created a new genus, naming it *Beauveria* after Beauverie and made *Botrytis bassiana* the type species; thus we end up with the present name of *Beauveria bassiana* (Balsamo) Vuillemin.

But what about the closely related fungus studied by Snow, Forbes, and others on the chinch bug? When these men worked with the chinch-bug fungus it was known as *Sporotrichum globuliferum* Spegazzini. However, it soon became clear that it was very closely related to, if not the same as, *Beauveria bassiana*. Accordingly, Picard transferred the species to the genus *Beauveria*, making it *Beauveria globulifera* (Spegazzini) Picard, distinguishing it from *B. bassiana* by minor characters, such as the more fluffy mycelium of *B. globulifera*. However, as indicated from the list we made of MacLeod's groupings, we find that *B. globulifera* is but a strain of *B. bassiana* and, in fact, they are identical species. Anyone trying to distinguish the differences between the fourteen species concerned at the time found that MacLeod's work was indeed a worthy North American contribution to nomenclatural clarification.

Having the nomenclature clarified gives us an understanding of one of the common denominators (the proper name, as it were) in our consideration of the causes of the diseases of many insects besides the silkworm and the chinch bug that are to be found infected with this white fungus. Also, instead of having to conceive of numerous (at least fourteen) species of white-muscardine fungi causing infection in over a hundred species of insects, one now has to be concerned only with whether it is *Beauveria bassiana* or *Beauveria tenella*. (Even at its worst, however, the nomenclatural tangle with species of *Beauveria* is as nothing compared with the still confusing situation that reigns among some of the Fungi Imperfecti and their Perfect [mostly Ascomycetes] counterparts, such as those that occur on scale insects infesting citrus trees.) And all of this concentration of terminology into two specific names revealed that actually a good deal more research on *Beauveria bassiana* (and its several aliases) was going on—both in this country and abroad—than was otherwise apparent. It continues to have an acknowledged role in nature as a regularly occurring enemy, unheralded and unsung, of a large number of destructive insects, including the chinch bug. In spite of the disappointing experiences in Kansas and Illinois, it was not entirely debarred from laboratory and field investigations, but the spotlight—now blurred—no longer focused on it as a sort of panacea responding to the handling of it by man in the artificial control of a serious insect pest.

Although probably a hundred species of North American insects have been recorded as hosts of *Beauveria bassiana* or one of its synonyms, one of its more notorious hosts is the European corn borer, *Pyrausta nubilalis*. The fact that the larva of the corn borer, a pest in North America as well as Europe where it appears to be somewhat more stabilized, is susceptible to infection by the fungus *B. bassiana* was first established in Europe by Metalnikov and Toumanoff (1928) while they were working with the International Corn Borer Investigation Committee, American financed but European based.

After Metalnikov's and Toumanoff's testimony that *Beauveria bassiana* was "very virulent" for the European corn borer, we find virtually no mention made of experimentation with this fungus in the remaining three volumes of the *Scientific Reports* of the investigations (Ellinger, 1928, 1929, 1930, 1931). However, we do find reports on experiments using other fungi including *Aspergillus flavus*, *Spicaria farinosa*, and especially the well-known *Metarrhizium anisopliae* (Wallengren and Johansson, 1929; Wallengren, 1930; Hergula, 1930, 1931; Vouk and Klas, 1931, 1932). We shall have more to say about this last-named fungus later, but it is worth quoting the last sentence in Hergula's 1930 paper as an illustration of how easily enthusiastic optimism could still be generated in spite of the discouraging results obtained in attempting to use a fungus to control the chinch bug. Hergula states, "The outcome of these experiments with *Metarrhizium anisopliae* gives us every right to expect that, after a large scale trial next year, we shall be able to offer an effective method of combatting the Corn Borer." Unfortunately, "next year" did not come in the sense expected by Hergula. (See chapter 5 for more details on the International Corn Borer Investigations.)

Although it was near the end of Roland Thaxter's leadership, the laboratories of cryptogamic botany at Harvard were still involved in work with entomogenous fungi. In this instance it was W. H. Weston who encouraged C. L. Lefebvre (1931*a*, *b*; 1934) to work on certain mycological and histopathological aspects of *Beauveria* attacking the corn borer. Lefebvre also collaborated with K. A. Bartlett (Bartlett and Lefebvre, 1934) of the Federal Bureau of Entomology in attempting to introduce *Beauveria bassiana* (proceeding on the basis of Lefebvre's earlier work and conclusion that *B. bassiana* and *B. globulifera* were distinct species) into areas of corn-borer infestation. They found that dusting infested corn stalks with spore powder did cause infection in the larvae and a reduction in their numbers, but they concluded the paper without definitely committing themselves to whether or not they had established the fungus in the areas where they had liberated the spores. In Canada, Dominion investigators (Stirrett, Beall, and Timonin, 1937; Beall, Stirrett, and Conners, 1939) found that *B. bassiana* was able to control the corn borer in field plots and that the time of applying the spores was of great importance in determining the promptness with which control could be effected. Incidentally, one of these workers (Timonin, 1939) found larvae of the Colorado potato beetle, *Leptinotarsa decemlineata*, to be susceptible to *B. bassiana* after finding the fungus on dead specimens in the field.

With all due credit to the few investigators involved, work on these fungi more or less limped along during the 1930s, 1940s, and into the 1950s. There were spurts of interest: Rex's (1940) hopes of using *B. bassiana* against adult Japanese beetles; McCoy's and Carver's (1941) method for obtaining spores of *B. bassiana* in quantity; Fawcett's (1944) defense of the potentialities of this fungus, as well as others, on the basis that what may be true regarding the practical use of one fungus is not necessarily the case with others and that in such cases as *Beauveria* and the chinch bug a "saturation point" of spores may be reached that obviates the need or practicality of distributing spores artificially; Dresner's (1949, 1950) confirmation of the distinct and wide-ranged pathogenicity of *B. bassiana* for insects and his report that the effects of the fungus, at least against certain insects, was manifested or aided by the production of a toxin by the fungus; Michelbacher and Middlekauff's (1950) findings on the role of *B. bassiana* as a natural controlling agency against overwintering stages of the codling moth; Jaynes and Marucci's (1947) use of the fungus against the same insect in laboratory and field tests; Rockwood's (1951) observations of the common occurrence of *B. bassiana* infecting insects in nature in the Pacific Northwest, which added to the comprehensive checklist of entomogenous fungi in North America compiled by Vera K. Charles (1941); the moderate, but not high, degree of susceptibility of a series of coleopterous stored-grain insects to *B. bassiana* and certain other microorganisms under conditions optimum for the growth and development of the insects (Steinhaus and Bell, 1953); plus other examples. We cannot discount the overall significance of these papers or their accumulative value. And the same can be said about most other Fungi Imperfecti (*Metarrhizium*, *Isaria*, *Spicaria*, *Cephalosporium*, *Sorosporella*, *Hirsutella*, and others), and even such classical groups as Entomophthorales and Chytridiales (Phycomycetes), *Septobasidium* (Basidiomycetes), and *Cordyceps* (Ascomycetes).

It would be impractical, and it is not my intention in this book, to give full chronological histories of each of the principal insect pathogens. And perhaps I have dwelt too long on *Beauveria* in using it as an example of the development of insect pathology as it relates to a well-known insect pathogen. However, I have yielded to the temptation to follow it a bit in detail because man's concern with it began in a rather dramatic, two-pronged manner: as a dread pathogen of the valuable silkworm and as a hope for the microbial control of such pests as the chinch bug and the corn borer and, as time went on, of a host of other noxious insects. On the one hand, great efforts (mostly overseas) were made to combat the disease (white muscardine), and on the other, the species and varieties of *Beauveria* are so distinctly pathogenic to so many pest insects and are so easily grown on artificial media that it seems reasonable to hope that if man can but learn enough to get his hands on the right "handles" he should be able to manipulate this organism to be a truly effective control agent. As this is written, in 1968, at least one commercial concern, Nutrilite Products, Inc., is still attempting to formulate the fungus into a marketable product. A paper from that company authored by Dunn and Mechalas (1963) concluded that *B. bassiana* met well the requirements of a "candidate" microbial insecticide for commercial production.

In spite of our desire not to recount the historical development of insect pathology according to the type of pathogen concerned, it would be inexcusable if we dropped our comments concerning fungi at this point. Indeed, as elsewhere in the world, considerable work—much of it basically excellent—was being done in North America with other groups of entomogenous fungi.

4. *Metarrhizium anisopliae* AND RELATED FUNGI

Although in some ways not as notorious as *Beauveria*, the steady investigation of *Metarrhizium anisopliae*,[2] the cause of green muscardine, has over the years been impressive. The reader will, perhaps, remember our reference to this fascinating species in the last chapter in which we told briefly of Elie Metchnikoff's report (1879) on the natural infection of the wheat cockchafer, *Anisoplia austriaca*, by this fungus, of Metchnikoff's vision of using it as a control agent against this and other insects such as the sugar beet curculio (*Cleonus*), of Krassilstschik's organization of a small spore-production laboratory near Kiev, and of the expectations held for it by certain European workers in the control of the European corn borer. Here again, the initial hopes that *Metarrhizium* would be an effective control agent were high, and we find that its history in this respect fairly well parallels that of *Beauveria*. Like *Beauveria*, it remains today a common natural enemy of numerous insect pests, and as with *Beauveria*, plodding investigators continue studying the fungus and the disease it causes, both from a basic standpoint and with the hope that someday, somehow, it might be turned to mankind's advantage on a massive and effective scale.

Following the time of Metchnikoff's work with *Metarrhizium anisopliae*, a considerable number of insect hosts (both from nature and in the laboratory) have been listed. The number of host species in North America presently approaches one hundred. Apparently the first record of the green muscardine fungus from insects in the United States was that made by Pettit in 1895. He found it infecting the wheat wireworm in New York State. Rorer (1910, 1913) conducted promising field experiments in Trinidad, using the fungus in attempts to control froghoppers. It had been noted in Trinidad since as early as 1890 by Hart and others. Koebele (1898, 1900) and Speare (1912) found it on the sugarcane borer in Hawaii. A year later it was reported from Mexico (Urich, 1913). And Stevenson (1916, 1918) found the fungus on twelve different species of insects in Puerto Rico after it had been introduced from Hawaii by D. L. Van Dine in 1911; however, Stevenson believed that actually the fungus was indigenous. Smyth (1916) described several aspects of green muscardine as it occurred there. According to Fawcett (1915), the fungus was introduced into Cuba.

The manner in which *M. anisopliae* infects insects is very similar to that of *Beauveria bassiana* and other muscardine fungi. Interestingly, among the first to work out some of the details of its pathogenesis in an insect was the American, R. W. Glaser (1926). He was brought to this study by the occurrence of green muscardine in his stock of silkworms. However, like that of others, Glaser's report did nothing to help settle the conflicting reports concerning the efficacy of the fungus in combatting insect pests.

Sporadic records of the occurrence of green muscardine among insects have been made in the intervening years. Closely related species (e.g., *Metarrhizium album* and *Metarrhizium brunneum*) were described, but the concentration of interest remained on *M. anisopliae*, and the host list for this fungus continued to increase. A significant change in emphasis came with the findings that *M. anisopliae* produces enzymes (Huber, 1958) and toxic substances (Kodaira, 1961; Tamura et al., 1964, 1965; Roberts, 1966*a*, *b*; 1969) that are involved in its pathogenic action in insects.

5. AND OF *Cordyceps*, *Septobasidium*, AND *Coelomomyces*

In chapter 1 we noted that because of their ready appearance and considerable size, the fungi now placed largely in the genus *Cordyceps* were the first to be recorded as associated with insects. We also considered briefly how fanciful conceptions of their nature (small trees) and uses (medicinal) made them a part of legend in old as well as modern civilizations. By the time almost three-quarters of the twentieth century had past, these fungi were fairly well understood from a taxonomic standpoint but from few other standpoints. Yet they continued to carry with them an historic flavor.

In the United States during the middle years of the twentieth century, most of the work—largely taxonomic—on the genus, and its close relatives, was accomplished by Edwin B. Mains (1890–1968), director of the University Herbarium at the University of Michigan at Ann Arbor. Although we exchanged correspondence on occasion, it was never my privilege to meet or to know Mains personally. However, his graciousness in sending me photographic prints of some of the species of *Cordyceps* he had described so that I could use them in my *Principles of Insect Pathology* indicated to me that he was a man of generous and understanding qualities.

He came to the University of Michigan as a mycologist engaged primarily in plant pathology. His successor as director of the herbarium, Alexander H. Smith, told me that Mains eventually became interested in *Cordyceps* partly at least because he found a population of scale insects near Au Train heavily infected with *Cordyceps clavulata*. Because of the challenge to find them and because he knew that Mains would study them, Smith, who was doing most of the mycological field work for the herbarium, began to collect *Cordyceps* more diligently. As a result, Mains made a major project of studying members of this genus, especially from a taxonomic standpoint.

Although, through our diagnostic service, I received only a moderate number of *Cordyceps*, some of these were very interesting. One specimen, collected by W. L. Jellison on ants in Burma, was forwarded to Mains (1948), who described it as a new species, *Stilbum burmense*, possibly the conidial stage of a species already found on ants. This experience initiated a genuine interest on my part in *Cordyceps*. However, except for the experience, in 1954, of collecting, with Mauro Martignoni, specimens of *Cordyceps clavulata* on scale insects (*Lecanium corni*) on young walnut trees in Switzerland, it was never my privilege to work directly or extensively with these fascinating pathogens of insects.

The tragedy, as concerns *Cordyceps* at the time of this writing, is that no major work is being done with these fungi anywhere in the world. Significantly, in 1962 when I was searching for someone to author a chapter on *Cordyceps* for a two-volume treatise on insect pathology I was then editing, there was no experienced authority to whom I could turn. Mains by then had retired and was not in good health. Fortunately for the treatise, Freeman L. McEwen (1963), although inexperienced with these fungi, as a gesture of friendly kindness, agreed to undertake a review of the literature of *Cordyceps* infection. Considering the circumstances, McEwen not only did a fine job with the assignment, but presented insect pathologists and mycologists with the only up-to-date review of its kind in the English language.

It would be quite unthinkable to consider the development of knowledge regarding entomogenous fungi without acknowledging the major work on the genus *Septobasidium* by John N. Couch of the University of North Carolina. Couch's work is highlighted by the volume titled *The Genus Septobasidium*, published in 1938 by the University of North Carolina Press. Even today these 480 pages remain a classic.

Couch was introduced to *Septobasidium* in 1920, when he was a graduate assistant in botany at the University of North

Carolina. His instructor, W. C. Coker, had him collect specimens of this genus. Moreover, Couch drew the probasidia, basidia, and spores for two species—little attention being given at that time to the associated scale insects. Coker's paper "Notes on the Lower Basidiomycetes of North Carolina" was published in the *Journal of the Elisha Mitchell Science Society* in 1920; it included Couch's drawings. Couch went on later not only to revise the genus and describe many new species but to discover that the fungi and scale insects live symbiotically (and mutually) at the expense of the host plant.

Couch's real interest in *Septobasidium* had begun in the summer of 1926 when he spent about two months in the rain forests of Jamaica. Although by this time others had reported the occurrence of scale insects (e.g., *Aspidiotus*) under the stroma of these fungi, no one had demonstrated any biological relationship between the fungi and the insects. Indeed, some observers thought that the presence of the insects was simply accidental or fortuitous, other investigators claimed that the fungus overgrew, parasitized, and exterminated whole colonies of scale insects, and some even suggested the use of *Septobasidium* as a method of biologically controlling the scale insects concerned. A few investigators suggested that the fungus was nourished by the excretions of the insects.

In a 1968 letter to me, John Couch wrote these cogent words:

> In the rain forests of Jamaica the fungus was quite abundant—several species occurring on a wide variety of trees and shrubs. I was impressed by the vast growth of the fungus *Septobasidium jamaicaense* on the tree tomato covering the stem from the ground upwards 6–10 feet with a felty mat as much as one centimeter thick and beneath which were countless numbers of scale insects. Here the discovery was first made that beneath the fungus there were healthy insects of all ages with many adult females giving birth to young. Also for the first time scale insects were found which were clearly parasitized by the fungus. Such insects were completely covered by a thick, hard fungal pad and though parasitized by the fungus were alive with their sap-sucking apparatus drinking in the juice of the tree. Also the remarkable haustoria in the hemocoel of the insects were recognized for the first time. It was obvious that between the fungus and the scale insects there was a far more beautiful and complex relationship than previous workers had even imagined.

> The following questions demanded answers: Why are certain insects parasitized while others remain free from infection? How does infection take place? How are new colonies started? How is the combination disseminated? Which came first, the fungus or the scale insect? Fortunately for this study I found an abundance of *Septobasidium burtii* (=*S. retiforme* according to Burt) on pin oak trees close to the botanical laboratory in Chapel Hill. It took about four years of intensive work to answer the answerable questions posed above and there are many more unanswered.

In spite of his outstanding work on *Septobasidium*, probably most pathologists know Couch for his studies of the genus *Coelomomyces* (order Blastocladiales). The first *Coelomomyces* infection was discovered by the British worker Keilin in 1921 in the mosquito *Aëdes albopictus* collected in the Federated Malay States. The next year Bogoyavlensky (1922) described a species from *Notonectra*, an hemipteran, collected in Moscow. Subsequently, other species were found independently in mosquitoes by Haddow (1924), Iyengar (1935), and Walker (1938). The first significant discovery of *Coelomomyces* in the United States was made by Couch (1945) and Couch and Dodge (1947). Couch's involvement with this genus began during World War II when diseased larvae collected from around military camps in southern Georgia were sent to him for identification of the parasites obvious in their coeloms.

Couch was the son of a Baptist minister, attended Trinity College (now Duke University), and earned his bachelor's, master's, and doctor's degrees from the University of North Carolina. After service in France in World War I, he returned as instructor of botany at the University of North Carolina where he subsequently became Kenan Professor in 1944, and was head of the department from 1943 to 1959. His career was bedecked with honors.

6. BACTERIA AS MICROBIAL CONTROL AGENTS

Prior to 1940 a number of bacterial and viral infections were isolated from insects and were used with varying success in microbial control. One bacterium that figured but briefly on the scene as a microbial agent was *Coccobacillus acridiorum.*

Since it was among the very first bacteria used as a microbial control agent it seems worthy of mention. D'Herelle (1911) in 1910 observed an epizootic raging among locusts (either *Schistocerca pallens* or *S. paranensis*) in the Yucatan, Mexico. Apparently the locust population was completely decimated by the occurrence of this epizootic, which d'Herelle attributed to an organism that he named *Coccobacillus acridiorum*, so that by 1912 locusts were no longer a problem in that part of Mexico. Later, he (1912) seemingly had great success in Argentina and Columbia controlling the locust populations and less in Algeria and Tunisia. D'Herelle's work excited a number of workers in different countries concerning the possibilities of using the organism as a control measure for insects. Although some workers were able to duplicate his successes to some degree, the majority could not; thus in time early hopes were dashed regarding the organism's use for microbial control.

Felix d'Herelle (1873–1949) was born in Montreal, Canada, of French-Dutch parentage. His father died when he was six years old so his mother took him back to France. He started his study of medicine in Paris and completed it at Leyden in Holland (Lépine, 1949). According to Lépine when d'Herelle graduated he was looking for a position and learned from a trade journal that the position as chief of the laboratory in the General Hospital of Guatemala was available. He applied, was accepted, and enroute on board ship he taught himself bacteriology, a branch of medicine that he was unfamiliar with at the time. When he arrived in Guatemala he was chief of the laboratory and also had a teaching position in the College of Medicine. However, Dean Alarcon of the Faculty of Medicine, University of San Carolo, stated (personal correspondance to Murray Sager in 1958) that d'Herelle never had any connection with the College of Medicine; and Pedro Arenales, with whom d'Herelle worked, stated (information enclosed in Dean Alarcon's letter) that d'Herelle came to Guatemala in 1904 and worked in the Central Chemistry Laboratory, headed by Master Professor Rene Guerin. Almost all of his work in Central America was on fermentation processes. He distilled a good quality of whiskey from discarded bananas, which received a gold medal from the United States Department of Agriculture (Arenales, 1958). Arenales was sorry the process was never published, since it might have proven to be a boon to the banana-growing countries. D'Herelle also discovered a fungus causing rapid death of coffee plants and suggested ways of combating it.

His interest in fermentation sent him to the Yucatan in Mexico to study the industrial utilization of waste pulp of *Agave rigida*. It was while he was here that he observed the locust epizootics and apparently also discovered bacteriophage about the same time. Thus d'Herelle appeared briefly but dramatically on the stage of insect pathology and then faded from the scene.

In light of the inadequate knowledge regarding host susceptibility and resistance, external conditions leading to the epizootics, and principles of bacteriology at that time, it is not surprising that contradictory results and claims were made regarding the efficacy of *C. acridiorum*. Of course, this does not say that d'Herelle's successes were all that he claimed them to be. It has been shown (Glaser, 1918) that of the number of strains that existed as *Coccobacillus acridiorum*, not all were equally pathogenic, some were not even the same organism d'Herelle claimed to have isolated, and pathogenicity was difficult to maintain on artificial media without frequent passage through susceptible locusts. Workers condemning the organism used it against locusts only distantly related to *Schistocerca* (which is cannibalistic and migratory). The case of *Coccobacillus acridiorum* was a clear example of the great need for more basic research before the organism as a microbial agent could be adequately judged.

D'Herelle's organism was not the only bacterium that was tried as a microbial agent and found wanting in the early twentieth century. A number of organisms were discovered—though with less fanfare no doubt—both in field and in insectory-raised insects, but none was ever effectively utilized in the United States to any extent. After Metalnikov and his associates found several spore-forming organisms, *Bacillus thuringiensis* among them, to be pathogenic for corn borer, pink bollworm, and gypsy moth, the results were encouraging enough that commercial preparations were prepared and distributed in France. Renewed studies on *B. thuringiensis*, first described by Berliner (1915), during the 1950s have rekindled interest in this organism as a microbial control agent. There was no widespread effort in the United States to use bacteria as control agents until the discovery of the milky diseases of the Japanese beetle.

It would appear that the earliest studies on the diseases of the Japanese beetle were made in 1926 and 1927 by George E. Spencer at the Japanese beetle laboratory set up by the Bureau of Entomology and Plant Quarantine at Moorestown, New Jersey. Spencer found larvae of the insect to be attacked by both bacteria and fungi. In addition, infectivity tests were run with the fungus *Beauveria tenella* (=*Isaria densa*) obtained by L. O. Howard from France, where it had been observed infecting *Melolontha*. About fifty percent of the test larvae became infected (Hawley and White, 1935).

In 1928, the study of the diseases of the Japanese-beetle grubs was continued by Henry Fox at the Moorestown laboratory in cooperation with R. W. Glaser then working for the New Jersey Department of Agriculture. Although they observed bacteria in the diseased larvae, their attention—particularly that of Glaser's—was drawn to a parasitic nematode, *Neoaplectana glaseri*, about which we shall have more to say later (Hawley and White, 1935).

Then, in September of 1933, an investigation of the diseases of the larvae of the Japanese beetle was renewed by I. M. Hawley and G. F. White (1935), also of the bureau. They separated diseased grubs found in the field into three groups: the black group, the white group, and the fungus group. The first two groups represented diseases caused by bacteria; of these the white group was destined to be of great practical and historical significance. As near as can be determined, the white group included those diseases that came to be known as the milky diseases of the Japanese beetle. From the first, it should be noted, the motivation for the study of the diseases of this insect was to find a microbial pathogen that could be used in the beetles' control. Interestingly, although *Popillia japonica* is a very destructive insect in the United States, it causes relatively little damage in its native Japan, preferring weeds to cultivated plants. Presumably it is held in check by its natural enemies and other ecological factors.

In the meantime, a young soil microbiologist at Rutgers, the state university of New Jersey, had selected the study of the diseases of the larvae of the Japanese beetle as his doctoral research problem. His dissertation on this subject was accepted by the university in May, 1937. Of singular significance was the fact that this young investigator, Samson R. Dutky, found that there were several diseases in the white group, of which "types A and B milky disease" were the most prevalent. In 1940, as an agent of the Division of Fruit Insect Investigations, Bureau of Entomology and Plant Quarantine, and as a research assistant in soil microbiology for the New Jersey Agricultural Experiment Station, Dutky described these milky-disease bacteria under the names *Bacillus popilliae* and *Bacillus lentimorbus. B. popilliae* the cause of the type A disease appeared to be the most promising of the two as a control agent and was to receive by far the most attention as a subject of research.

Dutky conducted some field experiments with the bacteria and showed that certain other scarabaeids were susceptible to them, but his principal interest remained with the bacteriology of the organisms, their physiology, and their effects on the host (e.g., see Dutky, 1963). Someone prepared an excellent exhibit of these two organisms for the American Association for Advancement of Science meeting in Columbus, Ohio, in 1938. I well remember being intrigued by this newly discovered organism, especially its refractile body.

From 1940 on, there developed several strong liaison programs between the federal government and a number of states in the area of Japanese-beetle infestation. Inasmuch as, at that time, a method of producing spores of *B. popilliae* and *B. lentimorbus* on artificial media had not been developed, they had to be produced by innoculating grubs with the pathogen, then harvesting the spores after they had developed in large numbers in the host insect. By thus producing spores that were incorporated into talc and other preparation, living insecticidal preparations could be provided agriculturists and others by the Moorestown laboratory, by state experiment stations, and by commercial manufacturers. Widespread use of such spore preparations is credited with much success in reducing the amount of damage that otherwise would have been caused by the grubs and adults of the Japanese beetle. A host of entomologists became involved in the use of these bacteria in combating the insect, among them, in alphabetical order, J. A. Adams, E. N. Cory, S. S. Easter, C. H. Hadley, G. S. Langford, P. J. McCabe, J. B. Polivka, E. H. Smith, E. H. Wheeler, and Ralph T. White. And there were, and are, references and publications not federal in origin. An example, is a bulletin authored by Raimon L. Beard and published by the Connecticut Agricultural Experiment Station at New Haven. This bulletin appeared in August, 1945, and at the time was an excellent report of various aspects of biological relationships between the bacteria and their hosts as well as an impressive summary and recapitulation of all the work on type A milky disease of Japanese beetle that had preceded it. A recent government bulletin by Fleming (1968) gave an excellent review, including a survey of many unpublished reports, on the total picture of the biological control of the Japanese beetle.

Considerable time and money have been spent on research on the milky diseases, certainly justified, since from the beginning of their discovery they were a promising means of controlling the beetle. Ever since 1939, when spore-dust mixture distribution by the United States Department of Agriculture was begun, reports of its success have been made. Marked reductions in the Japanese-beetle-grub population have occurred in all treated areas, and this spore dust constitutes one of the most effective means of bringing about a gradual decline of the beetle population. There is little doubt that up to the 1940s one of the most successful attempts to control an insect by microbial means was achieved in the use of the milky diseases, especially type A, against the Japanese beetle, *Popillia japonica.*

In discussing bacterial diseases of insects, the contributions of Gershom Franklin White (1873–1937) should not be overlooked. A man trained not only as a scientist, but as a physician (M.D. from George Washington University), he had an extremely broad knowledge of bacteriology as well as pathology and helminthology. Perhaps he is best known for his monumental observations on bee diseases establishing definitely the causes of the foulbroods. He also discovered and described a number of diseases, septicemias and protozoan infections, of insects other than bees. At the time of his death he was working on the Japanese-beetle diseases. In addition to studies on insect diseases, he collaborated with Kirby-Smith on determining the cause of creeping eruption (hookworm, *Ancylostoma braziliense*) in the Deep South, and in 1930–34 he participated in investigations of methods for producing and shipping sterile maggots for use of surgeons in the Baer method of treating osteomyelitis and other suppurating lesions in man. Dr. White served as a bacteriologist both at Cornell University and in the U.S.D.A. Bureau of Animal Industry, but he joined the Bureau of Entomology in 1907 and never left it except for a brief service in the Medical Corps in 1918. Insect pathology owes a debt of gratitude to this man who was such a thorough and painstaking investigator.

Rudolf William Glaser (1888–1947) perhaps came the closest to anyone in the United States to doing basic research on the problems of insect diseases, and without a doubt, he had the broadest interests. One only has to look over his bibliography to ascertain the wide areas of his research: wilt diseases of gypsy moth and other insects; bacterial diseases of caterpillars; the nature of polyhedral bodies in virus infections; intracellular bacteria; effect of food on length of life and reproduction of flies; cultivation of bacteriocytes of roaches; sterile culture of flies; immunity principles in insects; growth of insect blood cells in vitro; studies on inclusion bodies of silkworm jaundice; and study and culture of *Neoaplectana glaseri* for a microbial control agent for Japanese beetle. Glaser spent most of his professional life at Rockefeller Institute for Medical Research in the Department of Animal and Plant Pathology, acting on occasion as a consultant to the U.S.D.A. Bureau of Entomology in regard to a microbial control problem. He was able to pursue both basic research and the practical applications of insect pathology, a rare opportunity at that time. Glaser was one of my earliest mentors as my interest in the relationship between microorganisms and insects evolved into the main one of my life. It is sad that neither he nor Paillot lived to see the great burst of development in the field of pathology, since both had been true pioneers in the broadest concepts of insect pathology—that is, insect pathology was far more than microbial control. The work of Paillot, begun in 1913, unfortunately is still not appreciated, but there is no question but that the publication of Paillot's *L'Infection chez les insectes* (1933) and the establishment of the Laboratory of Invertebrate Pathology at Lyon were precursors of what was to come in North America after World War II.

7. SUMMARY

In surveying the development of insect pathology to 1940, one should not ignore the fact that there were workers who had the vision and aspirations and advocated an organized basic approach to the study of the diseases of insects, but for one reason or another could not implement the programs. As mentioned, both Forbes and Snow had recognized the need even in the nineteenth century for a better basic understanding. Harry Smith in a lecture given 4 December 1916 at the University of California, Berkeley, stressed the importance of control of insects by fungal, bacterial, and viral diseases. It was not until 1945 that he was able to establish a research laboratory encompassing this aspect of biological control. J. J. de Gryse of Canada was another one who had the vision, but the time had not come. In personal correspondence (1949), de Gryse wrote:

> In 1923 [in Canada] we had a general conference of all officers of the Division of Entomology. On that occasion, I suggested that we should undertake the study of insect diseases. . . . I was sat upon and ridiculed by all present. So I dropped the subject officially at least and tried to get co-operation from some bacteriologists and mycologists (on my own) whenever I came across diseased material. All I got out them was: Sorry, I've not time for this kind of work. In the meantime, I tried over and over again to interest my superior officers but their reply

invariably was: You can't do anything with disease and, moreover, the study of diseases is the function of the Plant Pathologist!

De Gryse was not to see his dreams fulfilled until the 1940s, when almost "subrosa" (so he says) he initiated the spruce-sawfly and spruce-budworm work. As a result of this work he was able to persuade the Canadian government that a laboratory to study insect diseases was necessary, and he was promised $150,000 to build and equip a laboratory. At the time of the finishing of the laboratory it cost more like $750,000. Regarding the aims of the laboratory, he wrote (personal correspondence, 1949): "I hope that, before long, we may achieve some practical results, but I am far more interested in the immediate developments of fundamental research. The rest will come in its own good time and will be all the more assured of success if the work is performed on a scientific basis." At long last, insect pathology was beginning to become established on a firm basis of scientific research.

There is no question that up to the 1940s, possibly with the exception of the Japanese-beetle work, most of the basic scientific principles of our knowledge of infectious diseases in insects had come to us from studies of the diseases of the silkworm and bees and that these studies contributed greatly to the early understanding of infectious diseases in man. Once the infectiousness of disease was established it was an easy step to become aware of the natural outbreaks of disease among noxious insects and thus seek ways to control them. In the main, the emphasis had been on the use of fungi, with only sporadic use of other microbial agents. Thus in 1940 the greatest need appeared to be not only more basic understanding of the principles of insect pathology but a unification and a bringing together of all the information available.

Certainly during the first half of the twentieth century there were American workers who were concerned with individual diseases and insects, but no one championed the idea of bringing all the scattered efforts together into the unified discipline of insect pathology. However, insect (and invertebrate) pathology did come into its own as World War II ended, becoming recognized as an honorable and distinct discipline. Not only in Canada and the United States but also in other parts of the world, new laboratories with staffs especially trained in phases of insect pathology were beginning to be established. There had developed a deeper appreciation of the necessity of accomplishing a greater amount of basic or fundamental research before the effects of disease on insect populations could be throughly understood. Lessons taught by past attempts to utilize microbial control methods were being absorbed. Insect pathologists had long since learned that a great deal more than the mere distribution of infectious agents is involved in the successful use of microbial control methods. Thus twenty-five years (1968) later as I write these words I am hard put to think of any significant number of major biological institutions in the United States that do not have staff members who are at least knowledgeable and appreciative of the field, even though that institution may not have a program or project dealing with insect or invertebrate pathology.

[EAS had not completed this chapter at the time of his death. As indicated in several places he had planned to include viral and nematode infections, and, no doubt, would have covered bacterial infections more fully. MCS.]

1. The term *muscardine* has been indiscriminately used in entomological literature to mean almost any type of fungal infection of an insect. However, as far as its modern, and perhaps more correct, usage is concerned, the word *muscardine* apparently originated in the Italian language with the word *moscardino*, meaning a musk comfit, grape, pear, and the like, or any of the various plants with musk-scented foliage or flowers. (*Musk*, incidentally, refers to the odorous substance from the abdomen of the male musk deer, used as a basis for perfumes.) The French have the words *muscadin*, meaning a musk lozenge, and *muscardin*, which, in addition to referring to the dormouse (*Muscardinus*), also means a comfit or bonbon. Because the bodies of insects (initially the silkworm) infected with certain fungi (e.g., *Beauveria bassiana*) are transferred into white (or with other fungi, into other colors) mummified specimens resembling in appearance comfits or bonbons, the natives of France referred to them as muscardin. It was apparently the French scientists who added a final *e* to the word and used it in referring to the fungus-covered insect. It has since been taken over as a bona fide English word, and English dictionaries and encyclopedias furnish us with at least three meanings or uses of the word: (1) as a noun meaning the fungus itself, (2) as a noun meaning the disease caused by the fungus, (3) as an adjective, *muscardined*, meaning infected and covered by the mycelium of the fungus. In Italy the disease itself, the white muscardine of the silkworm, was first called *mal del segno* ("the disease having a sign"); later, and at present, it is known as *calcino* ("calcium";

"white power"). In Japan the word *kokabyo* refers to virtually any type of fungal infection in insects, with *hakkyo-byo* referring more specifically to white muscardine. According to Aoki (1957) the Chinese equivalent of white muscardine is comprised of characters meaning, literally, "Buddha's bones and silkworm" (Hukuhara, 1968, personal correspondence).

2. Metchnikoff, who first noted and described the fungus, named it *Entomophthora anisopliae*. Sorokin described and named the genus, *Metarrhizium*, and he placed Metchnikoff's newly described fungus in this group. Rorer (1910) cited the Sorokin reference but I have never been able to find the original. The interlibrary-loan librarians of the University of California at both Berkeley and Irvine have been searching the world over for a source of this paper, which is cited by Rorer as "Zeit. der Kaiser. Land. Gesell. für Neurussland, Odessa, p. 280, 1879." I would appreciate hearing from anyone who has seen the original paper that described Metchnikoff's fungus as *Metarrhizium anisopliae*.

[Thanks to V. Cacese, coordinator of UCI branch libraries, University of California, Irvine; V. P. Pristavko, Kiev, USSR; and J. Weiser, Prague, Czechoslovakia, I think that the puzzle of the Sorokin reference has been solved. In 1910, J. B. Rorer erroneously cited the reference as given above, with the year of publication as 1879, and this error has been perpetuated in English literature. The correct reference is: Sorokin, N. 1883. Rastitelnye parazity cheloveka i zhivotnykh' kak' prichina zaraznykh' boleznei (Plant parasites causing infectious diseases of man and animals). Vyp. II. Izdanie glavnogo Voenno-Meditsinskago Upraveleneia. St. Petersburg, 544 pp. Pervoe prilozhenie k Voenno-Meditsinskomu Zhurnalu za 1883 g. (First supplement to the Journal of Military Medicine for the year 1883). pp. 268 (this is a printing error and should read 168–198). In this paper *Metarrhizium* is spelled throughout with one *r*. The Metchnikoff reference was also erroneously quoted by Rorer, and again the error has been perpetuated. The correct reference is: Metchnikoff, E. 1879. O boleznach litchinok khlebnogo zhuka (Diseases of *Anisoplia austriaca* larvae). Zapiski Imperatorskogo Obschestva sel' skogo khoziaistva Iuzhnoi Rossii. Odessa. pp. 21–50. The paper was published also as a reprint-pamphlet, and the text is identical. Metchnikoff, E. 1879. O boleznach litchinok chlebnogo shuka (Diseases of *Anisoplia austriaca* larvae). *Series title*, O vrednych dla zemledelija nasekomych (On insect pests of agriculture). Vyp. 3. Odessa. 32 pp. MCS.]

Two roads diverged in a wood, and I--
I took the one less traveled by,
And that has made all the difference.

Robert Frost

3

Insect Pathology in California

1. BEGINNINGS

I

It is appropriate to begin this chapter with a confession. I began writing this book attempting to compose it in passive voice and first person plural. I gradually found it necessary, at least partially, to use active voice more frequently and to abandon the editorial "we" because of the deadening, impersonal, and hollow ring that was creeping into what I was trying to say. Similarly, after two attempts to write an objective, impersonal account of California's role in the history of insect pathology in North America, I found the words having the sound of false modesty, shirked responsibility, and strained expression. It has always been my feeling that, as in the case at point, unless a man is a great or near-great man, it is highly presumptuous of him to write his autobiography—especially if it is to be published. I still feel this way, though I admit I would personally value having autobiographical accounts by my maternal Mayflower ancestor, Richard Warren, or my paternal German immigrant grandfather Erdmann. In any case, I now find it unavoidable to shift to what might be considered a sort of semiautobiographical form of composition. How can I possibly be truly objective about a segment of the subject of this book when I have been so intimately involved in it for more than a quarter of a century? To pretend such objectivity would indeed be presumptuous and would smack of hypocrisy and sophistry. On the other hand, I have no desire to make derogatory statements or hypercriticial analyses of my colleagues and contemporary workers for their shock value. I can but throw myself upon the mercy, understanding, and charity of the reader—be he "gentle" or not—and proceed as straightforwardly as I can and "as I remember it."[1]

Attention to the disease of insects in California appears to have begun with a wrangle between one S. M. Woodbridge on the one hand and A. J. Cook and Harry S. Smith on the other. The year was 1915. The scene was southern California, where the black scale, *Saissetia oleae*, was a serious pest of citrus and other plants. Woodbridge, apparently a Ph.D. "with 35 years experience in the news and periodical business" and a resident of San Bernardino, California, where it seems he maintained a laboratory of some kind, was a strong advocate for the use of a particular fungus[2] as a means of controlling the black scale in orchards. He insisted that he could prove that "my culture is a cure for the black scale pest as anti-toxin is a cure for Diphtheria." In all probability he acquired the idea from the observations and practices along similar lines being made in Florida. According to the *California Cultivator* for 18 February 1915 (see also the same periodical dated 14 January 1915), a statement made by the Los Angeles Horticulture Commission discredited this method of control and announced that the treatment would not prevent the serving of notice for cleaning up trees either by means of spraying chemical insecticides or of fumigation. The *Cultivator* then proceeded to publish an article (which also appeared about the same time in the monthly bulletin issued by the State Commission of Horticulture) by Harry S. Smith, who explained why Woodbridge's fungus treatment was suspect.

In light of what the future was to bring, it is beguilingly ironical that the principal discreditor of Woodbridge's claim was Harry Smith, who, at the time, was superintendent of the State Insectary at Sacramento, and, as we shall see, a man who was instrumental in making it possible to add the use of pathogens to that of entomophagous insects in the control of insect pests in California. His superior, and supporter, was A. J. Cook, California state commissioner of horticulture. However, it would appear that the warnings originating with Smith, and the dim view taken by him of Woodbridge's methods, were justified. At any rate, Woodbridge's fungus treatment did not enjoy a very long, very prosperous, or very general acceptance.

A climax to the contention between Woodbridge and Smith came with the private publication of a bizarre pamphlet assembled by Woodbridge in 1915 containing statements by Woodbridge defending his methods and a record of the exchange of correspondence between him and Smith, Cook, and others. Woodbridge included in the pamphlet testimonial letters to the efficacy of his methods, but the tract was devoid of data of the type required for conclusive scientific proof of any method of controlling insects. Because the back cover of the leaflet carried scandalous and perhaps libelous statements concerning Cook, this page carried a rubber-stamped line indicating that the bulletin was "FOR PRIVATE DISTRIBUTION." Thus, although copies of Woodbridge's document are rare, it is a fascinating memorabilia of California's first exposure to the subject of microbial disease in insects and the attempt to use microbial agents of disease to control noxious insects.

Having known Harry Smith well, I am convinced that he arrived at his conclusion concerning Woodbridge's control methods quite objectively and with the interests of California agriculture at heart. It is possible, however, that he had already developed a cautious skepticism of the use of entomogenous fungi because of the negating reports of Billings and Glenn (1911) concerning the artificial use of *Beauveria bassiana* in the control of the chinch bug in the Midwest, and of Morrill and Back (1912) concerning the employment of artificial means of enhancing the natural control of whiteflies in Florida. It is also possible that he was alerted to the pitfalls of the method by his father-in-law, Professor Lawrence Bruner, who was on the faculty of the University of Nebraska at Lincoln and who also served as state entomologist. L. O. Howard, U.S. Department of Agriculture entomologist, appearing in the *Yearbook of the Department of Agriculture* for 1901, reported that Bruner was employed by the Department of Agriculture during a portion of the summer of 1901, to investigate local outbreaks of grasshoppers in several of the western states and to pay special attention to the role of disease in the natural control of the insect populations. Bruner reported to Howard that "the whole matter relative to the killing of insect pests by means of fungus diseases is greatly overestimated, and that this is especially true of their use against destructive grasshoppers."

The records show that virtually no significant research was being done on insect pathogens in California at the time of the Woodbridge-Smith controversy. In a letter, quoted by Woodbridge (1915) and dated 26 April 1915 from H. J. Quayle to A. J. Cook, there is the following paragraph worthy of note:

> The Entomological Division of the Citrus Experiment Station has been carrying on some experiments with this [Woodbridge's?] fungus for the past few months. The experiments have gone far enough to warrant the general conclusion that this fungus is fairly efficient on the Black scale when confined in a moist chamber in the Laboratory, but the results in attempting to disseminate the fungus artificially in the field have thus far been wholly negative.

Of course, in the meantime considerable research was being conducted in Florida on the diseases of scale insects attacking citrus. Undoubtedly, the difference in attitude toward such research by the two states was the realization that the climate in Florida was naturally conducive to the development and use of entomogenous fungi, whereas the warm but dry summer climate of California was not.

In the 24 June 1915 issue of the *California Cultivator* is what appeared to be a definitive report on the subject. It is the report of a delegation of "well known scientists and experts" appointed by State Horticultural Commissioner A. J. Cook "to ascertain the results of inoculating trees with a fungus claimed to act injuriously on black scale insects." The official group consisted of H. J. Quayle, H. S. Fawcett, C. W. Beers, D. D. Sharpe, K. S. Knowlton, and G. P. Weldon. Accompanying them was another group of twelve individuals including Woodbridge and Commissioner Cook. (One cannot help wondering why H. S. Smith was not among them.) After visiting groves in the Glendale area and in the San Gabriel and San Fernando valleys, the committee reported that they could "find no evidence either in the treated or untreated groves that a fungus of any kind has entered into the control of this pest in these districts." Apparently the warnings issued by Smith had been warranted.

These early experiences of Smith's did not create a doubt in his mind as to the possible use of pathogens in biological control provided they were first carefully and scientifically studied and tested. There was no question but that Smith appreciated the important role played in nature by disease agents in the control of insects. As early as 1916, he revealed this appreciation by citing the natural-control role of diseases afflicting the gypsy moth and the chinch bug, as well as the disastrous effect of disease in the silkworm.

II

My first contact with Harry S. Smith was by correspondence. In April 1942 I received a letter from him requesting a reprint of a paper I had written for *Bacteriological Reviews* titled "The Microbiology of Insects." This very mediocre review had been written in 1939 (and published in 1940) while I was still a postdoctorate fellow, and I felt highly honored to receive this request from such a well-known entomologist. At the time I was not aware of the extent of his interest in the microbial diseases of insects. It was the first thing I had ever written on the relationships between insects and microorganisms—probably written more to educate myself than to elucidate the subject—and, for some reason, I did not expect any interest in the subject from entomologists. Inasmuch as the paper was concerned mostly with relationships other than that of microbial disease in insect, I am sure that "Prof Harry," as I was later to find him affectionately called, requested the reprint more from what he could expect by the title than from what was in it.

Before I was to become personally acquainted with Professor

Smith, a curious sequence of events and experiences was to take place that would bring me from what I thought was to be a career as a bacterial physiologist to one dedicated to learning what and how things "go wrong" in the life processes of invertebrate animals, especially insects. Little did I realize, when I received the reprint request from Professor Smith, how this man was to change the direction of my professional life; nor for that matter did I have the faintest notion that most of this professional life was to be lived at the University of California, where, without question, Prof Harry was one of the best-loved and most highly respected of men.

By the time I received the unobtrusive yet portentous reprint request I was involved in research on the microbiology and rickettsiology of ticks (later on arthropod-transmitted agents of diseases associated with World War II) at the Rocky Mountain Laboratory of the U.S. Public Health Service at Hamilton, Montana. By the spring of 1944, however, I was restless and uneasy. Beginning with the bombing of Pearl Harbor, several attempts to volunteer for the U.S. Army and Navy were denied, leaving me somewhat frustrated. To be sure, at the Rocky Mountain Laboratory, we had been busy on various war projects, including the manufacture of vaccine, and I had been serving as a Public Health Service member on the U.S. Army Bullis Fever Commission. Now that it was clear the war would be drawing to a close within another year, the bureaucratic control of the Division of Infectious Diseases (and perhaps other divisions) of the Public Health Service made remaining within the service less and less attractive. Fortunately, for my own peace of mind and relaxation I pursued during my off-hours in a minor way my deepening interests in insect pathology. Moreover, I was able to spend some of my evenings acquainting myself with the scattered, but fascinating, literature pertaining to all types of biological relationships between arthropods (mostly insects and ticks) and microorganisms. Actually, although done mostly on my own time, it was quite generous of the director to indulge me in this way, because he once expressed his opinion that though my interests in insect microbiology were understandable, "Who else would ever be interested?"

One spring day—I believe it was toward the end of May—matters boiled over more rapidly than I had anticipated. Morale at the Rocky Mountain Laboratory was generally low because of what many of us felt was undue federal regimentation and the condescending and restrictive manner in which Washington officials dictated policy, particularly as it pertained to freedom of research. Fortunately most of the work I was doing, and that which I was supervising for the director, R. R. Parker, met with little interference. Nonetheless, as an example, when the chief of our division in Washington, or his associates, refused to allow one of our outstanding parasitologists to publish a new and imaginative concept of the evolutionary development of fleas in relation to the development of their hosts, I suffered almost as much anguish as my associate. In my own case, the last straw—as frequently is the case in such situations—was really a very minor one. It had to do with the typing of a manuscript that had been repeatedly set aside in order to handle what I felt were routine housekeeping chores. When the secretary was once again ordered by the director's office to set it aside, I simply stepped in Dr. Parker's office and asked him why this had to be. The answer was very similar to those previously given: "The demands from Washington and a shortage of help made this necessary." Parker, although director of the laboratory, was himself a victim of the pressures from Washington and did not like the situation any better than the rest of us. Being the impatient young man that I was at the time, impulsively and on the spot, I decided to resign even though I had no other job in the offing. Parker chuckled as though I were joking about the matter.

I returned to my office, looked out the window awhile at the beautiful Bitterroot Mountains, then typed out my resignation. When I carried my written statement to Parker, he realized I was serious; he became quite upset and, of course, felt that my actions were most unreasonable. News traveled fast. Before the day was over several of my colleagues came by to ask if what they had heard was true, and in almost every case they congratulated me because they realized the generally unsatisfactory conditions that prevailed for a rash young man aspiring to be a scientist. I particularly remember that Dr.

Robert Cooley, the senior statesman of the laboratory and a man revered by the entire staff, came by and, after some sympathetic discussion, startled me with the comment that were he ten years younger he would do the same thing! The assurances of this great entomologist (ixodologist) meant a great deal to me. Out of the kindness of his heart and from his precious time he had taught me much about ticks that could not be found in books, including special techniques that enabled me to dissect these arachnids on a mass-production basis.

That evening I returned home just shortly after my wife, Mabry, had returned from a shopping trip to Missoula. I walked into the kitchen and rather abruptly told her of my resignation. Her reaction was typical of her. Without missing a single turn of her stirring spoon she replied, "Good, when do we leave?" Her response surprised me inasmuch as she, as well as I, loved the Bitterroot Valley and up until this time we had never seriously discussed the possibility of leaving. I had on occasion mentioned that under the circumstances prevailing in much of the Public Health Service, which at that time frequently placed research-oriented Ph.D.s under the jurisdiction of nonresearch-oriented M.D.s, perhaps I would be happier in an academic situation. In truth, the caste system of the service was not much in evidence at the Hamilton Laboratory, but one knew that it existed. I had to explain to my wife that I had resigned without having the faintest idea of where we would be going, and that since I had resigned as of July 1, this gave me only a month to wind up my work and find another position.

The next day the thought occurred to me that two recent visitors at the Rocky Mountain Laboratory, both of whom had shown some interest in the work I was doing, might be able to help me. Accordingly, I sent off wires to K. F. Meyer of the George Williams Hooper Foundation, University of California, San Francisco, and to Chauncey D. Leake, then vice-president of the University of Texas and in charge of the medical school at Galveston. Dr. Leake wired back advising me to contact Dr. Meyer, not knowing I had already done so. Unfortunately, Dr. Meyer was out of town and my wire had to wait until his return on June 15. But the agonizing wait ended happily when I received his wire saying he could use me immediately. A phone call and subsequent wire and letter confirmed my appointment to the faculty at Berkeley in the Department of Bacteriology, where I was to teach immunology, and to the Hooper Foundation, where I was to work and advise in the area of medical entomology (plague and encephalitis). I would also be permitted, as time allowed, to conduct research in general insect microbiology and pathology.

One is often tempted as he travels life's highway to contemplate what might have been had he taken another path. What if Dr. Meyer had not visited the Rocky Mountain Laboratory in April 1944, thus giving me the courage and audacity to contact him for a position, then getting a position, thus putting me at possibly the one university in the United States that had a man, Professor Smith, who was interested in and thinking of a person who might organize and devote his whole energies to a study of insects and their diseases, hence making the first laboratory of insect pathology in the United States possible. Would the dreams of a young scientist have reached fruition in some other way or in some other place? Who knows what chance happenings contribute to the fabric of one's life.

The rapidity with which I was able to find new employment surprised not only me but Dr. Parker as well. When he realized that I really was leaving, his irritation with me subsided and he was able to say to me that he had always suspected that I aspired to an academic life and that he had anticipated that some day I would leave the laboratory for a university or college. Until his death in 1949, we maintained the most cordial of relations including a very active exchange of correspondence relating to many matters of mutual interest. Some of this correspondence had to do with work I had participated in while at the laboratory but not yet published. This included a report on the first discovery of a spirochete in the tissues of hens in the United States (Steinhaus and Hughes, 1947) and an important bulletin published after Parker's death (Parker et al. 1951), which incorporated some of the most comprehensive, yet baffling, research in which I have ever been involved. Among other things it had to do with the mysterious regular occurrence of *Pasteurella tularensis* in creeks and streams. The life of R. R. Parker is a fascinating one; I owe him much—so does the Public Health Service, but I doubt that the hierarchy of this organization ever fully appreciated him and his contribution. Indeed, it is my opinion that except for his close associates, all too few in the world of medical science are fully aware of the fullness of the contributions, service, and dedication of this man.

On the first of July we piled our few belongings into a small "America" (Willys) and headed for California, arriving in Berkeley 4 July 1944. I reported to Dr. Meyer's office on the Berkeley campus early on the morning of 5 July. His secretary knew nothing of my coming; I explained to her what had transpired, and then sat down to wait. It was not long before Dr. Meyer arrived from the Hooper Foundation in San Francisco, where he had his principal office. He came over to Berkeley only to meet classes and to perform his duties as chairman of the Department of Bacteriology. He chaired the department in a rather perfunctory manner, relying largely on assistants to carry on the day-to-day operations. A. P. Krueger was the nominal chairman, but Meyer was carrying on as a wartime chairman. The only other faculty member on duty in the department was Michael Doudoroff, who handled the course in general bacteriology and conducted research in bacterial physiology. Doudoroff, a brilliant bacteriologist, did all he could to make me feel at home and was most friendly and cooperative. We shared a common interest in insect life, his being that of an amateur collector.

When Dr. Meyer arrived he greeted me in a friendly manner acting as though I had always been a member of his faculty. Without any ado he gave me my teaching assignment, which was to present most of the lectures in immunology (how thankful I was that one of the most competent of all my own teachers had been Professor W. A. Starin, immunologist at Ohio State University) and to otherwise assist Bernice Eddie, Meyer's right-hand professional assistant at the Hooper Foundation and Ruth Chesbro, an assistant in the Department of Bacteriology, who did most of the preparatory work for the class laboratories and participated in teaching the laboratory sessions. (Because of the wartime need for nurses and technicians, classes in bacteriology were being held on a year-round basis.). My duties at the Hooper Foundation were rather vague. I was to advise Meyer's graduate students working on sylvatic plague, and to be generally available as a consultant to William M. Hammon and William C. Reeves in their work on encephalitis as it related to mosquito transmission of the virus. Although I attempted to make my rounds faithfully, I doubt that I really earned my pay at the Hooper Foundation. I recall striking up a friendly relationship with A. L. Burroughs, then a graduate student working on the transmission of the plague bacterium by fleas. I had first met Al along with his professor, Harlow B. Mills, then at Montana State, when they had visited the Rocky Mountain Laboratory. Mills, as chief of the Illinois Natural History Survey, was later to host an important organizational meeting that was called to launch the *Annual Review of Entomology*.

K. F. Meyer is an exceptional man in many ways. How unfortunate for me and this book that it is not really germane to attempt to tell more about him. However, his impact upon bacteriology, epidemiology, and the pathology of higher animals is so great, and his life has been so rich with that which is unusual and exciting, that surely he will receive from more capable hands than mine the great biographical treatment he deserves. I last saw "K.F." in the spring of 1967—he was then eighty-three—when he visited me briefly on the Irvine campus of the University. Even then this man of many talents and interests (including philately), who perhaps most of all personified dynamism, was planning a trip around the world as a consultant for WHO. But let it be recorded that Meyer was very knowledgeable of the relationships between invertebrates and microorganisms, not only in the area of medical entomology but in that of mutualistic symbiosis. One can imagine my surprise when I discovered that this man so famous for all of his medically oriented research had in 1925 published a virtual monograph on the bacterial symbiosis of certain operculate land mollusks! Moreover, it was through his appreciation for the potentialities of the fields of insect microbiology and insect pathology, as well as his basic kindness as a human being, that he was soon to make it possible for me to pursue my basic interests. I have much to be grateful for from this man, who not only is a great scientist, but also a very human person under a gruff exterior.

I was assigned no laboratory research space until about September 1, when the Navy moved out of a small laboratory room belonging to the department. Although the room was devoid of equipment, I was soon able to sequester enough supplies and equipment to begin, almost surreptitiously, a study of the cecal bacteria of the harlequin bug, *Murgantia histrionica*. In the meantime, mindful of the interests of Dr. Meyer, I had submitted to him, on 14 July 1944, an outline of my proposed research activities. (Already it was clear to me that dividing my activities between Berkeley and San Francisco was not going to be practical over the long haul.) I suggested a project titled "A Study of the Rickettsiae and Rickettsial Diseases of California." I was careful to include in my proposal an investigation of the so-called nonpathogenic rickettsiae, hoping that this might give me an opportunity to explore still other types of relationships between insects and microorganisms of all kinds. I felt the proposal was a logical one because relatively little had been done in California on rickettsiae and the diseases they cause, and my experience with this group of agents was still fresh from the work I had been doing at the Rocky Mountain Laboratory. However, Meyer wisely did not approve of my working on rickettsiae or on the bacterium causing tularemia, because, unlike the situation at the Rocky Mountain Laboratory or the Hooper Foundation, there was insufficient protection against accidental infection of students and others frequenting the halls of academe. This denial turned out to be a fortunate one for me.

As our first months in the Bay Area passed by, we became adjusted to living in a metropolitan area. Without losing our affection for western Montana, we began to find the charm of living in California in the mid-1940s. Mabry put her training in bacteriology to good use by finding a "war job" at Cutter

Laboratories, a commercial concern in Berkeley, where she became a member of a team producing penicillin. At the university, I was enjoying the stimulation of teaching. At the same time, I was becoming acquainted with the entomologists who, for the most part, were then situated in two divisions of the College of Agriculture, The Division of Entomology and Parasitology located in Agriculture Hall (now called Wellman Hall) and the branch laboratory of the statewide Division of Beneficial Insect Investigations (later to become the Division of Biological Control) located at a station in Albany called Gill Tract, three miles from the campus. The headquarters of the latter division, chaired by Harry S. Smith, was at the university's Citrus Experiment Station at Riverside. Among the entomologists located on the Berkeley campus that I first came to know were E. O. Essig, then chairman of the division, Stanley B. Freeborn, A. E. Michelbacher, E. Gorton Linsley, and Ray F. Smith, then a graduate student of Michelbacher's. All of these, and others, were to play ancillary but significant roles in the development of insect pathology at the University of California.

III

It was a day in August 1944, that I first had the privilege of meeting Professor Harry S. Smith in person. I received a telephone call shortly after lunch from Professor Smith saying that he was in Berkeley on one of his frequent trips from Riverside and that he would like to visit with me. I was flattered to have his call and delighted to have the opportunity to meet him. I offered to come to wherever he might be. With dignified modesty, which I was to learn was typical of the man, he insisted that it would be simpler if he came to see me; moreover, he wanted to bring a friend of his with him.

Within the hour Smith arrived at my office in the Life Sciences Building; with him was Curtis P. Clausen of the Bureau of Entomology and Plant Quarantine of the U.S. Department of Agriculture. I had, of course, heard of Mr. Clausen and his work with entomophagous insects, so my pleasure was doubled. After a few amenities, we had an interesting chat pertaining in particular to a disease that the Division of Beneficial Insect Investigations was finding in its rearings of the potato tuberworm, *Gnorimoschema operculella*, at its insectary at the Gill Tract in Albany. Smith, apparently assuming I possessed greater knowledge than I did, asked me if I had the time and desire to study this disease for his division. Hopefully, if its cause could be determined, a way of suppressing the malady might be worked out. I eagerly agreed to undertake the project to the extent that my duties in bacteriology would permit. In my own mind, I was determined not to let anything interfere with this opportunity to demonstrate one of the applications of insect pathology. In retrospect, there is no question but that this visit and conversation with Harry Smith was a major turning point of my life. As someone has said, turning points of lives are not always great moments or dramatic events; they are often concealed in minor, sometimes almost trivial, occurrences. So it was that day in August 1944.

With the cooperation of Blair Bartlett, who was in charge of the insectary operation at that time, I obtained specimens of the diseased insects and soon completed a brief study of the disease. Three species of gram-negative nonsporeforming bacteria appeared to be involved; two were coliforms, but the principal cause of the troublesome mortality was caused by a strain of *Serratia marcescens*. (A note on these findings was published in the *Journal of Economic Entomology* in 1945.) By maintaining strict conditions of sanitation and by carefully regulating the temperature of the rearing rooms, the disease was brought under reasonable control. In the meantime I continued performing my instructional duties in the Department of Bacteriology and my consultative obligations at the Hooper Foundation. I also spent some of my evenings working on the symbiotes of *Murgantia* and, with the help of my wife, completed the index of a book, *Insect Microbiology*, I had begun in Hamilton.

Toward the end of 1944, during another of Harry Smith's visits to Berkeley, he hinted at the possibility of my engaging in cooperative work on a formal basis with his division on other problems pertaining to the diseases of insects. He explained that for some time he had been hoping that someday he might enlarge the acitivites of his division to include work on the diseases of insects with an eye to the possibility of using microbial pathogens in the control of insect pests. His division was already well known throughout the world for its work on entomophagous insects, and, only recently, in cooperation with the U.S. Department of Agriculture, he had initiated studies on the use of insects to control noxious weeds. To round out the work of his division, he felt that some work on the possible use of microorganisms in biological control of insects was necessary. Also, during this visit, Professor Smith suggested that I make a trip down to Riverside for the purpose of looking over some of the entomological problems of that area with the idea of ascertaining the possible role of insect pathogens in the control of citrus pests. He was particularly interested in whether or not there were any diseases that might affect the citricola scale, *Coccus pseudomagnoliarum.* I agreed with his suggestions in principal, but suggested that until there was a definite end of the war in sight, it was my duty to continue my instruction of technicians, nurses, navy personnel, and others taking bacteriology and immunology, and to help as I could in the medical research at the Hooper Foundation. With the shortages of certain chemical insecticides however, I was also philosophically aware of the necessity of developing alternate methods of controlling crop pests to ensure the country's food supply. Accordingly, with the inspiration of my few conversations with Professor Smith, I began in late 1944 to plan the establishment of a laboratory or institute either in the Agricultural Experiment Station or in the College of Letters and Science, which would undertake research in insect microbiology and insect pathology.

By the year's end there was growing confidence that the war would end during the coming year. This caused administrators and others within the university to begin making postwar plans. Professor Smith and I began to intensify our dialogue with regard to the possibilities of establishing a unit concerned with the study of disease in insects. By January 1945, it had become apparent that my interests, as well as those of Smith's, could best be served if I were to have an appointment in the Division of Beneficial Insect Investigations in what was then the Department (later the College) of Agriculture or in the Agricultural Experiment Station for the purpose of more directly administering the work I was already doing for Agriculture. In establishing such a unit it was understood that my duties would include the overall direction of such work in insect pathology as the university might do in the state of California. If this were to be done, however, several rather delicate matters had first to be resolved. One was that some agreement would have to be made with K. F. Meyer, who, after all, enabled me to come to the University of California and to whose department I owed considerable allegiance. I was not certain as to how firmly Meyer felt that I was committed to him, but I knew that I must continue teaching until Dr. Meyer could make arrangements for some one else to take my place. In order that my transfer to the Department of Agriculture might be accomplished, Professor Smith decided to hold conferences with C. B. Hutchison, then dean of the Division of Agriculture and director of the Agricultural Experiment Station.

I have never been exactly sure as to what gave Professor Smith confidence in me as being the one he would care to entrust with the development of insect pathology as he envisioned it in the university. My few publications pertaining to insect and tick microbiology were of such minor import that they could not have impressed him to any significant extent. Subjectively speaking, the only thing I might have had going for me was an almost unbridled enthusiasm regarding the potentialities of the field, which I felt was being sadly overlooked except by a few stalwart souls, most of whom we have already discussed in this volume. I do recall that during one of Smith's visits at Berkeley he politely, but quite intently, asked me concerning my background and training, and as to how I had become interested in the field of insect microbiology and pathology. Lead by Smith's penetrating questioning, I accounted myself somewhat as follows:

I was born during a blinding snow storm in the village of

Max, in central North Dakota, on 7 November 1914. My father, Arthur, of German ancestry, was a merchant-farmer who had come to this frontier area in 1906 with a group of Soo Line railroad surveyors from New Richland, Minnesota, where he was born. He, and several associates, decided to settle at this point to establish a town they thought would thrive because it was at a junction of the coming railroad. (I did not realize at the time, but my Aryan-German ancestry was a point in my favor as far as Professor Smith was concerned; for reasons never clear to me, he possessed, what I would consider to be, an excessive admiration for Aryan Germans, especially in the field of science.) I grew up much as would any boy in a small-town community (the railroad went thirty miles north to Minot and Max never became that big town my father and his friends had envisioned!) made up of about 500 rugged Scandinavians, Russians, Germans, and others, most of whom—except for the children—had recently immigrated from Europe. Being the oldest of four living sons, I enjoyed both the benefits and the disadvantages of being the eldest. After school and during the summers I worked not only in my father's store, which he operated with his brother Gust, but in the wheatfields of the farmlands they owned. High school was a rich experience for me, being an active participant in class politics, dramatics, debating, and in athletics of all kinds. Except possibly in athletics—our football and basketball teams were then among the best in their class in the state—my opportunity to participate in so many curricular and extracurricular activities was undoubtedly favored because of the relatively small number of competitors.

My interest in science derived from two sources. My mother was greatly impressed by what she considered the honest objectivity of scientists. She saw to it that I read reliable books in science. Because of its controversial nature—and probably also because of her enjoyment in arguing with local fundamentalist clergy—she made it a point that I read Darwin and Huxley. My father was a great admirer of inventions and inventors—new gadgets intrigued him—but he really wanted me to enter the business world, which by the time I had finished high school

I hated passionately, or a profession, such as law, where one had to be a persuasive speaker. Not until much later did he see much sense in studying science for what he felt was largely science's sake. The other source of encouragement I received to enter science, specifically biology, was my high school biology teacher, Miss Alice Paulson, whom I admired greatly and for whom no amount of study or laboratory work was too much. I am sure that the printing and publishing worlds have much to thank her for, because without her inspiration I undoubtedly would have maintained my aspirations to publish a newspaper.

I explained to Professor Smith that at the time I went to college at North Dakota Agricultural College (now North Dakota State University), I was undecided as to which subject, bacteriology or entomology, interested me more. Actually, I enrolled with the intention of majoring in entomology, and this determination was increased by the warm reception given me by J. A. Munro, head of the department. However, I also felt constrained to look into the possibility of majoring in bacteriology, and so I visited this department (just next door to the Department of Entomology) and found myself also warmly received by C. I. Nelson ("Cap" Nelson to his students), chairman of the department. Indeed, the kindnesses shown me, a tall, awkward, gangling, small-town boy of seventeen, by Professor Nelson and his associates—particularly Delaphine Rosa Wyckoff—probably had a great deal to do with my renewed fascination with microorgniasms, a fascination that had begun in my high school biology course and was stimulated by my spellbound reading of Paul de Kruif's histrionic *Microbe Hunters.* (Some years later when I chanced to meet de Kruif briefly, I better understood Heywood Broun's remark about him: "Never have I known a man who pursued knowledge with such gusto. . . . he presents it with such passion that the would-be heckler finds himself in the teeth of the hurricane.") When I returned to my room that afternoon I was totally confused as to whether I should major in entomology or bacteriology. When the time came for registration I declared my choice, and it turned out to be bacteriology.

I held to bacteriology (microbiology) as my major interest

throughout my four years of undergraduate work at North Dakota State and during my graduate work at Ohio State University, where I held a graduate assistantship from 1936 to 1939. During the last year of graduate work, I had the good fortune to enroll in a course in entomology given to summer students by Alvah Peterson. As part of this course, each student was requested to do a special problem somewhat of the nature of an investigational project. I chose to work on the bacterial flora of the alimentary tract of the large milkweed bug, *Oncopeltus fasciatus.* Professor Peterson showed considerable interest in my choice of subject and became particularly intrigued when I was able to demonstrate to him that unlike most Lygaeidae, this species did not possess the bacteria-filled gastric ceca. Thus began a long and cherished friendship with a great and genuinely good man who, even though I was not an entomology major, gave me much sound advice and steady encouragement. Once when I despaired that anyone would be interested in hiring a person who had the combined interest of a microbiologist and an entomologist, it was he who counseled me, "Don't become discouraged; regardless of what others may say or think, hold fast to your goals and aspirations—sometime, somewhere, there will be a need for an insect microbiologist." With such advice I would leave his office or laboratory with rekindled spirits, and dare to go to the library where I could read and admire the seemingly impractical yet, to me, captivating works of Buchner (*Tier und Pflanze in Intrazellularer Symbiose*) and Glasgow (*The Gastric Caeca and the Caecal Bacteria of the Heteroptera*).

By late 1938 I had completed all the requirements for a doctorate in bacteriology with a minor in entomology. Even though I had taken only one course in entomology for credit, I had audited several others, including C. H. Kennedy's course on the internal anatomy and histology of insects. My thesis was on certain physiological studies pertaining to the dynamics of the life and death of bacteria.[3] I had come to like biochemistry very much and had taken several courses in premed with the thought of enrolling in medical school if I could find a way of supporting myself and paying the tuition costs. At that time, assistantships and scholarships were not generally available for medical students. Needless to say I did not enroll. My major professor was Jorgen M. Birkeland, who, after I had finished my dissertation, detected and lent sympathetic support to my entomological-microbiological interests. He and N. Paul Hudson, chairman of the Department of Bacteriology, suggested that I apply for a postdoctorate Muellhaupt Scholarship (technically a fellowship) with the understanding that I would obtain my degree with the initiation of my work on the fellowship, should I be lucky enough to receive it. Professor Hudson was known to graduate students as a stern disciplinarian; however, after being graduated most of us realized that his strictness had been to our benefit. Moreover, those of us who had the good fortune to have him review our manuscripts learned much from him concerning precision in writing and respect for the written word. I have never forgotten his injunction to me, "Be careful what you write for publication; once printed, it is more permanent than marrige." To "N. Paul" and his faculty—especially J. M. Birkeland and W. A. Starin—I owe more than that which most students owe their professors; in addition I have been the beneficiary of years of counsel and information exchange through pleasant correspondence.

In applying for the Muellhaupt postdoctorate research fellowship, I submitted as a subject for investigation a study of the microbial flora of insects (see Steinhaus, 1941). Although by now deeply interested in this subject myself, I really could not envision a committee of judges approving such a proposal for one of two fellowships to be granted on a nationwide basis but to be served at Ohio State University. Accordingly, I was completely surprised when I was notified in May of 1939 that my application had been approved and that I would begin serving the fellowship 1 July of that year. In August 1939 I received my Ph.D. and continued to work on the fellowship.

In the meantime I had fallen in love with a charming, happy-spirited girl from Mississippi who was attending Ohio State to work for her master's degree in bacteriology. We both had been assigned to the same major professor, Jorgen Birkeland, and it turned out that Mabry was his first student to graduate with a master's degree and I was his first Ph.D. graduate. After Mabry received her degree in December 1938, she accepted a teaching position at, of all places, North Dakota State, the alma mater of both Professor Birkeland and myself; her chief there was the same "Cap" Nelson who had introduced both Birkeland and myself to undergraduate bacteriology. She held her position at North Dakota until I had completed my post-doctorate fellowship, and we were married in Auburn, Alabama, 14 June 1940.

The position with the U.S. Public Health Service that I was fortunate enough to secure had come my way primarily because Director R. R. Parker sought someone to work on the microbiology of ticks at the Rocky Mountain Laboratory and because one of my professors at Ohio State, Floyd S. Markham, who had been spending his summers at the laboratory, knew of my then bizarre combination of interests and recommended me. After borrowing $400 from kindly and trusting Professor W. A. Starin, I bought a second-hand car, and with $90 my wife had saved, we headed for western Montana, where I commenced my duties at the Rocky Mountain Laboratory in July 1940.

Professor Smith wondered how I had been able to continue my interests in insect microbiology while busily engaged in work on tularemia, several rickettsial diseases, vaccine, and other war-oriented research. I explained that some of the research—e.g., the microbiology of ticks—related directly to the research I was assigned by the government or that I was supervising for the director, R. R. Parker. However, I was able to engage in a great deal of "self-teaching" concerning the scattered literature on insect microbiology. Fortunately, the library at Montana State College, in Bozeman, was most generous about sending abstract journals and other publications, which made up for the deficiences of our own small library at the laboratory. Without initially conceiving of putting the information in book form, I began a literature survey of all aspects of insect microbiology by abstracting and compiling the information on cards, which soon filled several drawers of a filing cabinet. From this initial survey, a *Catalogue of Bacteria Associated Extracellularly with Insects and Ticks* was published in 1942. I continued studying the biological relationships of microorganisms and insects and ticks, and a substantial amount of information was gathered and eventually published under the title of *Insect Microbiology* in 1946.

Meanwhile, after serving as the Public Health Service member of a commission established by the army to study a disease (Bullis fever) occurring among troops stationed in Texas, and thought to be arthropod borne, I had thought that at the war's end I might seek a position that might give me more freedom to study along the lines of general insect microbiology or insect pathology. As explained earlier in this chapter, circumstances developed that brought about an earlier decision than I had contemplated.

This was the very rough and cursory manner in which I reviewed for Professor Smith my background, my training, and my experience, and my aspirations up to the time of my coming to Berkeley. (The reader may have noticed my frequent reference to Smith as "Professor Smith." Actually, most of his associates affectionately called him "Prof Harry," and the term was most appropriate for this kindly, gentle, and very informal man. However, probably because of the very high regard I held for him and the big difference in our ages, I could never directly address him other than "Professor Smith." Once he asked me why I did not call him "Prof Harry." I explained that it probably was a carry over from my childhood when I had been taught that to show respect for one's elders one used a formal title of address. He seemed to understand; and though I was always "Ed" to him, and properly so, I could never bring myself to address him directly as "Prof Harry," although I bow to no one in my respect, gratitude, and admiration for the man.)

During my first visit to Riverside in April 1945, I found it to be a wonderfully, pleasant, small city, with the Citrus Experiment Station endowed with an alluring informality in some contrast with Berkeley. Professor Smith and I exchanged many thoughts and discussed numerous ideas relating to how my interests in insect pathology might be incorporated into the objectives he had for his division. The trip had been a profitable one. On the train ride back to Berkeley (yes, we rode the train in those days!) I noticed that the flags were flying at half-mast in the small towns we were passing through. Shortly, the conductor came into the car, without comment he switched on the train's radio. Franklin Delano Roosevelt was dead. Except for the click of the wheels on the rails, all was silence. Then a male voice some seats back of me said, "My God, what will happen to us now with Truman as President!" And I, too, wondered. Who is this man Harry Truman, vice-president of the United States, who will now lead the country to the war's conclusion?

IV

The negotiations, to which I have already referred, between Professor Smith, Dean Hutchison, and Dr. Meyer were successful in every way with regard to my transfer to the College of Agriculture and the Agricultural Experiment Station. However, since Dr. Meyer still required my services as a teacher of immunology in his department for another semester, it was arranged that beginning 1 July 1945, I should hold a dual appointment serving bacteriology in a teaching capacity and agriculture in a research capacity.

Some weeks following this decision I was present at a meeting between Smith, Hutchison, and the assistant dean of the College of Agriculture, S. F. Freeborn. The meeting was called to discuss the budget, where the unit should best be located, a teaching program in insect pathology, and other matters. Among these was the question of what name or designation should be given to this new unit. It was to operate somewhat as an "institute" except that institutes were not in vogue in agriculture as they were throughout the rest of the university, for, in essence, the Experiment Station itself was equivalent to one large institute. After the name "Insect Microbiology Laboratory" was suggested and rejected, the name "Insect Pathology Laboratory" was agreed on. Two or three months later Professor Smith mentioned to me that the form "Laboratory of Insect Pathology" might sound better, so this was the name adopted and used until, years later, the unit became a separate department.

It is only fair to say that not everyone in the Division of Beneficial Insect Investigations was enamored with the establishment of such a distinct unit. One or two individuals mistakenly felt that the unit would parasitize the division's budget and that Smith—by his enthusiastic promotion—was favoring it. However, the reasons for separately designating the insect pathology setup were several and were practical. In the first place, since it was the only activity of the division located on the Berkeley campus proper, it would be confusing to refer publicly to the unit as the "Division of Beneficial Insect Investigations." We were not concerned with beneficial insects, and the northern headquarters for such work was at the Gill tract, Albany, three miles away from the Berkeley campus, and the main headquarters of the division was located at Riverside. It appeared logical to unify the work in insect pathology because, although pathogens are natural enemies of insect pests just as are entomophagous insects, the techniques, equipment, and entire approach to the study of microorganisms were generally different from those used in studies of insect parasites and predators. Also, from their conversation, I am sure that the dean and Professor Smith felt that the prestige of the college and division could be enhanced if attention were drawn to the fact that work on the diseases of insects (in addition to that being done on bees on the Davis campus) was being conducted at the University of California. The name "Laboratory of Insect Pathology" on letterheads, and through other uses, would presumably help to accomplish this. They also believed that since, except for individual workers and projects, ours would be at that time the only general research and teaching unit of its kind in the United States, it would be of considerable help to outsiders to have some name for the unit that could be easily remembered, and to which inquiries and diseased insect specimens could be sent directly. Perhaps the most practical reason Hutchison and Smith agreed with my proposals to establish the insect pathology work as an autonomous unit, to be directed and supervised as a unit, was to facilitate its administration, secure its budget support, and the like. In any case, the foregoing reasons for organizing the work as a unit and for giving it the name "Laboratory of Insect Pathology" were the principal ones in my thinking and were, without hesitation, supported by Hutchison and Smith.

Toward the end of 1944, and during the beginning months of 1945, while details of my transfer to Agriculture were being worked out, interesting ideas were forming in the minds of a number of people as to just what the main thrust of the work of the new laboratory was to be. It was clear that Harry Smith wanted me to work on the diseases of citrus insects and on the diseases of insects being reared in the insectary. Hutchison wanted me to clear up the research initiated prior to America's entrance into the war by Sokoloff and Klotz (1941, 1942) at Riverside on a *Bacillus* "*C*," which they had isolated from the soil and from the California citrus red scale, *Aonidiella aurantii.* These workers had reported that the bacillus was capable of invading and destroying the adult scale insects on lemons under laboratory conditions, and Hutchison felt he was under some pressure to have these findings applied to infested citrus groves. Although not enthusiastic about the possibilities involved, Smith concurred with Hutchison—it was a clear example to the dean of the practicability of having an insect pathologist about. The U.S. Department of Agriculture forestry entomologists, stationed at Berkeley, had consulted with me concerning some of the epizootics of disease they had observed in certain forest insects in California and were anxious to have a study made of the diseased material they could provide. Finally, during the winter months of 1944–45, I became intrigued by a disease occurring widely in the alfalfa caterpillar, *Colias eurytheme*, because of what A. E. Michelbacher and his graduate student Ray F. Smith had related on several occasions (Michelbacher and Smith, 1943). It was clear that I should try to become acquainted as rapidly as possible with the insect problems of California, and to outline some sort of plan to follow in organizing the research of the laboratory.

Accordingly, on 26 February 1945, I submitted to Dr. Meyer a brief sketch of those aspects of insect pathology that I thought might appeal to his scientific interest and thus retain his good will and keep up his interest in the proposal he was agreeing to in his negotiations with Dean Hutchison. I emphasized the long-range and more fundamental aspects in my memo to him. It was pointed out how epizootics of disease among insects could be studied in a manner that would contribute to our understanding of epizootics and epidemics in higher life forms. Because Meyer's original training was as a pathologist in veterinary medicine, my discussions with him about disease phenomena in invertebrates were received with sympathetic understanding. However, several weeks later, in presenting a similar outline to Smith I did not feel on as solid ground and was not at all confident as to its reception. I felt very strongly that he should know at the outset that I believed that there was little hope for the use of microbial agents in biological control unless it had its foundation on basic and fundamental research. However, I knew that practical results would be expected by some of my colleagues, and in short order. Moreover, I realized that the funding the laboratory received would, to a large extent, depend on the practical results that might stem from our research. As it turned out, I had nothing to fear. Harry Smith assured me that he understood my concern regarding the need for basic research and that he would support me in this approach, and this he did. Concerning the descriptive outline I presented him, he responded with a note saying, among other things, that he found nothing in it "with which I cannot agree heartily." This understanding attitude on Smith's part was a great psychological relief and convinced me beyond all doubt that I wanted to devote my life to teaching and conducting research in insect pathology.

Although in retrospect my proposal was somewhat naïve and overly ambitious, at least it cleared the air for all concerned as to what some of our initial goals were. Among the general subjects that I suggested we explore (agreeing, of course, to look into the citrus insect problems that were of immediate concern to Smith and Hutchison) were the following: (1) the general biological relationships between insects and microorganisms not excluding the phenomenon of intracellular symbiosis; (2) the study of all types of disease, noninfectious as well as infectious, and all types of microbial pathogens of insects; (3) the importance of microbial pathogens in the *natural* control of insects; (4) the importance of pathogens in the *artificial* control of insects; (5) the gross pathologies and histopathologies of insects; (6) the phenomenon of immunity in insects; (7) the principles of epizootiology of insect diseases; (8) the geographical incidence of insect diseases; (9) the diseases of insects reared in the laboratory or insectary, and their control; and (10) the teaching of insect pathology to advanced undergraduate and graduate students.

Of fundamental importance in the philosophy behind the 1945 proposal was the belief that insect pathology, that is, "what goes wrong" with insect life, is a fundamental, legitimate branch of entomology just as is insect taxonomy, insect physiology, insect morphology, insect ecology, and the rest. If a particular institution did not have an entomology department, insect pathology was a legitimate branch of invertebrate pathology, which is a legitimate part of comparative pathology, which is part of the biological science known as pathology in the broad sense. The applications of insect pathology, be they the use of microorganisms in the control of pest insects, or the suppression of disease in beneficial insects, or whatever, are not to be confused with the basic science of insect pathology itself. Of course, as the entomological sciences were then organized within the University of California, it was only sensible that the basic and applied aspects be under the jurisdiction of the laboratory unit set up for that purpose. It was my conviction that the primary initial contribution of our laboratory was to bring together the loose ends lying in plant pathology, entomology, and the various disciplines of biology generally—to bring them together into a well-rounded, coherent, and distinct discipline of Insect Pathology in which all manner of diseases and injuries known to occur in insects could at least theoretically be included.

This, then, was to be our main goal—the crystallization of fragments into a single, readily recognizable field of endeavor, complete with its teaching, research, and public service components. Although it may have sounded a bit pedantic when I proposed this goal to Professor Smith way back in 1945, his reactions did not give the impression that he thought so. Indeed, just the opposite—he heartily endorsed it, ending his comments of approval with an emphatic statement to the effect that it was time that entomologists recognized such a field as theirs rather than letting it go by default to plant pathologists because they, and mycologists, had done most of the investigation into the diseases of insects for the reason that most of the well-known diseases up to that time happened to be caused by fungi. Also, Smith had no objection to my feeling that from a practical standpoint, insect pathology should include certain aspects of the general field of insect microbiology and certain of the biological relationships existing between insects and microorganisms not pathogenic to them. He acknowledged the truth of the statement that no where was insect pathology being recognized as a distinct discipline, that virtually all approaches to the study of disease in insects had heretofore been made on the basis of specific problems (e.g., the diseases of the silkworm, honey bee, and Japanese beetle), or the long-time interests of individual scientists (e.g., Metalnikov, Paillot, Glaser), or a combination of both. Never before had an institution, as the University of California was about to do, established a unit devoted to all aspects of teaching and research in insect pathology—a genuine no-holds-barred carte blanche! How fortunate for me that I was to be associated with a man who had vision and was willing to support one's dreams. No doubt, in the beginning, Smith had only thought in terms of immediate biological control of insect pests, but it is to his credit, and my eternal gratitude, that he wholeheartedly supported the new Laboratory of Insect Pathology in all of its aspirations.

In the early stages of my discussions with Professor Smith, it was undecided as to whether the laboratory was to be located at Riverside, Berkeley, or Albany. Smith and I had agreed that Riverside might be the most appropriate location; my wife and I were quite willing to move to Riverside because at that time it was a beautiful, small city and had not yet become the victim of a population explosion with its accompanying problem of smog. However, Dean Hutchison overruled our decision pointing out that in order to conduct classes in insect pathology it would be preferrable to have the Laboratory and its staff on the Berkeley campus. (At that time the Riverside campus did not offer class instruction; it consisted essentially of the Citrus Experiment Station.) He suggested that we consult with Professor E. O. Essig, then chairman of the Division of Entomology and Parasitology on the Berkeley campus, to see if some space could not be obtained from this division. Because of the war-caused low enrollment in the university, the Division of Entomology and Parasitology had several rooms not in use. Through the kindness and generosity of Essig, we were assigned a large laboratory (Room 209) in the northeast corner of the second floor of Agriculture Hall together with an adjoining office (Room 210). Soon after we surrendered the office to a member of entomology's own increasingly growing staff so that really at the beginning Room 209 constituted the entire spatial facilities for our laboratory, except for part of a green house we borrowed at the Gill Tract for insect-rearing purposes. Dean Hutchison assured me, however, that as soon as appropriate funds became available, additional and more adequate facilities would be found. The room assigned for the laboratory previously had been used by graduate students. After considerable scrubbing and cleaning and rearranging of the desks and benches in the room, a comfortable and brightly lit laboratory emerged. I gradually moved in during the summer months; by late August 1945 it was my sole headquarters on the campus. I had left bacteriology and Dr. Meyer with a touch of sadness, but also with gratitude to him for having served so generously as my entree into the University of California.

In the meantime, I had been asked by Smith to submit a budget of sorts for our first year of operation. Realizing the limitations imposed by the meager funds that would be available, I was acutely aware that the laboratory's beginnings would have to be of a modest nature. (It must be remembered that this was before the era of federal grants-in-aid. Of course, the Experiment Station was supported with federal funds, but these were then limited and still are.) Accordingly, for personnel, I proposed that in addition to myself, the assistance of one full-time technical assistant was necessary for the efficient functioning of the laboratory. (It was my unexpressed hope and intention that within two or three years we would be able to add one or two professionally trained men to the staff, together with adequate technical assistance for each of them.) I also asked for the part-time services of a typist or secretary. In order to discourage neither Smith nor Hutchison with the probably real cost of maintaining the laboratory, I purposely submitted a very low figure for the amount of money required for supplies and equipment. My total request came to $3,500. This low figure was dictated also by the knowledge that there was very little money in the current university budget that could be diverted to setting up the laboratory at that time. (The organization of the laboratory had not been anticipated in time to have the request for funds incorporated in the previous year's annual budget request.)

Apparently my request for funds was greater than had been expected by either Smith or Hutchison. However, with characteristic innovation in problem solving, Professor Smith made available from his Riverside laboratory a compound microscope and a microtome. In addition, I was given $2,000 to "get things started." Because teaching would be involved, the Division of Entomology and Parasitology, which had jurisdiction over the curriculum in the entomological sciences, kindly paid for bringing in high-pressure steam lines to accommodate a small autoclave that had to come out of the $2,000. Remembering what happens to little acorns, I did not allow myself to be dismayed. I was doing the kind of work I loved best, and also I had my thoughts on the course that I was to teach in insect pathology. The Divisions of Entomology and of Beneficial Insect Investigations at Riverside gave no undergraduate course work, but they did accept graduate students. Most of the undergraduate courses in entomology in the university were given at Berkeley with a few at Davis. Aware of the fact that no textbook existed for the proposed course in insect pathology, I elected to defer giving the course as long as possible, in the meantime working on my lecture notes and trying to decide whether or not to attempt to do a "crash job" in writing a reference text. Eventually I decided to attempt it; I had the manuscript ready around the middle of 1948, although the publisher asked me to date the Preface as of February 1949. *Principles of Insect Pathology*, such as it was, was available for classes.

In spite of spending most of the summer teaching immunology and bacteriology and preparing the laboratory in Agriculture Hall, the spring, summer, and fall of 1945 were eventful seasons for my research as well. In January of 1945, after the completion of the investigation for Professor Smith on the bacterial disease of the potato tuberworm, *Gnorimoschema operculella*, I happened to find a microsporidian (*Nosema*) in the hymenopterous parasite *Macrocentrus ancylivorus* being reared on the same insect. These parasites were being reared for distribution in California to aid in the control of the oriental fruit moth, *Grapholitha molesta*. We also found a *Nosema* present in the insectary host insect (*Gnorimoschema*), as had been reported by Allen and Brunson (1945). In 1947 more detailed studies of this *Nosema* as well as another microsporidian, a *Plistophora*, occurring in the tuberworm were made and found to be new species (Steinhaus and Hughes, 1949). The reason for mentioning the finding of microsporidia in our insectary stocks in 1945 is that it served to accelerate, at a rather critical time, Smith's interest in getting our laboratory started. He indicated this in a letter on the subject that he wrote me on 5 February 1945.

The summer of 1945 was eventful for another reason associated with insect pathology. I have already mentioned being intrigued with a disease of the alfalfa caterpillar, *Colias eurytheme*, at that time the most destructive pest of alfalfa, which is one of the more important crops of California. This disease was called to my attention by A. E. Michelbacher and Ray F. Smith, who were working on the ecology and natural enemies of the insect. The cause of the disease had been identified for them by one of the university bacteriologists as being bacterial in nature, as had Wildermuth (1914, 1922) and Brown (1930) earlier. Ray Smith took me along on one of his field trips into alfalfa-growing areas in northern California where I could see epizootics of the disease among caterpillar populations. Upon examining specimens of the diseased insects it was soon evident that the wide-spread malady was caused by a nuclear-polyhedrosis virus. These findings were in accordance with the suspicions of Chapman and Glaser (1915) and the conclusions of Dean and Smith (1935), who had observed the diseased caterpillars in Kansas. We were later to make a more thorough study of the nuclear polyhedrosis in *Colias*; the significance of learning about it in 1944 and 1945 was the effect it had in determining the disease or diseases we should concentrate on to bring about practical results in the fastest manner possible so as to convince the powers that be that insect pathology was worth not only the moral and monetary support it had been promised, but to get the amount of this support substantially increased. Although I was quite willing to study the disease of scale insects, as Harry Smith and Dean Hutchison wanted, I really was not confident that we could come up with enough to create the necessary confidence and "positive thinking" necessary to generate funds for the kind of work we wanted to do.

In a talk before the Entomological Club of southern California, at Alhambra on 1 June 1945, I took advantage of the opportunity (with Professor Smith and other entomologists from the university in the audience) to stress the great need for basic research in the field of insect pathology; I de-emphasized—mostly by omission—the potential use of fungi and adlibbed the fact that the climatic differences between California and Florida were such that even if the friendly fungi were effective in Florida against scale insects, there was not the same likelihood that they would be effective in California, at least not until after a great deal more fundamental research were done. I went a step further by emphasizing the potentialities of the possible use of such diseases (polyhedroses) as those we had observed causing striking epizootics in the western hemlock looper, *Lambdina fiscellaria somniaria* (called to my attention by R. L. Furniss) and the alfalfa caterpillar in California.

After this talk Professor Smith indicated that he understood why I had emphasized the need for basic research before practical results could be expected and why I apparently wanted to initiate our research program on a broader base than just the diseases of scale insects. As he had done for others in similar situations, Smith gave thoughtful comments that provided me with the reassurance I needed and sent me back to Berkeley a happy man. As I rode the train back, I was determined that as soon as possible I would concentrate on the virus disease of the alfalfa caterpillar to which Ray Smith had introduced me and would use it to attempt to demonstrate that the insect pathogens could be manipulated to control harmful pests by methods similar to those used to control insects with chemical insecticides.

In keeping with my promise to Dean Hutchison, soon after becoming established in "Ag Hall" I began to concern myself with the so-called *Bacillus C* infection in the California citrus red scale, *Aonidiella aurantii*, previously studied by Sokoloff and Klotz of the Citrus Experiment Station and reported by them in 1941 and 1942. Politically, I found this to be a rather delicate problem on which to work, inasmuch as the claims for the possible control use of this sporeforming bacterium by Sokoloff had been vigorously questioned by some of the entomologists at the Citrus Experiment Station. Others at the Station were strong supporters of Sokoloff's claims, and the dean had received letters from individuals, not on the staff, accusing him of suppressing or not adequately supporting Sokoloff's work. I was somewhat reluctant to serve as a "referee" of sorts in this controversy, but I could not very well refuse to reinvestigate the problem since both Dean Hutchison and Professor Smith, as well as others at the Riverside Station, were anxious that the matter be clarified. Having obtained a culture of the original *Bacillus C* from Klotz, who was most cooperative (Sokoloff had by this time taken a position elsewhere), I soon ascertained that the organism was actually a strain of *Bacillus cereus*, one of the most common sporeformers occurring in nature. My efforts (and later those of an assistant, Karl Snyder) failed to repeat the results obtained by Sokoloff and Klotz, and we could not demonstrate the reported invasive properties of the bacillus. Lethal effects were obtained with broth cultures of the organism, but these were shown not to be associated with any real invasion of the insect by the bacillus. These results were reported to Smith and Hutchison, who seemed convinced that further experimentation was unnecessary. The controversy gradually died.

In the meantime I was increasing the amount of time I could spend on studying the polyhedrosis of the alfalfa caterpillar. Making trips to the alfalfa fields incidentally gave me an opportunity to study an outbreak of bacterial septicemia in grasshoppers. It is doubtful that it was caused by the same organism that d'Herelle attributed to be the cause of epizootics in Yucatan, Mexico, in 1911. Projects of various types were being requested or suggested from numerous sources, healthy signs of interest in our budding Laboratory of Insect Pathology. In addition to citrus insects, there were crop pests (e.g., corn earworm, the California oakworm, aphids, mealybugs, the Fuller rose beetle, armyworms, and cutworms) that were the special concern of growers of particular crops. These growers, particularly when organized into groups, were sometimes quite emphatic in their requests that we work on the diseases of the insects of their concern because, unfortunately, the use of microbial control agents appeared as a panacea to some of them. For this reason, in publications and whenever I spoke to grower groups, I would explain why we had intentionally decided on a course of doing a great deal of basic research on which to base the applied work we expected to do, and that in any case we were not expecting panaceas. One of the proposed projects we attempted to initiate was a study of the microbial disease of wireworms. In order to carry it off, however, we felt that we needed the cooperation of workers in other states, particularly those in the U.S. Department of Agriculture. Unfortunately, probably because we also needed financial support, for one reason or another our plans were never executed, and work on the problem never did get underway.

In early December 1945, Professor Smith informed me that it was necessary for him to make up the budget estimates for the fiscal year 1946–47. In line with our previous discussions, he agreed to ask for a laboratory technician to assist me—I was still the only member of the staff of the Laboratory of Insect Pathology—in addition to a request for $1,200 for supplies and equipment. Particularly significant was his decision to arrange matters so that the insect pathology unit would have it own budget, and be independent of any other funds in the division. Up to this time the pathology work had been supported from excess or unused funds originally allocated to a large project on the oriental fruit moth being investigated at our insectary in Albany. I was naturally very pleased with this development and encouraged by what was, under the circumstances of the times, truly generous support rendered by Harry Smith in response to my requests. Although I knew we needed more than $1,200 for supplies and equipment if the laboratory were going to get anywhere, I knew Smith was doing his best for the project and wanted me to feel that the funds he allotted were adequate. In a letter dated 5 December 1945, he comments: "I am also requesting $1200. . . . This seems rather large, but I know that there are still several items of equipment needed to make your facilities reasonably adequate, and it will, of course, be necessary for you to do some field work, for which we will need to provide funds." (This $1,200 plus the initial $2,000 for equipment seem pretty paltry when compared with the $750,000 that Canada spent to build and equip its first Laboratory of Insect Pathology at Sault Sainte Marie [personal correspondence, de Gryse, 1949].)

Of significance was the fact that even at this early period in the development of our work, considerable interest in what the University of California planned to do in the field of insect pathology was being engendered in various places. I particularly recall the letter of inquiry written to Professor Smith by A. B. Baird, chief, Biological Control Investigations, Division of Entomology, Canadian Department of Agriculture. Mr. Baird had indicated his organization's interest in the subject and wished to know what was being planned at the university. Also, I was visited by O. C. Woolpert, who made me a very attractive offer to join his federal laboratory in the eastern United States as head of a research unit concerned with various biological relations between microorganisms and insects, including certain phases of insect pathology. Although the salary offered was considerably more than I was getting at the University of California, I did not feel that the opportunities for research were any greater, or perhaps as great, as those being offered to me at the University of California, nor did his organization have the atmosphere I cherished, especially just recently having come from a federal agency. His offer, however, did indicate to me the awakening interest in the basic relationships between insects and microorganisms. Similar evidences of interest were being indicated by invitations to appear on entomological programs of various sorts, requests for special scientific or semiscientific articles on the work we were doing or planned to do, and newspaper and magazine articles pertaining, unfortunately, to the more dramatic aspects of our work. Insect pathology, as a separate and distinct discipline, was definitely in the process of being born. What role was our laboratory to play in nurturing this newborn child of science and in aiding its development?

2. RESEARCH AT THE LABORATORY OF INSECT PATHOLOGY

Here is an excellent field
for experimentation.

Harry Smith

I

After the end of hostilities in Europe on 8 May and in the Pacific on 14 August 1945, the administration in the University of California generated a surprising attitude of liberality in increasing the amount of research devoted to the problems of agriculture. Agriculture is not only one of California's principal industries, but California leads the entire United States in the number and amount of marketable crops it produces. Of this number of crops—in the neighborhood of 225—virtually every one of them is attacked by insect or disease of one type or another. And, although DDT and other new chemical insecticides were becoming available in 1945 to farmers and growers, new and better ways of controlling pests by strictly biological means were being sought. The rest of the United States had lagged in this area of applied science; it is to the credit of California, and such leaders as Harry S. Smith, that California boldly embarked not only on a program encompassing the use of entomophagous insects in the control of harmful species but (in cooperation with the federal government) on a program of using insects to destroy harmful weeds, and now, it had established the "third leg" of the stool, the use of microorganisms (including viruses) to control insect pests. But this third part of the triad could not rest on the efforts of one man. Not only did I need assistance, but I was determined that as soon as I could convince my administrative superiors, I would ask that there be a gradual increase in the number of professional staff employed in the Laboratory of Insect Pathology.

In June 1946, final approval for the appointment of a senior laboratory technician was received. Some weeks previously I had already made contact with Kenneth M. Hughes and had offered him the position, which was set up essentially to assist me in the general investigational work and to study pathological changes in diseased insects, particularly the histopathological changes. In the meantime, the university was preparing a supplemental budget for 1946–47, in which we included a request for a second laboratory assistant for the purpose of assisting in the production of mass culture of insects and disease organisms at the Gill Tract laboratory in Albany. It had already become apparent that in order for us to expand our production and testing work adequately, we would have to obtain facilities in addition to those we had in Agriculture Hall on the Berkeley campus. There were no facilities available there unless space were to be taken from someone else. The only other facilities controlled by Smith's division in northern California were those at the Gill Tract in Albany, where, as I have already mentioned, work on entomophagous insects and on insect parasites of the Klamath weed was being conducted. A request for a professional position, a junior insect pathologist in the Experiment

Station, was also made in this supplemental budget. This request, however, was not approved, although Dean Hutchison was very favorable to the idea; so the request was made again in the biennial budget for 1947–49. It was some months later that funds were provided to hire a graduate student as a technician who was to assist in the rearing of insects for experimental purposes—and later, hopefully, for the purpose of mass-producing insect pathogens.

The supplemental budget also included an additional $1,700 for supplies and equipment for insect pathology. Therefore, as matters stood, at the end of the 1945–46 fiscal year, the Laboratory of Insect Pathology had a total supplies and equipment budget of $2,900, plus the provision for two laboratory assistants and a good indication that a junior insect pathologist would be provided in 1947. At this point it is interesting to recall that whereas six months earlier Smith considered $1,200 to be a "rather large" sum for our work, he now felt that the amount of $2,900 was rather small, since, as he said, "I am sure you will have considerable difficulty equipping the laboratory at the Gill Tract (space had been found for us there) and caring for all your other expenses and equipment with this meager sum." My aspirations for an expansion of the laboratory and its activities seemed to have the possibility of gradually materializing.

When Ken Hughes reported for duty on 1 July 1946, I was not sure as to just how his services as a technician might best be put to use for the rapid and sound development of our work. It was obvious that with his zoology-laboratory technical background, he would not be interested in field applications—and he frankly told me so. However, since it might be some time before field tests could be arranged, I felt that in the meantime we should concentrate on the basic research necessary as well as on the development of methods of mass-producing insect pathogens in quantities adequate to make significant field tests. I envisioned two initial types of approach to the problem of the field application of microbial pathogens of insects: (1) the perodic treatment of an infested crop with large quantities of the pathogen applied as sprays and dusts, in the manner used with chemical insecticides, and (2) the introduction of microbial pathogens into populations of insects in specific areas with the hope that they would maintain themselves on their own accord or with intermittent introductions of the pathogens.

One of the first things Hughes and I set about finishing up was the microsporidian infections, mentioned in the preceeding section, as afflicting the potato tuberworm and other insects being reared in the insectary at the Gill Tract. We found two kinds of Microsporidia involved. After working out their life histories, both were found to be new species, *Nosema destructor* and *Plistophora californica.* Since we had a number of projects going and this was slow and somewhat tedious work, it was not finished actually until early 1948 (Steinhaus and Hughes, 1949). At the time Hughes "came aboard" I was deeply involved with my investigation of the polyhedrosis of the alfalfa caterpillar, *Colias eurytheme*, which had intrigued me since late 1944. I had just begun a study of the histopathology of the disease, and was anxious to begin observations with the electron microscope to see if I could find within the polyhedral inclusion bodies the spherical particles described by Glaser and Stanley (1943) from the polyhedra of diseased silkworms or the "virus aggregates" reported by Bergold (1943).

Although I had had some experience with operating an electron microscope while I was in the Public Health Service (using the microscope at the Armour laboratories in Chicago, I had attempted to ascertain what happened to the rickettsiae of Rocky Mountain spotted fever and typhus fever after they had been made into vaccines) and had done some work on the instrument at Berkeley, I found that the long hours of sitting at the instrument required more patience than I had. Accordingly, I asked Ken if he would be interested in trying his hand at it with regard to the polyhedrosis of the alfalfa caterpillar. He said he would. At the time the only electron microscope on the Berkeley campus was in the Department of Physics. We were given permission to use it, but as was frequently the case in those days, the instrument was used by such a variety of personnel that it was constantly breaking down. Nevertheless, Hughes quickly learned the necessary techniques and was on the way to becoming our principal electron microscopist.

In 1947 and 1948, in correspondence with Bergold, in Germany, I had learned that instead of the spherical virus of Glaser and Stanley or the virus aggregates he had previously observed, Bergold had found that the true whole virus particles of the silkworm and the nun moth polyhedroses were shaped in the form of rods approximately 40 by 300 millimicrons in size. Using techniques similar to those used by Bergold, Hughes found similar rod-shaped particles in the polyhedra characteristic of the disease that I had been studying in the alfalfa caterpillar. Hughes's very first electron micrographs of this insect virus appeared in a paper concerned with a general description of the polyhedrosis (Steinhaus, 1948).

In February 1947, while examining some obviously diseased specimens of variegated cutworm, *Peridroma margaritosa*, I observed in the cytoplasm of the cells of the fat tissue large numbers of small granular inclusions (Steinhaus, 1947). I was not certain as to their nature, but having read Paillot's work on what he called *pseudo-grasserie* I thought the disease might be somewhat similar, but the particles themselves were too large to be virus particles. Surmising, as with polyhedra, that the granules were inclusion bodies and the virus particles lay within the granules, Hughes, Mrs. Harriette Wasser (another technician recently granted by the university) and I discussed the idea. It was agreed that the thing to do was to purify (wash) the granules, and then attempt to dissolve away the protein covering with weak alkali in order to reveal the virus particles we hoped were contained within the granule. However, I stressed the point with Hughes and Wasser that since certainly no one else in the world could be working on such a problem, there was no hurry in view of all the other matters we were attempting to accomplish. After all, did not Lord Chesterfield say to his son, "Whoever is in a hurry, shows that the thing he is about to do is too big for him"?

It just so happened that at the time, W. F. Sellers was stationed at Fontana, near Riverside, carrying out liaison work between the biological control people in the U.S. Department of Agriculture and the university, and those in the Commonwealth Institute (then Imperial Bureau) of Biological Control—by whom he, an American, was employed. Although his primary concern was entomophagous insects, he had become acquainted with

Gernot Bergold in Germany and knew of the latter's interest in insect viruses. One day, during the fall of 1948, Sellers came to the lab to show me some photographs he had received from Bergold. As he spread the electron micrographs out before me, I was speechless; they showed rod-shaped particles virtually identical to those we, by this time, were finding in the granules from the variegated cutworm! Bergold's virus was one he had found in the pine-roller, *Cacoecia murinana*, in Europe. And I had told Hughes and Wasser that no one else could possibly be working on such a disease! I immediately asked Sellers if Bergold had published his results. "No," he replied, "but I believe he has prepared a paper on his findings and it may be in press." I took the liberty of writing Bergold about the photographs Sellers had shown me; Bergold generously sent a copy of his manuscript as well as other photographs. In the meantime, I had busied myself putting our findings down on paper, and in November we submitted the manuscript to the *Journal of Bacteriology*, where it was published in the February 1949 issue (Steinhaus, Hughes, and Wasser, 1949). Bergold's paper, "Über die Kapselvirus-Krankheit," appeared in the *Zeitschrift fur Naturforschung* (3*b*, 338–42) in 1948, but after we had submitted our own manuscript. Thus Bergold, fairly and squarely, had established the viral nature of the granules, although had we not been taking our time we could have had priority. There is a lesson in this incident somewhere for those concerned with priority, and sometimes, for the good of the scientist and his institution, priority is important.[4]

Bergold's magnificent discovery was marred only by a statement in his paper, based upon bad advise from colleagues in Germany, suggesting a kinship between the granules and Microsporidia. Also, Bergold did not see his observations as confirmation of the suspected virus nature of Paillot's "pseudo-grasserie" particles, which the latter investigator had discovered and described in 1926, 1934, and 1937, and to which I had referred in my 1947 paper. It was only two or three years later, when I showed Bergold some of Paillot's original histology slides which Madame Paillot had sent me, that he acknowledged the gross similarity.

There was another interesting aspect to this work; it had

to do with terminology. Because histopathological examinations of the diseased tissue cells, whether on fixed slides or in freshly dissected wet-mounts, showed the presence of characteristic granules in large numbers, we referred lightheartedly to the disease in the variegated cutworm as a *granulosis.* This was purely laboratory vernacular. When the journal paper, and about the same time the section on this group of diseases in my *Principles* was written, I could not think of a more precise and a more scientific-sounding word or term. In literary desperation the paper was submitted using the term *granulosis* but in apprehension lest the editor demand something better. Actually, inasmuch as the appearance of the diseased tissue through the optical microscope was distinctly granular, and inasmuch as the objects themselves appeared as minute granules, the term was descriptive of the cellular pathology of the disease. In any case, with the publication of the paper, the term caught on, and what had been a frivolous laboratory term became the name of a large and important group of virus diseases. Bergold had referred to the granules as *capsules*, a term we were reluctant to use because it might cause confusion with the well-known polysaccharide capsules of pneumococci. Yet, admittedly, although the word *granulosis* (plural *granuloses*) might be satisfactory as the name of the disease, there was nothing very descriptive about calling these minute objects "granules." After considerable correspondence between Bergold and his associates and our group, we amicably agreed on the terms *granulosis* for the disease and *capsule* for the granule or inclusion body. And just as one would say "psittacosis virus" or "poliomyelitis virus," so by combining the name of the disease with the word *virus* one would say "granulosis virus" (see Hughes, 1958).

Incidentally, during the writing of *Principles of Insect Pathology*, I became quite conscious of the matter of terminology and was dismayed at the careless and imprecise use of many words and terms by authors of papers writing on the diseases of insects. Of course, who was I to set myself up as a judge, but to this day I cringe when I see or use certain expressions. However, one usage I was determined to do something about, not only in the publications emanating from our laboratory but in our teaching, was that pertaining to the virus diseases of insects. Other than the granulosis-capsule affair just described, the diseases characterized by the presence of polyhedral bodies (polyhedra) in their tissues were in sad terminological disarray. Early workers (e.g., Reiff, 1911; Howard and Fiske, 1911) in the United States referred to the well-known disease in the gypsy moth, *Porthetria dispar*, as *wilt*, *wilt disease*, or *flacherie.* The terms wilt and wilt disease were apparently carried over from similar terminology used for certain plant diseases. The disease in caterpillars of the nun moth, *Lymantria monacha*, was designated by the Germans as *Wipfelkrankheit.* And most leading languages had their own common name for the disease in the silkworm, *Bombyx mori*; thus, it was called *grasserie* in French, *giallume* in Italian, *Gelbsucht* in German, and *jaundice* in English. And the Asiatic languages had their equivalents. I had nothing against the use of common names, if meaningful. After all, the same phenomenon occurs with regard to the common names of human diseases, of animals, and of plants. However, I did object to the use of such terms as polyhedral disease or polyhedral virus because these combinations were saying many-sided disease and many-sided virus, which usually was not what was at all meant. Accordingly, after having made this same mistake myself in *Insect Microbiology*, I began (in 1948) using the word polyhedrosis (plural -es) for the disease, and the term polyhedrosis virus to designate the virus, just as one says tuberculosis and tuberculosis bacillus. (*Webster's New International Dictionary* followed suit, as they also did with the use of the word *symbiote* versus *symbiont.* See chapter 6.) Later, when it was found that in some diseases the polyhedra originated in the nucleus and in other diseases they originated in the cytoplasm of the infected cell, the adjectival forms *nuclear polyhedrosis* and *cytoplasmic polyhedrosis* came into use. The words *nucleopolyhedrosis* and *cytopolyhedrosis* are also used to indicate these two types of disease. My use of the words *epizootic*, *enzootic*, and *epizootiology*, as used by those studying the diseases of invertebrate animals, was also challenged by many entomologists, including Prof Harry, who liked to refer to outbreaks of disease among insects as *epidemics* even though the word literally means "on the people." However, by constantly making clear my meaning, by late 1954 (Steinhaus, 1954), most insect pathologists agreed that we should follow what is standard terminology among most animal pathologists. Besides, it was a bit ludicrous to end up with a sentence saying something about "an epidemic among an epidemic of insects." It was because of the unnecessary variation of terms in the literature, the confusing terminology used by different authors, and the considerable amount of incorrect usage etymologically speaking that we finally attempted to provide a glossary of terms to assist students and others writing of disease in insects (Steinhaus and Martignoni, 1962, 1967, [1970]). [Many of these terms also appear in the new *Dictionary of Comparative Pathology and Experimental Biology* by Leader and Leader, 1971. MCS.]

In any event, terminology aside, it was the fascinating observations we were making on the virus diseases of the alfalfa caterpillar and the variegated cutworm that introduced us to the diseases known as the polyhedroses and the granuloses. And all of it was exciting and fun. In the words of Ernest Goodpasture, the vertebrate pathologist, we looked "at science as a career like fishing. If you can make a living out of it, it's a fine thing. You are making a living out of a pleasure—the satisfaction of your curiosity."

By this time, Ken Hughes was developing into one of the field's finest electron microscopists. Although he did not possess the "union card" of a Ph.D., he was doing the work of a professional. Suddenly, in 1949, we were given a grant to purchase our own electron microscope; because we lacked sufficient space for it, and the Department of Plant Pathology had the space, we collaborated with that department in using the instrument. Hughes represented our laboratory in this collaborative arrangement. Later, with the understanding that Hughes, who held a master's degree, would complete his Ph.D. in two more years, I was able to get the dean to agree to promote him from his position of a laboratory technician to that of junior specialist, then to assistant specialist, then associate, which was a type of semiprofessional position in the Experiment Station.

Once he had his doctorate, I was confident that Ken could be advanced to full professional status and be completely on his own. With this anticipation, I urged him to use his talents with the electron microscope to investigate the precise nature of the development of insect viruses within the cells of their hosts. In his slow, methodical, thoroughly excellent manner he proceeded to do so (see Hughes, 1952, 1953, 1958). But almost from the time I asked Hughes to complete his work on his degree, I could tell that something was troubling him.

I believe it was sometime in 1954 that Hughes first indicated that he did not feel that he wanted a career in science and that he would like to escape the "crowds and traffic of the city." He assured me that he was completely content with his position at the laboratory and that he appreciated the freedom to do research that had been given him; nevertheless, he had decided not to attempt to finish his Ph.D., and that when he could find a place to get away from it all, he would like to do so. Hoping this was but a passing psychological phase, I urged him to stay on because of the obvious outstanding future he had at the university. Ken possessed more than superb technical skills. With his fine, intelligent, and philosophical mind Hughes would have made an excellent university instructor, and it was my hope that eventually he could take over some of the instructional duties so that this part of our activities could be expanded. However, Ken did not share this confidence in himself, and after several conversations with his wife, Luree, I knew that it was just a matter of time until Ken would, indeed, have to "get away from it all." This he did in the spring of 1956, resigning on 1 May to go to Haines, Alaska, where as a sort of arctic Thoreau he was able to go to the wilderness, and by taking a job in a supply warehouse he was also able to "get out of science." There was no question but that Ken was hearing a different drummer and marching accordingly. Liked by everyone, everyone wished him well.

The summer and fall of 1946 turned out to be a period which had convinced Harry Smith that my aspiration to build insect pathology on a broad base—while recognizing that it had to be kept within the bounds forced upon us by the limited funds, limited space, and limited personnel—was a logical one. My philosophy, for both teaching and research—that insect pathology is a discipline that legitimately is concerned with whatever "goes wrong" with life processes in insects—was now accepted by Smith. Moreover, he understood the reasons for almost simultaneous probes into the various types of disease, caused by different types of microbial pathogens (protozoa, fungi, viruses, bacteria, nematodes, etc.), in different types of insect hosts. Thus, while our reports and published papers gave the appearance of not being able to settle on a single, definite project, we were maintaining unified forward thrust of insect pathology as a whole. We were unable to spread as widely as we would have liked—for example, we had to neglect, for the time being, the noninfectious diseases, such as diseases caused by deranged physiology and metabolism, nutritional diseases, genetic diseases, teratologies, physical and chemical injuries, etc.—but we resisted becoming insect virologists, insect bacteriologists, insect mycologists, insect protozoologists, and the like. As overly ambitious as it may sound, we attempted to be insect pathologists, willing to tackle anything.

In the midst of working on a coccidian disease of *Anagasta* and *Plodia* (Steinhaus, 1947), a bacterial disease of grasshoppers (Steinhaus, 1951a), and the nuclear polyhedrosis of the alfalfa caterpillar (1948), among other things, I was pleasantly surprised when Harry Smith asked if it might be possible to distinguish two physiologically (but not morphologically) different strains of the mealybug *Pseudococcus maritimus* by comparing their mycetomes and intracellular bacterialike symbiotes of the two strains (one of which feeds on grape and is attacked by a chalcid parasite, and the other on citrus, pear, and apple, and is not attacked by the chalcid) with each other and with the type form of the insect found on *Eriogonum* near Santa Cruz, California. We found there were differences in the symbiote picture, but no conclusion was reached as to the stability of these differences. The importance of this was that Professor Smith now was willing also to permit us to include aspects of nonpathological insect microbiology in our purview so long as it did not alter the main thrust of our work in pathology.

But these varied activities also meant something else. We needed a third professional worker. A request for a junior insect pathologist was honored by Smith but postponed by the dean until it could be supported out of next year's budget. Nonetheless, I needed help in the field on the alfalfa caterpillar project. After explaining my needs to the dean, he responded, "All you need is an extra pair of hands." Sensing that he was about to offer some type of help, I agreed, trusting that I would be lucky enough to find a good head coming along with the pair of hands. It was arranged that enough funds would be available to hire some student help. This might be all that was needed for the present, and I had in mind precisely the student who could give that help. He was a young paratroop veteran who had impressed me with his dedication by commenting that, so help him, he was so interested in the subject that he was just going to have to be an insect pathologist. The young man, who was to become my first Ph.D. graduate, was Clarence G. Thompson—for some inscrutable reason christened "Hank" by his wife and called by this appellation by all his friends. Hank accepted the job as a student helper.

It was not long before it became apparent that Hank Thompson had that good head even though all we could pay him for was his hands. To encourage him, and because he deserved it, I included his name in the authorship of our papers (e.g., Steinhaus and Thompson, 1949*a*, *b*, *c*) when we came to publish our work. When it came time to select a research problem to satisfy his Ph.D. requirements, I offered to permit Thompson to use that part of the research in which he was most involved as a basis for his thesis. He accepted the arrangement, and through his natural field skills and a great deal of hard labor we completed the first successful experiments in the control of a field crop pest through the artificial distribution of an insect virus (Steinhaus and Thompson, 1949*b*; Thompson and Steinhaus, 1950*a*, *b*).[5]

In 1947, as Thompson was completing his Ph.D. requirements, we received authorization for the junior insect pathologist position that we had requested; after considering several candidates, the appointment was offered to Hank. It was not long before he was promoted to assistant insect pathologist. Although quiet and unassuming by nature, Thompson was well liked by his colleagues. He was the willing recipient of, and participant in, practical jokes—such as the time his friends filled his tobacco pouch with caterpillar frass, which he attempted to smoke for some time before he suspected that something was awry. His "quiet" nature has, unfortunately, extended to the area of scientific publication. Apparently, this gifted and highly competent insect pathologist hates to write. As a result, although he has accomplished a great deal of important research, he has rarely published his findings except when he has been "forced" to co-author with a colleague. Fortunately, Hank attends meetings, presents papers, and is easy to communicate with, thus most American insect pathologists have an idea of the amount of work he has accomplished but deplore his reluctance to stop researching long enough to write up his results for publication. Certainly no one can accuse Thompson of being publication happy.

One of the several problems on which Thompson worked while with the laboratory helped us out politically. Almost from the time we initiated the program in insect pathology, the forest entomologists—particularly those with the U.S. Forest Service stationed on the Berkeley campus—had been importuning us to work on the diseases of certain insect pests of forests. This pressure became intensified as the Canadian Forestry Service initiated striking projects in insect pathology. F. P. Keen, C. B. Easton, and especially George R. Struble, all of the federal agency, were particularly interested in forming some sort of collaborative approach to the study of disease in forest insects in California. Apparently one of the federal administrators, P. A. Annand, discussed the possibility of a joint department with Professor Smith, who, as did I, did not react positively to the suggestion. However, we did collaborate to the extent that we could, and always responded to their requests for diagnoses of any diseased or dead insects sent or brought to our laboratory. Fortunately, Hank Thompson became interested in virus-caused epizootics in populations of white fir sawfly, a member of the *Neodiprion abietis* complex in the Sierra

Nevada in northern California. This enabled our laboratory to demonstrate its genuine interest in the diseases of forest insects and willingness to cooperate with the federal Division of Forest Insects to the extent we were able.

In the early spring of 1953 Fred C. Bishopp, then assistant chief in charge of research of the federal Bureau of Entomology and Plant Quarantine, came for a visit. He explained that his agency was anxious to expand their research in insect pathology (they already had going projects on the diseases of bees and the Japanese beetle) and that they had received authorization to establish an appropriate laboratory, and wondered if I would be interested in heading it up. The offer was a very attractive one, and I was honored to be considered, but having once worked for another federal agency, I could rather quickly respond that I preferred the academic life. He then asked if I could recommend someone for the job. Obviously the best man then available was our own Hank Thompson. In all fairness to Hank, I could not withhold this information from him or Bishopp.

Later, I described the position to Hank and suggested that if he were interested (though secretly I hoped he would not be) he should write J. E. Hambleton, in whose division the laboratory would be located. Hank found himself in a quandary because, as he said, he was quite happy to remain with the University of California. He requested some weeks to make his decision, finally deciding that the challenge of initiating his own unit was an opportunity he should not deny himself; in the meantime I had rationalized that if Hank decided to accept the position with the U.S. Department of Agriculture, at least we would have one of our own men in an important central position as far as the development of insect pathology in the United States was concerned. Thus it was that, a year later, in May of 1954, Hank Thompson joined the Section of Bee Culture and Biological Control (later called the Beekeeping and Insect Pathology Section, and still later the Insect Pathology Laboratory, and currently the Insect Pathology Pioneering Laboratory) of the U.S. Department of Agriculture's Entomology Research Branch at Beltsville, Maryland, where he remained as principal insect pathologist until he transferred to the Forestry Sciences Laboratory at Corvallis, Oregon. We shall have occasion to examine his role in these agencies in a subsequent chapter. [Unfortunately the chapter "Insect Pathology in States Other Than California" was never written. There these agencies would have been discussed. MCS.]

II

On 1 July 1947, the name of the Division of Beneficial Insect Investigations was changed to Division of Biological Control. During staff-meeting discussions on the subject of choosing a more appropriate name, I expressed my regret that of the several choices (e.g., "Division of Applied Insect Ecology" was also considered) all reflected only the applied aspects of what we were doing. I again tried to emphasize that the use of microorganisms in the control of insects (i.e., microbial control) was only one of the applications of insect pathology, and that in any case the applications should not be equated with the basic science, insect pathology. However, from the standpoint of the entire division it was only logical that the name Division of Biological Control be chosen as the most appropriate. Moreover, the name Laboratory of Insect Pathology still remained to indicate that it was an autonomous unit within the division.

By 1947 our research program had developed to a point that it was necessary to conform to the Experiment Station practice of having definite, formal projects that were reported on quarterly (later semiannually) and annually. To avoid pinning down the research too specifically, we gave broad, umbrella titles to most of our six original projects. These were:

Project No. 1306. Virus diseases of insects.
Project No. 1307. Protozoan diseases of insects.
Project No. 1308. Fungus diseases of insects.
Project No. 1309. Bacterial diseases of insects.
Project No. 1310. The pathological and microbiological examination of insect specimens submitted for identification of the possible disease-producing agents present. (In

the laboratory this was known as the "Diagnostic Service Project.")

Project No. 1311. Miscellaneous research and experimentation.

Late in 1947 it appeared that the laboratory might be the beneficiary of some special federal Experiment Station funds made available through the Flannagan-Hope Act. This windfall was realized by February 1948, and we made plans accordingly. I shall never know why, but apparently Assistant Dean Freeborn, who in his capacity as assistant director of the Experiment Station, carefully read all Experiment Station research reports, and had been impressed with those from our laboratory. He had spoken in a very complimentary fashion to Professor Smith regarding our work. I am reasonably confident that this was a factor in our being assigned a major portion of the Flannagan-Hope funds. Also, because of the promise shown by our work on insect viruses as possible agents of insect control, we had been able to convince the officials of the need for more basic research on viruses. As a result of receiving the Flannagan-Hope funds, we were directed to establish a special project to add to those just listed: Project No. 1333. The nature and properties of insect viruses.

Much of the responsibility for conducting this project was to be Ken Hughes's, and it was at this time that we were given funds to employ Harriette Block Wasser. (Mrs. Wasser included her maiden name, Block, which occasionally caused her name to be cited in literature as "Blockwasser.") Thus, at this moment in time, Hughes, Thompson, and I found ourselves blessed with the help of three laboratory technicians: Karl Snyder, graduate student (who was rearing most of our experimental insects, with advice from Glenn Finney, who operated the insectary at Gill Tract), Eunice Crapuchettes (a former student in my bacteriology and immunology classes, who was married in December, 1948, and replaced by another former student from the same class, Helen Owsley), and Harriette Wasser, who came experienced primarily in biochemistry, a field we anticipated a great need of as our work continued with the insect viruses.

While on the subject of technicians, I wish to divert slightly from the subject at hand to pay testimony to the technicians, assistants, stenographers, and secretaries—undignifiedly termed "nonacademic employees" by the university—these unsung heroes who served our laboratory so well. Personally, I never had an assistant, technician, or a secretary I did not like (though I do not claim that this feeling was mutual)—I have been fortunate indeed! In virtually all cases they terminated either because they married and moved away or because they began to raise families—an activity unfortunately usually incompatible with continuing a career—or in the case of graduate students they finished their degrees and moved on to other jobs. For an administrator, having a competent secretary is of the highest order of importance. Without a doubt, the most underpaid person on any staff is an efficient secretary, and the most overpaid individual is an inefficient secretary! As my mind recalls these alter egos, my gratitude, esteem, and approbation goes out to each and every one of them. During the nineteen years I was on the Berkeley campus, there were Eleanor Sneddon, Natalie Herring, Leona Elsken, Hazel Flick, Grace McNaulty, Thelma Young Andriese, Jeanne Hennecke, Pat Jenna, and Grace Lee—all members, at one time or another of our stenographic-secretary staff. Then, in addition to Hughes, Clark, Snyder, Thompson, Wasser, Crapuchettes, and Owsley, already mentioned, those who served with me as technicians or assistants (though at times there had to be some reassignments because of changes in work-load demands) were Jean Miller, Catherine Boerke, Helen Court, Gordon Marsh, Mariece Batey, Richie Bell, Elizabeth Jerrel, Joyce Dineen, Nancy Scherer, William Whitehead, Ruth Leutenegger, and Henry Scott. Others who were necessary because of their talents in rearing experimental insects for the entire laboratory were Robert Langston, J. J. Alcedo, Jr., James Milstead, and Carl Reiner. Bob and Jim later became technicians for Mauro Martignoni. In addition to the above mentioned personnel there were also Grace Chang, David McMullen, A. M. Tanabe, R. J. Scallion, and W. R. Kane, who were technicians for Tanada, Martignoni, and Falcon. Daphne Stern, Stella Quam, and Josephine Foley were secretaries at Gill Tract. Competent, loyal, dedicated colleagues all!

Like Hank Thompson and Ed Clark, another of our technicians who began working for us while doing his graduate work was Irvin M. Hall—but more about Irv later. Last but not least, there was Lorenne Sisson, one in a million as far as the ultraefficient handling of overall divisional affairs was concerned. It was she as secretary, and later administrative assistant, to the chairman of the division at Riverside who ultimately handled such matters as correspondence, reports, and budget requests. She contributed mightly to our welfare during all the years we were a part of Biological Control headquartered in Riverside.

During those early years of the Laboratory of Insect Pathology there was a great esprit de corps plus unlimited enthusiasm on the part of all of those who were working there. Here was a new field, so much to be done, and seemingly so few to do it. Not only did the group work hard, but there always seemed to be time for a little fun, too, during the course of sailing the "Good Ship Inpath" as we referred to ourselves on occasion. I am indebted mostly to Ken Hughes for refreshing my memory with some of the following stories of those who were the pioneers of insect pathology at Berkeley.

Hank Thompson seemed to have more than his share of unnerving experiences while he was at Berkeley. There was the time when he had a jug of *Colias* virus "goop" blow up in his rented room, making him most unpopular with his landlady. Probably the most disconcerting time for him was when he lost the front of his pants on one of Berkeley's busiest street corners. He had just gotten off the bus at University and Shattuck when he became aware of an unusual amount of ventilation. Looking down he found that the entire front of his pants had disintegrated from acid no doubt. The rest of the trip was made with his jacket draped around his waist! He and Irv Hall managed to blow up a fungus spore-collecting apparatus, Irv getting a bad gash on his face that required ten stitches. Hank could also often turn a joke to his advantage too. After Ken, Irv, and Harriette Wassar had doctored his

Pepsi by using anything but Pepsi, he announced that "it surely didn't taste right" but, on second thought, "the carbonation must have been left out." He got even by loading up Ken's pipe tobacco with oak-moth frass, which Ken admitted made a most peculiar-tasting smoke.

Ken and some of the others had a great laugh out of the blown-up composite picture, which appeared one morning on my desk, that Ken had made of me with a nude Marilyn Monroe perched on the front of my desk. (This was in the days before topless dancers!) They weren't above pulling my leg about a diagnosis either—Ken made a slide of insect fragments and polystyrene latex and I diagnosed it as granulosis!

While Art Heimpel was with us, we all learned of his great talent for telling a story, but he also admitted to a knowledge of palmistry. It seems that he had learned the rudiments from an aunt who was an accomplished soothsayer. On more than one occasion he demonstrated his skill in this art.

Even though Karl Snyder did not receive his Ph.D. in insect pathology, he did work with us during the time he was obtaining his degree in entomology. There always seemed to be lots of stories circulating about him. He, too, was an accomplished story teller. Karl always seemed to do things differently than most people--which he used to attribute to the fact that he was left-handed. One day he came in apparently somewhat agitated. Then it came out what was troubling him—he had dreamed that he had granulosis, and he couldn't help thinking about what a mess he would be in if he really had it!

Those who knew John Briggs as a graduate student remember that he was an accomplished junk collector ("you never know when it will come in handy"), all of which he kept under his work table because his wife, Lou, wouldn't let him bring any of it home. The peak of his junk collecting came when he found a workable TV in the garbage can back of his apartment. John used that TV until he left Berkeley.

Mauro Martignoni during the short time he was a student in Berkeley made quite an impression, especially on the girls. When he first came his English was precise but not always understandable. (When he was coming through customs, the inspector rapidly and brusquely asked him a number of questions which confused Mauro causing him to lapse into his native tongue, whereupon the inspector most causticly said, "Why don't you stay home until you learn how to speak English?") After he had gotten over his initial shyness, he began to pick up American slang, which mixed with his Italian accent was something! He had been in Berkeley only a few weeks when he was present at a party. After a certain amount of good wine, the group got into a songfest, and to everyone's surprise, Mauro knew more verses to "She'll be comin' 'round the mountain" than anybody.

Ed Clark was driving back to Berkeley from Truckee in a university car one time when a deer jumped off a high bank onto the highway directly in front of him. Ed couldn't avoid hitting it. The impact caved in the front end of the car and completely ruined the deer. Ed stopped, as did the car behind him, which proved to be a deputy sheriff of that county. There was no question but that it had been an accident so the deputy and Ed loaded the deer carcass into the deputy's car, supposedly for delivery to some institution. It occurred to Ed that he should have the word of a witness as to how it happened when he delivered the car back to the university, so he asked the deputy to write him a note about the accident. He scribbled something and handed it to Ed, who tucked the paper into his pocket without looking at it. Fortunately he did not need the witness for the university, for the paper simply said, "I saw it happen" with the deputy's signature.

Although the Flannagan-Hope funds allowed us to expand our activities and personnel, an immediate crisis occurred concerning space. And let it be recorded here that from this time on, throughout my career with the University of California at Berkeley and most of my time at Irvine, never did I, or the operation for which I was administratively responsible, ever have adequate space! The need for *Lebensraum* was to be the bane of my existence and was frequently the, or one of the, determining factors in basic decisions I had to make. In the present case, however, the need for more space in two different locations became intertwined. With the understanding and cooperation of James K. Holloway, a well-known U.S. Department of Agriculture worker, who was then directing the cooperative work with the university (as represented by Carl B. Huffaker) on the use of insects to control weeds, particularly the Klamath weed, we acquired more greenhouse and headhouse space for insect-rearing purposes and a benched laboratory for research purposes. Similarly, by a directive from Dean Hutchison and the kind cooperation of E. O. Essig, chairman of the Division of Entomology and Parasitology, we doubled our space in Agriculture Hall. At first we were assigned, and were glad to accept, a suite of rooms in the basement of Ag Hall, providing filters were placed in the windows to keep out contaminating dust. However, Professor Essig suggested, although apparently he later forgot, to the dean that we remain on the second floor, and that toxicology, "with all its clutter and smells," be placed in the basement. Fortunately, I kept a record of this transaction, because some years later we were accused of rejecting the space in the basement for what was more attractive space on the second floor.

Although this expansion of space gave us more room in which to conduct our research and our graduate teaching, it split us virtually in half between the campus location and the Gill Tract, three miles away. At times, these three miles were as difficult to bridge as 300 miles would have been. As director, I found it most difficult to give full and adequate attention to the work at the Gill Tract while my office, secretary, assistants, and operating base were on the main campus in Berkeley. Similarly, for seminars, to assist in teaching, and to conduct certain business, our pathologists at the Gill Tract found the campus to be "an awful long way away." Even parking for them was a problem on the Berkeley campus. In order to make sure that my necessary absences from the Gill Tract did not allow a feeling of neglect or nondirection to set in, I unofficially placed Thompson in charge of the insect pathology operations there. This facilitated the conduct of research, but unavoidably strained the feeling of the unity of the laboratory. A similar strain existed when, later, we established a branch of the laboratory on the Riverside campus. There was nothing any person could do to completely alleviate these situations—even Harry

Smith, who seemed to be constantly traveling between Riverside and Albany-Berkeley, could not keep a sort of underground north-south tension from developing within the statewide division as a whole. Eventually, the spatial problems between Riverside and Albany-Berkeley were to bring about a complete administrative separation of the Biological Control and Insect Pathology groups in the north from those in the south. In any case, as far as the Laboratory of Insect Pathology was concerned, the Berkeley and Albany branches maintained as good communications as possible under the circumstances, and both parts continued to be highly productive. Gradually, most of the Experiment Station research was conducted at the Gill Tract while my own research, and that of most of our graduate students, was done on the campus.

By 1949 our work on insect viruses, protozoa, and some bacteria and fungi developed to such proportions that we abandoned our research on the diseases (mostly fungus) of scale insects. From time to time, as I had begun to do in 1944, we examined and briefly studied the internal symbiotes of a number of common species of these insects (Steinhaus, 1951*a*); we also made a brief study of *Myiophagus* found in purple scale sent to us by Y. Tanada from Hawaii and considered by Fisher in Florida to be the cause of the mortality seen in the purple scale in that state. Speaking of Miss Fisher, in November 1948, she wrote to Harry Smith enquiring as to whether or not there was evidence of the chytridiosis (*Myiophagus*) occurring in California. (He referred this part of the letter to me to answer and arranged to have Charles Fleschner of the Riverside division send her scale-infested lemon leaves from southern California.) She also asked his opinion concerning a statement attributed by H. S. Fawcett to R. S. Woglum to the effect that Bordeaux mixture or sulfur coating does not cause any significant scale build-up in citrus groves in California. The reader will remember from chapter 2 that this became a much debated matter in Florida during the 1940s and after. Accordingly, Smith's reply (8 November 1948) is of interest:

> In regard to the residue matter, I think that Mr. Woglum's statement which you quoted from *The Citrus Industry* would

> represent the observations of most if not all our entomologists. Citrus trees are commonly sprayed with a Bordeaux mixture for brown rot, the lower half of the trees only being sprayed, and so far as I know no one has observed that the scale has built up more on the sprayed half of the trees than on the unsprayed half. On the other hand, it is a common observation here that roadside trees which are generally covered with dust have heavier infestations than trees farther toward the interior of the grove. This has generally been attributed to the effect of the dust on the build-up of the scale although the exact mechanism is unknown. Some of our entomologists have demonstrated that inert dust is lethal to various predatory and parasitic insects and that the build-up is due to the destruction of the enemies. Others feel that the presence of dust provides more favorable substratum which improves the chances of the crawlers to become established.
>
> Since there is probably a great deal more dust on the foliage of citrus trees in southern California on account of the absence of summer rains, it may be that additional residues such as would take place when a grove is sprayed with Bordeaux would not have any noticeable effect. At the moment this is the only explanation I can think of.

With regard to the purple-scale-infested lemon leaves sent to Fisher by Fleschner in the winter of 1949, the Florida scientist found that at least two species of scale fungi do occur in California. (She had been under the apparently correct impression that no entomogenous fungi of purple-scale were known in California.) In a 1950 manuscript prepared for *Citrograph*, Fisher stated that two specimens of the red-headed fungus (*Microcera* sp.) and several specimens of an unknown brown fungus were found growing on purple scales that were apparently dead when the samples were taken. She stated further, "As far as is known, these two fungi are not scale-killers, and are not harmful or beneficial to the grower." She concluded that comparatively dry summer weather of California does not necessarily mean that entomogenous fungi cannot survive and grow in this state, even though the two species she found do not occur in great numbers. About the same time, our own observations generally confirmed her findings, except that in the case of some scale insects, for example the black scale, entomogenous fungi have been found on them in districts near the coast, even in southern California. This fact was known in the early years of the century and was referred to in an article in the *California Cultivator* of 24 June 1915 (p. 703).

III

By 1950 it was clear that by the proper administration of the nuclear-polyhedrosis virus of the alfalfa caterpillar, this insect could be suppressed to a level where it caused no substantial damage to the alfalfa crop, and made the use of chemical insecticides unnecessary. However, it had two disadvantages: the virus required a five- to seven-day incubation period before it would kill the insect pest, and larvae dying of the virosis usually disintegrated on the plant in such a way as to make it less palatable as hay for cattle. Of course, this disintegration of the diseased caterpillars also helped disseminate the infectious material over a greater area of the plant. Furthermore, from a commercial standpoint at that time, the virus could only be produced in living insects, which yielded a product that required considerable cleaning up before it could be packaged; moreover, as Thompson showed, enough virus-diseased caterpillars could be collected by the grower himself to provide him with an adequate supply of virus to disseminate the following year. However, of these several disadvantages, the one that troubled me the most was the length of time required from the initial infestation by the virus until the death of the insect. Only by luck or by using Ray F. Smith's methods of predicting a destructive outbreak of the caterpillar on a particular field could one apply the virus and have it act in sufficient time to kill the pest and thus protect the crop. Of course, chemical control methods were effective within a few hours, but most of these left undesirable or harmful residues on a crop used as fodder for the cattle of California's huge dairy industry. How could the incubation period of the disease and the time for the effectiveness of the virus be reduced?

In 1945 I had isolated an interesting strain of a sporeforming bacillus from diseased larvae of the Indian-meal moth, *Plodia interpunctella*. Later, in 1949, a similar strain was isolated from diseased larvae of *Aphomia gularis*. Both appeared to me to be strains of *Bacillus cereus*, a very common, widely distributed bacterium found in soil, dust, and other materials. Yet, there was something different about these strains and *B. cereus*; classical strains of the latter did not kill test larvae when fed to the insects in the same dramatic way that our strains did. At that time, there was one outstanding authority on spore-forming bacilli, so without further study on our part, we shipped our cultures off to Nathan R. Smith, who confirmed the identifications. Sometime later, 1950, he recognized a similarity between these strains and one known as *Bacillus thuringiensis*. In the case of all three strains, Smith observed that the spores lay obliquely within the sporangium; he indicated (in correspondence) that perhaps the virulence we were finding might somehow be associated with this characteristic.

One day while worrying about the relative slowness with which the polyhedrosis virus killed its caterpillar host, it occurred to me that inasmuch as bacterial diseases generally have relatively short incubation periods in susceptible insects (there are exceptions such as *Bacillus popilliae* in Japanese beetle), why not try a bacterium against the alfalfa caterpillar. The correspondence with Nathan Smith on the sporeformers caused me to remember that we had been holding a strain of *Bacillus thuringiensis* in one of our refrigerators for a number of years. In fact, I had obtained it from Smith in 1942, and he had received it from J. R. Porter in 1940. Porter had secured it in 1936 from Otto Mattes (1927), who isolated it in Germany from diseased larvae of the Mediterranean flour moth, *Anagasta kühniella*. Berliner (1915) had first isolated it from the same insect in 1911. Why not give it a try? Then serendipity entered the picture. It so happened that we had a large tray holding several hundred alfalfa caterpillars handy in the laboratory. They were all healthy and feeding on cuttings of alfalfa. But it would take two or three days to have fresh cultures of the *Bacillus*, and by that time the larvae would be pupating. So, I made fresh transfers from the old cultures, then washed the old growth off. Using an ordinary clean atomizer type insecticide sprayer, I sprayed the old spore suspension over the foliage in the rearing tray. As I did so, I thought "how foolish," for this culture had been transferred on artificial media since its isolation by Mattes in 1927, so surely any pathogenicity or virulence it may have had for insects had long since been lost.

When we returned to the laboratory the next morning most of the caterpillars had ceased feeding and were, indeed, sick and dying. How could this be? Perhaps the sprayer had not been clean after all, and somehow it had retained the residue of a chemical insecticide. However, an examination of the blood of the ailing caterpillars showed the presence of *Bacillus* rods but not in convincing enough numbers to explain the amount of obvious morbidity. That evening, when enough growth had appeared on the new agar slants, I sprayed a new batch of caterpillars with the fresh cultures, making certain that there was no possibility of the presence of a chemical insecticide. Hurrying back the next morning, I was met with disappointment in the tray that had been sprayed with the fresh cultures of *Bacillus thuringiensis*, for the caterpillars appeared to be unaffected and were feeding on the alfalfa in a normal manner. In the tray that had previously been sprayed with the old, long-held culture, most of the larvae were now dead. Why the difference? Why were the larvae in the tray which had been sprayed with an old, presumably effete, culture killed so dramatically, whereas when sprayed with a fresh, vigorous culture, the larvae appeared to be unaffected?

With subsequent testing it was soon determined that it was not a matter of contamination of equipment with chemical insecticides. It was indeed a fact that the lethal effectiveness of *B. thuringiensis* became manifest only after the bacteria had sporulated. Accordingly, without making any special study of the bacillus or its sporangium, we proceeded—with the help of Hank Thompson—to conduct field trials of spore suspensions of the Mattes strain of *Bacillus thuringiensis*. By trial and error, we determined the type of media for spore production, and I spent long night hours counting the living and dead larvae brought in by Hank from the test plots, microscopically (and by culturing) diagnosing all the diseased larvae, making sure whether or not the larvae had died from the bacillus infection.

We had noticed (Steinhaus, 1951*b*; Steinhaus and Jerrel, 1954) the peculiar crystalline inclusions associated with the spores, originally described by Berliner and Mattes. However, we did not have the laboratory personnel to study both the precise manner in which the organism killed its host and the applied aspects as well. (We were really violating one of my basic precepts in not undertaking a study of the nature of the microbial organism before it was used as a control agent.) Inasmuch as this bacillus acted so much more rapidly in destroying noxious Lepidoptera (in some cases it caused the insect to stop feeding within a few hours) than had the nuclear-polyhedrosis virus, we decided to concentrate on this aspect of *B. thuringiensis* for the time being. Fortunately, Canadian workers were also interested in *B. thuringiensis* and other crystalliferous bacteria, and it was they who firmly established the association of one of the toxic reactions of the bacillus with the crystal. Gradually it became clear—as other North American workers as well as Europeans became interested—that these crystalliferous bacteria were to provide one of the most dramatic and most interesting challenges to insect pathology and one of the greatest opportunities for microbial control. And really, the story begins not with our work, or the work of the Canadians, or that of Berliner and Mattes, but with that of the Japanese in 1902, and possibly even with that of Pasteur during his studies of what was called *flacherie* in the silkworm between 1865 and 1870. (For a brief resume of the history of *B. thuringiensis*, see Steinhaus, 1960*a*.)

IV

Worth mentioning at this point is the frustration I felt in attempting to interest industry and insecticide concerns in the possibilities of microbial control agents (not only bacteria but fungi, protozoa, nematodes, and especially viruses). While it must be admitted that most of the activity of our laboratory was still dedicated to the basic research we felt was necessary to properly undergird the applications of insect pathology, we realized that, since our research was being supported by the Experiment Station, we would have to come up with something of practical significance to justify this support. (Federal grants for basic research had not yet become our mainstay.) However, putting it bluntly, the insecticide industry just was not interested in nonchemical insecticides—except a few of them kept their eyes on what we were doing so that our findings would not suddenly emerge as a threat to them. Apparently, since they did not feel they had to fight us, they had little interest in joining us. I well remember how, during the early 1950s this state of affairs made me do a "slow burn." Finally, in exasperation, and with subdued anger, I decided to write an article that would hopefully be read by people in industry and that would point out that manufacturers might be missing an opportunity by ignoring the potentialities of microbial agents as insecticidal agents. The journal, *Agricultural and Food Chemistry*, was good enough to publish the article (Steinhaus, 1956). Much to my surprise the tactic worked. Inquiries arrived from virtually every major insecticide manufacturer in the United States. A dozen or so sent personal representatives.[6] But now what? I suddenly realized that I knew nothing of how to deal with commercial businesses, to talk the language of their representatives, or how to convince them to make the investments necessary to give our discoveries a commercial trial.

Thus began a new experience for me, one that was to last the next five years while I continued my basic research on problems unrelated to *B. thuringiensis* (by now everyone was getting into the act) and maintained my teaching and administrative responsibilities. Since the time I had had to work in my father's country store, I had disliked being involved in any business or commercial venture. In the university's Experiment Station, our obligation, in addition to conducting basic research, was to attempt to help solve the agricultural problems of the state by making field trials to test our experimental batches of microbial insecticides. It was not necessarily our function to maintain control of a given pest on a specific crop for a particular group of growers or farmers. We could be of technical assistance to industrial concerns, but as members of the Division of Agricultural Sciences we were not permitted to serve as paid consultants, as were our colleagues throughout the rest of the university. In general, after Experiment Station scientists developed a product, a method, or a treatment, it was up to industry, or some segment of our free enterprise system, to further develop the product or idea into a marketable form. Thus, when commercial interests took over a project, we usually turned our attention to other problems of agriculture and public service. However, in the case of controlling insects by methods of biological control, especially as it concerned the use of insect parasites and predators, it was usually left to the Experiment Station or the government to carry on. There was not much profit in liberating a predator or parasite that would establish itself and more or less permanently control a pest. This is also true with certain microbial control agents, but--and this was one of the most difficult points on which I had to try to convince commercial interests--some, perhaps most, microbial agents could be marketed in a manner analogous to that of chemical insecticides. We attempted to show them that over the long haul, marketing such products could be commercially feasible. But, as we were so often told, we were only university professors and experimental scientists, who never had to meet a payroll; besides, it was not our money we were risking when we recommended such a project! I had sense enough to realize that there was some truth in this philosophy, but I was naïve enough to believe that somewhere in that great, profit-making, industrial-giant world was somebody, some manufacturer, willing to listen and to take a chance. It turned out I was right. Bassi's hope of 1836, and LeConte's of 1873, were not barren speculations.

As far as I was personally concerned, the American commercialization[7] of microbial control agents began on 27 September 1956, with a visit with Robert A. Fisher, then Director of Research and Development of Pacific Yeast Products, Inc. (later Bioferm Corporation, which was still later sold to International Minerals and Chemical Corporation) of Wasco, California. His visit, as were those of other industrial representatives, was the direct result of the article addressed to industry in the *Agricultural and Food Chemistry* magazine. Pacific Yeast Products was a small company, but they had developed techniques of mass-producing yeast and bacteria for commercial products, such as vitamins, and upon reading that article and another appearing almost simultaneously in *Scientific American* (1956*b*), they envisioned the use of these same techniques in the production of great quantities of *Bacillus thuringiensis.*[8] After ascertaining the growth requirements of a microorganism in pilot-plant equipment, they could transfer their production to large fermentation tanks of 40,000 liter capacity. (A general account of their methods of mass-production was presented by Briggs, 1963; see also Martignoni, 1964.)

Fisher and his associates, George Gelman, president, and J. M. Sudarsky, vice-president and general manager of the company, authorized their research microbiologist H. L. Wolin and later their research and development manager, John C. Megna, to prepare trial "batches" of *B. thuringiensis* and *B. sotto*, which we had provided them. On 5 November 1956, we received from Wolin three "samples of spore and crystal material" for testing purposes inasmuch as the firm had no facilities for rearing insects. Using the imported cabbageworm, *Pieris rapae*, as the test insect we found that of the three samples, the one labeled S6-91 to be as highly potent as any of our own spore-crystal preparations. Having in my hands this first commercial preparation of "*B.t.*," as the bacillus had become nicknamed, was akin emotionally, I imagined, to the time when Paul Ehrlich tested the first commercially produced (but more famous) "606." I was happy to be able to report back to the company that it appeared that they had something.

Apparently the results of their pilot plant batches were encouraging enough that the company decided to proceed with attempts to make a commercial product of the material. Their story is not, unfortunately, one of instant and overnight success. Indeed, this pioneering venture was a rough one in a number of respects, and it was they and one of their competitors, Nutrilite Products, Inc., I had in mind when in 1966, at a plenary symposium of the Entomological Society of America, I said, "It takes courage to invest in such a venture, and courage to stay with it until eventual success is attained—sometimes at an initial costly effort." However, they persevered, and of this writing I am told that the microbial insecticide section of the Bioferm Division of International Minerals and Chemical Corporation is running in the black, at least so far as their product *Thuricide*, containing *B. thuringiensis*, is concerned.

The problems involved in manufacturing a living insecticide were not all technological in nature. To be sure, all manufacturers (and initially there were several) had to experiment with methods of increasing the yield of spores and crystals, had to devise methods of standardizing the quality and potency of their products, had to perfect methods of mass production that would retain the beneficial attributes of the product when manufactured in the pilot plant, and had to develop modifications and formulations of their products for use against different insects. In addition, the manufacturer had to satisfy the state and federal laws and regulations with regard to safety and utility. Because their product was one of the first ever presented to the government as a general insecticide (dusts containing spores of *Bacillus popilliae* to control the Japanese beetle were approved on a different basis), the Bioferm people had to pioneer once again in working their way successfully through the maze of trials and tests required by government agencies, specifically the U.S. Department of Agriculture and the Food and Drug Administration. Although questions of bacterial mutation and the general safety of microbial insecticides were of concern and received comment (Steinhaus, 1957, 1959), one of the pioneering papers relating attempts to satisfy Federal agencies as to the safety of *B. thuringiensis* to life other than that of certain insects was that by Fisher and Rosner (1959). Fisher (1963) also knowledgeably reviewed the important matter of conducting adequate and proper bioassay of microbial pesticides. Added to all this, of course, is the fact that manufacturers and commercial distributors have to develop methods of marketing the product so that it can be considered an economic success by the boards of directors or the stockholders.

During my early dialogues with representatives of a number of different industrial concerns, I tried to interest them in the production of certain "promising" insect viruses. Certainly, the virulence with which certain nuclear-polyhedrosis viruses destroyed large populations of their hosts (e.g., the alfalfa caterpillar, the western yellowstriped armyworm, the gypsy moth, and the bollworm or corn earworm) appeared to me to indicate that properly manufactured insect viruses would make ideal insecticides. Especially did this appear likely in instances in which a reasonable incubation period for the viruses could be tolerated. Unfortunately, I had little success because, at the time, clean-cut production methods (such as growing the virus in tissue culture or embryonated hen's eggs or in tanks of culture media) were not available. Because the mass production of antibiotics and vitamins gave the fermentation industry a know-how they could readily adapt to entomogenous bacteria and fungi, they looked to the possibilities inherent in these organisms first. Even our own efficient insectary methods of mass-producing the appropriate insect host, which could then be mass infected as one would a tissue culture, was met with skepticism. As an academician and university scientist I had to realize that in the world of commerce, more caution than I displayed had to be taken when matters of economic risk were involved.

This matter of the applicability of insect viruses, as well as other insect pathogens, came up for discussion during a very interesting meeting called by Merck Sharp & Dohme, manufacturers of chemicals and biological products, for 20 and 21 January 1959, at their Branchburg Farm near Rahway, New Jersey. This "Merck Symposium on Microbial Control of Insects" consisted of panel discussions between representatives of Merck & Company (the parent company) and fifteen or sixteen interested entomologists and insect pathologists. Again I gained new insights into all that is involved in developing a new type of product by a commercial concern. Merck Sharp & Dohme's serious interest in the possibilities of microbial control were first communicated to me by a letter (23 August 1957) and subsequent visit (17 September 1957) from Dr. Richard F. Phillips of their research laboratories. We worked with the Merck people in much the same way as we did with those at Bioferm; being a public, tax-supported institution, we had to offer our services freely and equally to all, and especially to individuals or to companies that were located in, or had representative branches in, California. However, the greatness of the University of California was such that we could and did consider ourselves a world university, faithful to the interests of the state which gave us our basic support, but free to address ourselves to problems on a world scale or any place in the world we were permitted to enter. In addition to being primarily interested in *Bacillus thuringiensis*, however, Dr. Phillips accepted my enthusiasm concerning insect viruses in spite of the inherent difficulties their mass production involved. Accordingly, we furnished him not only with cultures of crystalliferous bacteria, but with virus preparations of four nuclear polyhedroses and three granuloses. He, and their virologists, were particularly interested in the polyhedrosis virus we had gathered from the corn earworm, *Heliothis zea.* An excellent choice, as we shall note later.[9]

As a result of the symposium, during which both the advantages and disadvantages of the use of microbial insecticides were discussed, Merck & Co. decided to escalate their activity in the field, especially as far as *B. thuringiensis* was concerned. The company's serious interest in microbial insecticides was evident by the fact that they were desirous of hiring any one of three of my former graduate students, but none was available. John Garber, manager of the company's Industrial Organic and Agricultural Chemicals section, and J. M. Merritt, manager of the Plant Chemical Section of Product Development, became busily engaged in initial production efforts. Their approach was cautious, scientific, and thorough. Matters appeared to be going along well, formulations were developed and field-tested, and by May 1959 they had devised a trade name (*Agritrol*), the label under which they hoped to market their product once they had it in mass production. Then came a crushing blow to the aspirations of the Merck scientists we have mentioned. On 11 November 1959, I received a telephone call from Merritt saying that the board of directors (or the equivalent powers that be) had decided to eliminate much of their activity in the development and manufacture of agricultural chemicals and biologicals, and with this decision the entire program in microbial insecticides was "being shoved down the drain." He assured me that this decision had been based on general principles, that it in no way reflected on their work on, or hopes for, microbial control products, and that it was just one of those things that happens in industrial situations when apparently for economic reasons, magnates of management make a decision that, as in this case, wiped out programs and jobs. Alas, I had thought Merck & Co. would, because of their size and financial strength, be the company to exploit the use of insect viruses for man's benefit; at least their men had listened to our plea and were thinking about it.

In the meantime, another California manufacturer—in addition to Bioferm—was progressing with its interest and work in the production of microbial insecticides. I have already mentioned Nutrilite Products, Inc., of Buena Park, California. H. F. Beckerdite, farm manager of this company first contacted the university through I. M. Hall, who headed up what, at that time, was the southern branch of the Laboratory of Insect Pathology on the Riverside campus. Inasmuch as this firm manufactured a vitamin-mineral product partially constitued from alfalfa, they were interested in nonchemical means of controlling alfalfa pests. If some such method could be found they then might also be interested in marketing it commercially. Hall acquainted Beckerdite with our work with *Bacillus thuringiensis*, who, on behalf of his company wrote to us (24 October 1956) in Berkeley requesting a culture of the bacterium. This we were pleased to supply. In March of 1957, Beckerdite sent us a sample of material, apparently a bran mash culture, of the *B. thuringiensis*; as we had for Bioferm, we tested the product for Nutrilite on larvae of *Pieris* and *Colias*. In mid June of 1957 Mr. Beckerdite, accompanied by Dr. Stefan Tenkoff, then vice-president in charge of manufacturing for Nutrilite Products, visited the Berkeley campus. Apparently as a result of our conference, a decision was made to proceed further in attempts to develop a marketable living insecticide containing *B. thuringiensis*; at any rate, soon thereafter I received copies of letters they had drafted to the U.S. Department of Agriculture and to the California State Department of Agriculture "requesting registration." In October 1957, Paul Dunn, who had been a student of, and an assistant to, Hall at Riverside, joined the staff at Nutrilite to "follow through with the development of the commercial aspects of *Bacillus thuringiensis.*" Toward the end of July 1959, I received a letter from Tenkoff saying, "Dr. [Thomas H.] Jukes and his associates are gradually taking over the work relative to the further development of our *Bacillus thuringiensis* product as well as our intended research program in the general field of biological insect control. . . . " Jukes had come from Lederle Laboratories to be named vice-president in charge of research at Nutrilite. One of Jukes' associates was Robert White-Stevens, who served as Nutrilite's director of Agricultural Research and Development, and under whom Dunn continued his work on their product, which had now been given the name *Biotrol.* (They subsequently used the name *Bacticide* for another formulation.) It looked as if Nutrilite was preparing to proceed on a wide front in biological control.

Apparently Jukes and White-Stevens did not find at Nutrilite the resources commensurate with their expectations or aspirations; hence they parted company before the year was up. However, Nutrilite continued its work on microbial insecticides. In addition to experimenting with the commercial production of *Bacillus thuringiensis* (e.g., Dunn, 1960; Mechalas and Dunn, 1964), it also investigated the commercial possibilities of fungi, such as *Beauveria bassiana* (Dunn and Mechalas, 1963) and a number of insect viruses (Chauthani, 1968).

But let us return to the activities of the Bioferm Corporation from which we became diverted a few pages back. Although the university could not assume a special or vested interest in any one company, there remained an involvement that should be recorded since Bioferm employed one of our Ph.D. graduates, John D. Briggs. The significance of his sojourn with Bioferm lies not only in what he accomplished for that company but in the fact that because of his academic orientation as a graduate student and as an insect pathologist at the Illinois State Natural History Survey Division, then later as a member of the faculty and administration at The Ohio State University, he served as an important liaison between industry and insect pathologists in academic institutions and government research laboratories both in America and overseas. Having actively participated in the trials and tribulations, the pluses and the minuses of commercialism, he was singularly competent in communicating these incomprehensibles to academicians and researchers outside of industry.

Philosophically, Briggs was sympathetic to the idea of manufacturing insecticides of which the active ingredients were insect viruses and other pathogens in addition to *B.t.* However, he also appreciated his company's concern to prove they could succeed with a relatively easily produced bacterial product before they tried agents with greater production risks associated with them. Fortunately, coincidental with Briggs's acceptance of a professorship at the Ohio State University in the fall of 1962 and his replacement by Carlo Ignoffo, Bioferm undertook the production of viral insecticides, concentrating on the nuclear polyhedrosis of the bollworm or corn earworm, *Heliothis zea*, one of the nation's most serious pests.

Prior to joining Bioferm as director of entomology, Ignoffo had impressively entered the insect pathology scene through a prodigious output of research on insect pathogens, especially viruses, while stationed at the Brownsville, Texas, laboratories of the Entomology Research Division of the U.S. Department of Agriculture's Agricultural Research Service. Many of his publications appeared in the *Journal of Insect Pathology* and its descendent the *Journal of Invertebrate Pathology*. Particularly notable is the series of papers (Ignoffo and associates, 1965 et seq.) on the propagation, bioassay, and field use of a nuclear-polyhedrosis virus which attacked *Heliothis virescens* and *H. zea*. (See also a general review on the production of insect viruses written by Ignoffo, 1966.) With the background of knowledge obtained from his studies in Texas, Ignoffo brought a new level of excellence and potentiality to the industrial production of viral insecticides. As these words are written, there is an apparent hesitation and reluctance on the part of federal agencies to decide on and approve toxicological testing procedures of insect-virus preparations for registration purposes. Let us hope that by the time these words are read, the governmental agencies concerned will have resolved the difficulties involved, and that formulations of insect viruses will be on the market serving the interests of mankind safely and effectively—this, in accordance with the possibilities we dreamed of back in 1945 as we stood in the alfalfa fields of California observing the manner in which natural virus-caused epizootics (as well as those later artificially initiated) suppressed destructive populations of the alfalfa caterpillar. [EAS would have been gratified to know that International Minerals and Chemical Corporation was granted temporary exemption from tolerance in December 1970 and final clearance in May 1973 by the federal government for the commercial production of the *Heliothis* virus. MCS.]

In addition to the inherent potentials apparent in the use of insect viruses to control noxious insects, one of the reasons for urging industry to consider these and other agents was the fact that it appeared that all interested commercial companies were concentrating on *Bacillus thuringiensis*. If all companies succeeded in developing products of which the active ingredient was this bacillus and its toxins, it seemed obvious that the competition would cause some to fail and thus discourage the whole idea of microbial control. Indeed, to some extent this did happen. Virtually all pharmaceutical houses manufacturing fermentation products showed interest and toyed with the idea of manufacturing the microorganisms, which would then be distributed by insecticide companies, in a manner similar to that in which Stauffer Chemical Company marketed Bioferm's Thuricide. Well-known chemical companies, such as DuPont, Dow, Velsicol, and Industrial Biochemicals also indicated serious interest. Some companies, such as the Wallerstein Company and the Pfizer Company did some pilot plant work with *B. thuringiensis*. Rohm & Haas Company actually developed a product—which tested out very well—in a manner similar to that produced by Merck & Company. The Rohm & Haas product was given the trade name *Bakthane*. The Grain Processing Corporation in Iowa was another organization that gave *B.t.* a try. In two or three instances the bacillus was ignored and insect-virus preparations were made with commercial intent, but apparently these efforts did not long survive.

In the meantime, the mass production and "commercialization" of microbial insecticides had been taken up by European governmental laboratories and research stations, particularly in Germany, France, and Czechoslovakia. Through no fault of the scientists involved, reports such as that appearing in the British *New Scientist* for 24 March 1960 and that carried in the *New York Times* for 3 April 1960, implying originality in these countries of the development of microbial insecticides "that can be produced on a commercial scale" were inaccurate as far as the timing was concerned. The same applies to an account that appeared in *Time* magazine for 2 March 1962 with regard to very creditable experimental work done in the USSR.

V

As Hank Thompson assumed his new duties with the U.S. Department of Agriculture in Beltsville and Ken Hughes headed for the "peace and quiet" of Alaska, changes took place in the professional personnel of the laboratory. In the meantime, one of the more significant developments and expansions of the laboratory was taking place on the Riverside campus. Almost from the beginning of the laboratory's existence it was apparent that we could scarcely do justice to pathology problems of southern California working solely out of our Berkeley laboratories. Although I made occasional trips to southern California in an attempt to operate on a statewide basis, it was clear that just as the division needed its northern branch, so did the laboratory, although headquartered at Berkeley, need its southern branch. Fortunately, we were able to obtain a position for an insect pathologist and some space for a research laboratory about the time Irv Hall obtained his Ph.D. with us in 1952.

Irvin (let no one call him Irwin or Irvine) Monroe Hall, Jr., had served as a senior laboratory technician from 1948 to 1952 at the Gill Tract, where, for the most part, he assisted Thompson and conducted research on a problem he presented for his doctoral dissertation. It concerned the use of microorganisms in the control of the sod webworm, *Crambus bonifactellus* (Hall, 1954). He assumed his duties on the Riverside campus in March of 1953, and the opening sentence of a letter from him dated 1 April 1953 reads, "The southern branch of Insect Pathology is at last a going concern." Irv's pleasant, outgoing, sociable personality, and his talent of always being able to sustain his part of a conversation, enabled him to integrate himself and his work rapidly with his colleagues and the Riverside group. Officially we were able to initiate the opening for Hall's position on the basis of the apparent need for developing a type of microbial control for the western grapeleaf skeletonizer, *Harrisina brillians*; however, real success in this regard was not forthcoming because of inadequate field populations on which to run conclusive tests. By May he had commenced field and laboratory studies of a granulosis and a nuclear polyhedrosis of the alfalfa looper, *Trichoplusia ni*, and a fungus (*Beauveria bassiana*) against curculionids infesting alfalfa. All of this, and other possibilities, made for a lively exchange of correspondence between Hall and myself and gave us much to discuss whenever we visited at either end of the laboratory's axis.

By 1956, Hall's program at Riverside was entering a maturation phase and was also becoming enmeshed in some of the stresses and strains beginning to appear in the statewide structure of the Department of Biological Control. A significant change in the nature of the research of Irv and his assistants was their increased interest in entomogenous Entomophthoraceae. Up to this time they had been primarily concerned with insect viruses and with *B. thuringiensis*. Irv, as was my aspiration for all of my graduate students, was educationally trained to be a "compleat" insect pathologist with competency, active or latent, in all aspects of disease of insects. Accordingly, in his role as supervisor of the Riverside branch of the laboratory, he could confidently confront virtually any problem of insect disease that came his way.

By the mid 1950s, the alfalfa caterpillar was losing its distinction as being the most serious insect pest of alfalfa—the crop with the largest acreage of any in California. A new threat to this important forage crop loomed in the form of the spotted alfalfa aphid, *Therioaphis maculata.* This insect, common in Mediterranean countries, was first found in the United States in New Mexico in 1954; by 1955 it had spread through eleven states of the southwest, including California. Its spread occurred at one of the greatest rates of any pest ever introduced into this country; by 1957 it had been reported from thirty states in a more or less continuous belt from Virginia through the Gulf States to California. The severe injury it caused alfalfa, and certain related plants, created the nearest thing to entomological panic I have personally witnessed. What did all this have to do with the laboratory's activities, and in particular with Hall's research program? Although chemical insecticide programs were instituted in attempts to control the insect, great interest was had in its natural enemies, especially because alfalfa was a forage crop for cattle, and the fewer chemicals used the better. Among these natural enemies were several species of entomophthoraceous fungi which caused impressive epizootics in populations of the aphid when atmospheric conditions were right for its growth, development, and spread. Undoubtedly, the importance of these fungi and his work on them fired Hall's (Hall and Dunn, 1957, 1958) interest in Entomophthorales generally.

Although at Berkeley we had begun to receive fungus-diseased specimens of the aphid as early as 1955 (Steinhaus and Marsh, 1962), it was clear that because of the insect's initial ravages in central and southern California, Hall, with his interest in entomophthoraceous fungi, was the man to "carry the ball" on this project. With limited assistance, he did so in an impressive manner; and while the control effects of the artificially disseminated fungi, the principal one being *Entomophthora exitialis*, were not as effective or as dependable overall as other types of control, the work led by Hall created an appreciation of the possibilities inherent in insect pathogens by some hitherto skeptical entomologists. The highly valuable by-product of this episode was that it opened the doors of justification to permit Hall to proceed with his basic research interests in the Entomophthoraceae.

In December of 1956 I received from Irv a letter reflecting discouragement on his part with regard to policy set by the Experiment Station, which directed that insect pathology was to continue to be centered at Berkeley and that most of any future growth in the laboratory was to take place at Berkeley, with Riverside augmenting its activities only to the degree necessary to take care of pathology problems in southern California. Actually, this policy worried me before it did Hall, because with a state the size of California, I could see no reason why Hall's operations in the south should not be permitted to expand to whatever size required. My philosophy was, and always has been, that there is so much to be done in insect pathology, and invertebrate pathology generally, that the greater number involved the better. Moreover, I had my hands full with the problems in northern and central California—so the more problems that Hall could take over, the more desirable it would be. Irv had interpreted the directive to mean that his "unit is not intended to grow into a workable balance between basic and applied research." I, of course, wrote Irv assuring him of my support and my confidence in him; I also let him know of my efforts to have the university provide his laboratory with another professional insect pathologist as well as with more technical help. Irv responded by saying that there seemed to be no problem after all. This was typical of our relationship for we seemed to be able to resolve any problem that arose.

Earlier, I explained how our well-liked, pipe-smoking associate C. G. Thompson accepted a challenging opportunity to develop a program in insect pathology for the U.S. Department of Agriculture. At that time, inasmuch as we were still the only university training graduate students in insect pathology, finding an accomplished replacement for Hank was certain to be difficult. Indeed, because of the shortage of fully trained professionals, we adopted the expedient of hiring one of our talented graduate students, who was already serving us as a senior laboratory technician and was finishing his work on his doctorate. On 1 December 1953, we obtained permission to appoint Edwin Cook Clark as acting junior insect pathologist in the Agricultural Experiment Station. As soon as he had fulfilled the requirements for his Ph.D. he was given a regular appointment (February 1954) that was advanced to assistant insect pathologist on 1 July 1955. Ed Clark's dissertation was on the microbial diseases of the Great Basin tent caterpillar, *Malacosoma fragilis*, which periodically defoliates large areas of bitterbrush, a highly valued browse plant utilized by cattle, sheep, and deer in western states.

Upon joining the staff, Clark continued his work on the diseases of tent caterpillars. His primary entomological interest had always been forest insects, and he was encouraged to pursue this interest with us. While doing so, he cooperated with and provided guidance to the U.S. Forest Service on the use of a nuclear-polyhedrosis virus in the control of the Great Basin tent caterpillar. As of 30 June 1956, he resigned his position with us to accept one in the College of Forestry at the University of Idaho in Moscow, Idaho.

VI

With Ed Clark's departure for Idaho imminent, I immediately began screening possible candidates for his replacement. Although in principle I was very much against institutional inbreeding, it was obvious that one of the best on-coming insect pathologists was one of my own former students, Yoshinori Tanada. Informally, he is called "Joe"—the only reason for this moniker that I was ever able to discover was that his wife, Edna, preferred it to the longer, but certainly more elegant, "Yoshinori." Joe, born in Puuloa, Oahu, in 1917, received his Ph.D. from the University of California in 1953; his dissertation had to do with the infectious diseases, especially the virus diseases, of the imported cabbageworm, *Pieris rapae* (Tanada, 1953). At the same time he was holding the position of junior entomologist in the Agricultural Experiment Station at the University of Hawaii. By the time we sought his services, in 1956, he had been advanced to the rank of assistant professor and had also received regential approval for a promotion to associate professor as of 1 July 1956. On 30 March 1956, we offered Tanada the position of assistant insect pathologist (Step II) in the Experiment Station, somewhat doubting that our offer would attract him. His duties were to include the conduct of basic research, but, at least at first, he was to be primarily involved in applied aspects of insect pathology.

Fortunately for us, because of the opportunities he felt were possible at the Laboratory of Insect Pathology, Joe accepted our offer on 11 May 1956, to begin 1 September 1956, although he actually was unable to arrive until mid September. Thus there came to Berkeley a man who was to become one of America's truly leading insect pathologists, a man whose publications were once described to me by another of pathology's greats as "perfect, inviolate, every one!" Joe, a man small in stature, but great of heart, kind, considerate, and loyal to the point of self-effacement—but not necessarily meek, as assistants or secretaries of his can testify, when things were not done as they should be. Gentle, Joe, a talented, highly productive scientist to whom I had to hand disappointing administrative decisions on numerous occasions but with whom I never had a full-blown argument let alone a quarrel, remained one of the steadfast central pillars of insect pathology at Berkeley even when, eventually, the discipline (as an administrative unit) was dissolved.

It would be foolhardy of me to attempt to review here Tanada's principal contributions to insect pathology. All anyone need do is to consult any literature summary in virtually any area of insect pathology and the name of Y. Tanada will be there. Although mindful of his obligations with regard to research in applied insect pathology, one of Joe's first research projects (actually initiated at the University of Hawaii) was concerned with synergism among insect viruses—a point worth mentioning because, even though through the years he concerned himself with a large number of subjects, he always retained this as one dear to his heart, and hence from time to time researched it and closely allied phenomena related to stress and immunology. Tanada's laboratory research was of the highest quality, and his approach to much of his work in insect pathology was ecological in nature, as his work and writings on the epizootiology of insect diseases testify. This approach is also evident in his research on, and contributions to, the microbial control of insects.

I shall have occasion later in this treatise to speak further of Joe Tanada. I cannot go beyond this point, however, without briefly mentioning that there were times Joe showed me a loyalty that was truly touching. I only wish to record here that during some of the most trying days on the Berkeley campus, when the independence of our Laboratory (and later Department) of Insect Pathology was being challenged, I was not insensitive to how Joe, quietly but resolutely, stood by my side through thick and thin. This can be exemplified by a note I received from him at a critical time in 1960, a note which included the words, "If the Administration is going to condemn you for [insisting on our independence and integrity], then I for one should also be condemned." This statement succinctly indicates the fidelity and character of Yoshinori Tanada.

Almost coincidentally with the addition of Tanada to our staff, we were fortunate also to add Mauro Emilio Martignoni. Mauro had previously, from June 1951 to February 1952, been a graduate student but was called home to Switzerland when his father died quite suddenly. Martignoni (pronounced *martin-yoni*—as I embarrasingly discovered after I had in ignorance been attempting to pronounce the *g* in his name for several days after he first arrived) was born 30 October 1926, in Lugano, Switzerland. His father, Dr. Angiolo Martignoni, was a leading Lugano lawyer and government official. Proud of his Italian-Swiss heritage, Mauro showed a superb grasp of the potentialities of insect pathology for Europe, and especially as it related to forest insects in his beloved Switzerland—this, even as a university student in Switzerland and before receiving specialized training in the subject. I was further impressed with Mauro when I visited him (and his beautiful recent bride, Lu) in Zurich in 1954 and saw the excellence of the research he was conducting both in the laboratory and in the field (in the Engadine) where he was studying a granulosis of the tortricid *Eucosma griseana*, on which he did his doctoral thesis (Martignoni, 1954).

With the departure of Ken Hughes for Alaska in the spring of 1956, I recalled passages from letters Martignoni had written.

The essence of these was that if ever we had an opening in our professional staff after he received his Ph.D., he would like to be considered for it. Of course, such a possibility was in my mind while I was his guest in Switzerland in 1954. He and Hughes were very close friends during the time Mauro was a student in this country, so it is a bit ironical that it was Hughes's departure that created the opening that I hoped we could now have Martignoni fill. Unfortunately, Martignoni had not quite finished his doctorate, and with the high regard that he was held by his colleagues and professors (e.g., Professor O. Schneider-Orelli and Professor P. Bovey) in Switzerland, could we induce him to leave when undoubtedly they would extend to him competing offers?

To America's everlasting good fortune, our approach to Martignoni was received with interest and encouragement for us. Indeed, outside of a brief red-tape delay in obtaining his visa (we had to provide the American Consulate General in Zurich with a certificate of good conduct from the Police Department of Berkeley) no serious problems developed. Mrs. Martignoni was already a citizen of this country. We circumvented the momentary hindrance of Martignoni's not having a doctor's degree by requesting his initial appointment as assistant specialist, which did not require the union card, effective 1 October 1956. By the following 1 July, Mauro had officially received his documents, at which time his position at the University of California became assistant insect pathologist in the Agricultural Experiment Station.

With Mauro's arrival in October, one of the most fascinating and precious relationships I ever was to have with another human being began. No scientific colleague has ever caused me more frustration or given me greater satisfaction than has Mauro Emilio Martignoni. It is to our mutual credit, and to that of our respective wives, that over the years we have been able to be blisteringly direct and frank with each other and yet retain mutual affection and respect. Tall of stature and physically vigorous, exceptionally talented and brilliant of mind, methodical and precise, strictly courteous and polite (what a struggle it was to get him to call me "Ed"!), friendly and possessed of an insatiable and superb sense of humor, Martignoni was, and is, nonetheless intolerant of inefficiency and stupidity, stubborn until convinced ("But just try to convince me!"), overly concerned lest he has inadvertently offended someone whose respect he desires, and basically shy and at times unduly timid.

As with Tanada so with Martignoni, it would be superfluous for me to attempt to review his publications and contributions. He stands with others with whom it has been my privilege to be associated as one of the true greats of modern insect pathology. But there is this to be said about what Martignoni has seen fit to publish: It is not excessive. Mauro is as painstakingly meticulous about his publications as he is about his research. He publishes only when he is good and ready, and when he feels that what he has to say is ready to be said. Although he could competently handle any aspect of insect pathology, his first and greatest love has been the insect viruses, whether they be in insects in the field, in test tubes in the laboratory, or in insect tissue cultures—a technique he was among the first to master, but, because of the modesty and caution of which I have spoken, he has not received the credit he deserves.

Mauro and Joe worked independently on their research, cooperating and assisting each other when the need arose, but they rarely collaborated. Together they constituted a pair which gave the laboratory its thurst for excellence and caused me unbounded pride. It was because of this separate, but nonetheless synergistic, dynamism that I could, as director of the Laboratory of Insect Pathology, administer to their needs and yet retain some semblance of sanity.

VII

In the meantime, the research activities of our laboratory took on renewed vigor. The principal thrusts maintained much of what we had been doing and included projects on the nature and properties of insect viruses and their diseases, the bacterial diseases of insects, the diagnosis of diseases of invertebrates, and some work on certain fungus and protozoan diseases of insects. Martignoni, Tanada, and I were involved in most of these programs. Considering that during most of the existence of the insect pathology unit in Berkeley not more than three professional workers were active at any one time, and realizing that a considerable amount of good work was accomplished but for one reason or another not published, the productivity of the laboratory (and department) was as Joe Tanada (1968, unpublished manuscript) says, "truly remarkable." According to Tanada, over 175 publications—some of them large papers and books—were published between 1945 and 1965.

Breaking down just one of the subject headings given in the preceding paragraph a bit further and relying on information summarized by Tanada, it should be recorded that most of the work on, for example, insect viruses was concerned with (1) the structure of viruses and their inclusion bodies, (2) the histopathology of a number of virus diseases, (3) studies in pathophysiology (or physiopathology, if you prefer), (4) synergism and interference between insect viruses, (5) the resistance of insects to virus infections, (6) the cross-transmission of insect viruses, (7) the trans-ovum transmission of insect viruses, (8) insect tissue culture and virus infection in vitro, (9) the effect of stressors on insect viruses (and certain other pathogens), (10) the diagnosis of numerous virus diseases and the discovery of many new viruses, and (11) the annotated listing and the assembling of a comprehensive bibliography of insects reported to have virus diseases. Among the papers published on these subjects are those of Clark (1956), Hall (1953), Hughes (1950, 1952, 1953*a*, *b*, 1957, 1958), Hughes and Thompson (1951), Leutenegger (1964), Martignoni (1957, 1958, 1964), Martignoni and Langston (1960), Martignoni and Milstead (1962, 1964, 1966*a*, *b*), Martignoni and Scallion (1961*a*, *b*), Martignoni, Zitcer, and Wagner (1958), Sager (1960), Smith, Hughes, Dunn, and Hall (1956), Steinhaus (1948*a*, *b*, 1949*c*, 1952*c*, 1953*a*, 1954*b*, 1959*c*, 1960*f*), Steinhaus and Dineen (1959, 1960*a*), Steinhaus and Hughes (1952), Steinhaus, Hughes, and Wasser (1949), Steinhaus and Leutenegger (1963), Steinhaus and Marsh (1960), Steinhaus and Thompson (1949*c*), Tanada (1954, 1956, 1957, 1959, 1960*a*, 1961, 1964, 1965), Tanada and Chang (1960, 1962*a*, *b*, 1964, 1968), Tanada and Leutenegger (1965), Tanada and Tanabe (1964, 1965), Tanada, Tanabe, and Reiner (1964), Wasser (1952), Wasser and Steinhaus (1951), and Wittig, Steinhaus, and Dineen (1960).

Among the unit's authors of papers on bacterial diseases of insects and bacteria associated with these and other arthropods were Hall and Arakawa (1959), Hall and Dunn (1958), Steinhaus (1945*b*, 1947*c*, 1951*a*, *b*, 1952*b*, 1957*b*, 1959*b*, *e*, 1961), Steinhaus, Batey, and Boerke (1956), Steinhaus and Brinley, (1957), and Steinhaus and Jerrel (1954). Papers on entomophilic protozoa and fungi were published by Abdel-Malek and Steinhaus (1948), Brooks (1968), Hall (1952*a*, *b*, 1959), Hall and Bell (1960, 1961, 1962, 1963*a*,*b*,*c*), Hall and Dunn (1957, 1959), Hall and Halfhill (1959), Lipa and Martignoni (1960), Lipa and Steinhaus (1959), Steinhaus (1947*a*), Steinhaus and Hughes (1949), and Tanada (1955, 1964).

The diagnostic service performed by the laboratory and department was one of the most rewarding as far as having the satisfaction of serving other scientists, mostly entomologists, and agriculturists was concerned. The number of accessions received and processed numbered in the thousands. Inasmuch as this work has been summarized in the literature (Steinhaus, 1951*b*, 1955*b*, 1957*a*, 1963*c*; Steinhaus and Marsh, 1962, 1967) more need not be said about it here. In 1963, the diagnostic service was carried on by Martignoni and Marsh, and since 1964 by George O. Poinar and G. M. Thomas.

Epizootiology, or the ecology of insect diseases, entered into much of the work on microbial control. Some of the staff's publications reflect an emphasis on the principles involved: Clark (1955), Martignoni and Auer (1957), Martignoni and Schmid (1961), Steinhaus (1954*c*, 1958*b*, 1960*g*), Tanada (1959, 1960*b*, 1961, 1967), and Tanada, Tanabe, and Reiner (1964).

For a while, directly after the laboratory was made the department, all work on the use of microorganisms to control pest insects was assigned to the Department of Biological Control; while our unit was constituted administratively as a laboratory (and later when it became a division), a moderate number of our publications had to do with microbial control. It was our intention to keep microbial control on an experimental basis while other entomologists and growers and industrial concerns practiced most of the actual application of the microbial insecticides we developed or with which we worked. Among our papers were those by Clark and Thompson (1954), Grigarick and Tanada (1959), Hall (1955, 1957*a*,*b*), Hall and Andres (1959), Hall and Dunn (1958), Hall, Hale, Storey, and Arakawa (1961), Hall and Stern (1962), Martignoni (1964), Rabb, Steinhaus, and Guthrie (1957), Steinhaus (1945*a*, 1947*b*, 1956*a*,*b*,*c*, 1957*d*, 1959*a*, 1960*a*, 1967*b*,*d*), Steinhaus and Thompson (1949*a*, *b*), Tanada (1959, 1967), Tanada and Beardsley (1957), Tanada and Reiner (1960, 1962), and Thompson and Steinhaus (1950*a*, *b*).

A special project on the diseases of gastropods (snails and slugs) was initiated by Wayne Brooks, one of our graduate students, who received an appointment in the Experiment Station. The title of his dissertation was "The role of the holotrichous ciliates, *Tetrahymena limacis* (Warren) and *Tetrahymena rostrata* (Kahl), as parasites of the gray field slug, *Deroceras reticulatum* (Müller)" (Brooks, 1968).

The parade of citations to which the reader has been subjected in the preceding paragraphs—if he has cared to go through them—was frankly a sampling of the principal publications made by the small group of workers assembled into what was the Laboratory of Insect Pathology and had become the Department of Insect Pathology. It does *not* include some of the very interesting and significant contributions made by graduate students specializing in insect pathology, and about which we shall have more to say in our section on the teaching of insect (invertebrate) pathology. [Not written. MCS.] Another reason for listing the citations to some of the literature our group produced is to indicate that the department which came into being in 1960 was, even though small, a vibrant, healthy organism, and that what was to happen next was in no way meant to diminish this robustness. Indeed, with the promise of additional federal funds to enable us to expand our operations to study some of the noninfectious diseases of invertebrates, things could not have looked better.

[The following note was found with this section: "Also add Halls's work at Riverside after we split off as a department and he joined the group at Riverside, and Kellen's work at Fresno because it ended up under the university's jurisdiction." MCS.]

3. A PERIOD OF TRANSITIONS

Catch, then, oh catch the transient hour;
Improve each moment as it flies!

Samuel Johnson

I

In spite of all the lip service given to basic research—science for science's sake or research done primarily to further man's understanding of nature—rarely is a new science, or new branch of science, allowed to come into full being and allowed to flourish merely because it satisfies man's curiosity. Somewhere, lurking in the background, there usually has had to be at least a hoped-for practical justification for it, if it is to attract material support. Not always, but usually. This is not necessarily an evil motivation—most scientists, with any social consciousness at all, desire that their labors have some meaning to life and some benefit to mankind, even if these are to be in the remote future. On the other hand, most scientists in the process of making a discovery are, at that particular point in time, excited, encouraged, and satisfied by the new knowledge just because it is new and revealing. The fact remains that most financial support of scientific research is based on at least the hope that "something will come of it." So it was with insect pathology. The earliest work on the diseases of the honey bee and the silkworm was because of the economic losses incurred. So on up to the present day.

Insect pathology at the University of California—although heavily engaged in basic research—was, aside from federal funds and grants-in-aid, supported by, and had its start and found its home in, not the College of Letters and Science but the College of Agriculture—or, to properly include the Agricultural Experiment Station—the Division of Agricultural Sciences. Nor did it originate in a department of basic entomology but rather in a department concerned with one of the great applications of entomology—biological control. Indeed, were the truth to be told, the original Division of Beneficial Insects (later the Department of Biological Control) had to be administratively separated from other departments of entomology in the university system primarily to give it breathing space and a chance to grow and to protect it from being overwhelmed by an already great application of entomology—chemical control.

To be sure, much basic research was accomplished in the Department of Biological Control, but from the beginning, consciously or subconsciously, its raison d'être was to use living agents to control other living agents—primarily to use insects to control other insects. Harry Smith, the father of biological control in California, was himself known as much for his classic contributions to population dynamics and population ecology as for his notable achievements in introducing and colonizing beneficial insects for the biological control of harmful species. But biological control per se is an application of ecology or entomology—or more strictly when it applies to insects—of insect parasitology. It is not in itself a science in the sense of possessing the meaningful -ology (or -logy). Instead, it represents a field of application and technology based on the sciences of entomology, parasitology, ecology, and others. (This is why another perhaps more appropriate name for the field would be "applied ecology" except that the latter has wider connotations.) It has to do with the phenomena *in these sciences.* Wherever there are phenomena, there can be a science to describe and explain those phenomena; phenomena breed sciences. Some might choose to call biological control an applied science, but this merely extends the tangle of semantics involved.

Thus, although insect pathology at the University of California (as it has at most other institutions) found its birth and was initially nurtured by the administrative division called biological control, it is basically not a part of this technology. That application of insect pathology known as microbial control *is* legitimately a branch or part of biological control. (Insect pathology has other applications of interest to agriculture, such as the suppression of disease in beneficial insects of all types including the honey bee and the silkworm.) But the science, the -logy, of pathology no more belongs to the field of biological control than other branches of entomology do—insect physiology, insect ecology, insect morphology, and the rest. Certainly there is nothing wrong with insect pathology being pursued in a department or other administrative unit designated as "biological control," but the science of insect pathology (a branch of entomology, or of invertebrate pathology, or of comparative pathology, or of pathobiology, or—best of all—of pathology when the latter is properly defined in its broad sense as the study of disease) must not be equated with any of its applications. Insect pathology is not synonymous with microbial control!

The philosophical point on which I have been dwelling was fundamental to an administratively unpopular position I felt it necessary to take in the late 1950s. As the year 1960 approached, I was convinced that, acknowledging our everlasting debt to Professor Harry Smith and to the Department of Biological Control in which the Laboratory of Insect Pathology was a unit, the science would eventually be smothered by not only its own application but that known as biological control generally. Moreover, and this was even more jarring to our administration, I did not feel that agriculture stopped at the seashore as far as our unit was concerned or that the diseases of marine and terrestrial invertebrates were less important than those of insects and mites.

From the practical standpoint, it was not too difficult to convince my administrative superiors that we should broaden the scope of our activities to include studies on the diseases of snails, slugs, earthworms, and other terrestrial invertebrates of agricultural importance. But they balked whenever I brought up the idea of marine invertebrates. This, they said, would be the concern of the Institute of Marine Resources on the San Diego campus, or of marine biologists, or of some part of marine stations generally. (Curiously, in spite of the opportunities for the study of disease in marine invertebrates, few marine stations had pathology units at this time.) Clearly, therefore, I was not being at all successful in selling the idea that insect pathology (and certainly invertebrate pathology—the very name of which they feared would worry the zoologists in the College of Letters and Science) was a science worthy of investigation for its own sake. They acknowledged, but seemed unmoved by, the old truth that if we launched a broadened program of basic research, the practical applications would flow from the results naturally. Moreover, inasmuch as our laboratory was in Agriculture, the applications in, and contributions to, such fields as medical and veterinary research were not appropriate "at this stage of your Labortory's development." (Fortunately, the U.S. Public Health Service, from which we had been receiving much of our federal grant support, had not been of a like mind, even with regard to those diseases of insects which had obvious agricultural applications.)

Although as early as 1949, I revealed my expectations with regard to insect pathology's being the forerunner, and even the "backbone," of invertebrate pathology (e.g. see Steinhaus, 1949, p. 701), only as 1960 approached did I feel it imperative to take a stand on the matter so far as our university administration was concerned. (Had I known in 1949 what I had learned by 1965 about the workings of university administrative procedures, I would have proceeded earlier and at a higher level—the fear and conservatism that sometimes prevail at the level of department chairmen, deans, directors, vice-chancellors, and the variety of vice-presidents and assistants to presidents can be appalling.) If we were to be confined to the study of diseases and abnormalities of insects alone, then properly insect pathology should be included in the Department of Entomology along with insect physiology, insect ecology, and the rest. The application microbial control could, in any case, in accordance with the administrative setup prevailing in the university at that time, be properly assigned to the Department of Biological Control. But, it appeared to me that to limit our work solely to the diseases of insects was definitely clipping our wings, limiting our potential, and assuming gun-barrel vision of an horizon of important possibilities. All invertebrates should be within our purview. Compared to the research being conducted on the diseases of vertebrates (man and other higher animals) and on plants, that being done on invertebrates, even including the burgeoning of insect pathology, was meager.

In the strict sense, pathology is a biological science, and as such could perhaps not be fully tolerated or accommodated in a division or college of agricultural sciences. Yet, plant pathology (sometimes submerged in plant or crop protection) was able to be accommodated. Knowing the entanglement of red tape that would be involved in attempting to move insect or invertebrate pathology out of Agriculture, and conscious of the debt our particular unit owed to Agriculture in giving us our start and the birthrights and opportunities which followed, I knew it was useless to make such a request. (Interestingly enough, some years later I learned from one of the leading zoologists in Letters and Science that the Department of Zoology had been watching our developing basic research program and, particularly our teaching program, and had given some thought to requesting, through the appropriate committees of the Academic Senate, our transfer to that College.) In any case, I was never really concerned with moving out of Agriculture; rather, my concern was with just how to go about convincing my associates and the administration that we should broaden our work at least to include all invertebrates of importance to agriculture—then, having accomplished this, how to obtain blanket approval to work on the diseases of invertebrates of any and all kinds.

Changes were imminent in the administration of the Department of Biological Control, so now seemed the time to face the higher administration with the "facts of life" as I saw them pertaining to invertebrate pathology. I had the moral support of the members of the laboratory staff, but the entire Biological Control Department was involved with its own problems, so I realized that if I were to make a fight, I would have to make it largely alone. This gave me pause. I was not afraid of a good fight, particularly when, as in this case, I felt I was on the side of the angels, but doubted my ability to convince an ostensibly friendly but hard-headed, reputedly conservative group of administrators of the merits of my case, especially when I knew they would have strong pressure for an opposite viewpoint from certain influential colleagues outside of the Laboratory of Insect Pathology. As I saw it at the time, three accomplishments were necessary to bring this about: a new distinct department or institute, the words "invertebrate pathology" incorporated in the new name, and more financial support and staff. The opportunity to realize the first and last of these came more rapidly than I had expected.

II

In 1951 our mentor, Professor Harry S. Smith, reached the university's compulsory retirement age of sixty-seven. Fortunately, through his abiding interest in biological control and in the statewide department he had built and his affection for the members of its staff, we were to enjoy his company and wise counsel until his sudden death from a cerebral hemorrhage on Thanksgiving Day, 1957. Prof Harry was ably succeeded by his illustrious and long-time friend Curtis Paul Clausen, who, by taking advantage of the relatively early retirement time allowed by the U.S. Department of Agriculture, was able to leave that organization to assume the administrative duties relinquished by Smith.

Curtis was a graduate of the University of California (Berkeley), and from 1916 until he enlisted in the Army (1918) for World War I, he was assistant superintendent of the State Insectary at Sacramento while Smith was superintendent. Returning from the war, Curtis joined the federal Bureau of Entomology, where in 1934 he was placed in charge of the Division of Foreign Parasite Introduction and, later, also of the Division of Control Investigations. Under Professor Clausen, the Department of Biological Control, statewide, continued to advance scientifically and prestigiously. So did the Laboratory of Insect Pathology, which Clausen supported enthusiastically (e.g., see Clausen, 1954) and without the bias that some had feared, for he was a famous and pioneering "parasite man" and had written the classic *Entomophagous Insects* in 1940. From a personal standpoint, I found the change of guard quite in order and quite a natural one—I had never forgotten with what quiet but intense interest Curtis Clausen sat through my first face-to-face meeting with Harry Smith back in 1944.

Trouble had been brewing beneath the surface of the Department of Biological Control for some time. Separation of the north and south branches seemed inevitable, for it simply was not practical to have a chairman, no matter how efficient, located more than 400 miles away. Statewide and intercampus departments were decreasing, and there was a developing spirit of campus autonomy within the university system. This desire was clearly seen by President Clark Kerr, who is generally credited with having instituted the degree of administrative decentralization which gave each of the campuses (at this writing there are nine) their own academic integrity and relative autonomy—unhappily the same degree of autonomy was not to be as readily forthcoming with respect to the business and finance area of the university operations.

The stresses and strains in the statewide structure of the Department of Biological Control arose from a feeling on the part of those working at Albany-Berkeley that they were second-class citizens within the department, that they suffered the disadvantages of an outpost or colony, and that with the department chairman located at Riverside, the northern group suffered from lack of appreciation and attention. Just how real these grievances were, was difficult to assess. Since the insect pathology unit was autonomous and headquartered in the north, we did not feel this discrimination, imagined or real, as much as did the men concerned with insect parasites and predators; it is not unlikely, however, that Hall might have felt a similar inconvenience at Riverside since he was the branch of our laboratory there. The fact that the group in the north had a vice-chairman, Richard L. Doutt, as an administrative head did not help matters a great deal. When Doutt had asked to be relieved of his administrative responsibilities, Clausen, who had been most considerate of the needs of insect pathology, asked me to assume the vice-chairmanship with the hope of stabilizing matters. From my standpoint it was a successful year of departmental administrative activity, but I was under no illusions that I, or anyone else, could stem the undertow of what was not so much discontent with a chairman or vice-chairman but really desire for independence from Riverside. Beginning 1 January 1959, I too, asked to be relieved of administrative duties for the department, preferring to concentrate on the welfare of insect pathology without having to lean over backwards not to favor the work with pathogens over that with insect parasites and predators.

A separatist movement, formally initiated with a letter, in 1957, through Professor Clausen's office to vice-president of Agricultural Sciences Harry R. Wellman, was still simmering despite its administrative rejection. The letter had formally requested that, in keeping with the general autonomy being given separate campuses, "the northern branch of the Department of Biological Control now be given *de jure status* as a separate department." Thus matters simmered along through 1958 and 1959, with both Hall and the pathology group at Berkeley remaining as aloof as possible from the discontent in the minds of some of the men in the remainder of the Department of Biological Control, even though we had supported them in their letter of 1957.

By the spring of 1959, it was clear that Curtis Clausen was considering retiring as chairman of the Department of Biological Control. Being aware that a few of the parasite men in the department looked askance at the autonomy enjoyed by the laboratory, he surprised me one day as we were returning from lunch by the direct questions, "Ed, would you like to have insect pathology as a separate department, and not have to put up with this nipping at your heels by one or two malcontents?" My response was a quick, "No, I like things as they are." But in answering Clausen so quickly I was not fully anticipating what might lie ahead after he resigned as chairman of the department as well as the eventual consummation of the desire on the part of the staff of Biological Control in the north to separate themselves administratively from that in the south. But I thought on the matter. The separatist movement most likely would split apart the northern and southern branches of the Laboratory of Insect Pathology as well. Such a separation was only natural in that the philosophy pertaining to the department as a whole would also apply to the statewide administration of the laboratory. Moreover, it was to Irv Hall's advantage that he cast his lot with his associates at Riverside.

III

On 19 May 1959, I received a memo from Clausen stating that he was retiring from the university on 30 June and that Charles Anthony Fleschner had been appointed as the new chairman. While reflecting not the slightest on Fleschner, this appointment came as a surprise: it had been understood, but never formally promised, that the next chairmanship would rotate between the southern and the northern branches of the department. Inasmuch as Smith, Clausen, and now Fleschner, were all "southerners," it is only fair to say that feelings of independence were again provided incendiary fuel, although by this time those of us in the Laboratory of Insect Pathology were having aspirations of our own for at least enough independence to broaden the scope of our activities beyond insects. In spite of this situation, we in the Laboratory of Insect Pathology and the people at Albany all generally agreed to give Charlie a chance. Moreover, inasmuch as Fleschner assumed the chairmanship with the highest of motives and the best of intentions and goodwill, it was only fair that he be afforded an appropriate honeymoon period. This included abandoning, at least temporarily, any idea of promoting complete separation of the Laboratory of Insect Pathology from the Department of Biological Control. We were already, by regential action, a "nondepartmental laboratory" which automatically gave us autonomy as a distinct budgetary research unit.

Because there appeared to be a difference in understanding as to our administrative status between the chancellor of the Berkeley campus, the university dean of Agriculture, and director of the Experiment Station, I wrote university Dean Daniel G. Aldrich, Jr., on 25 May 1959, asking for clarification, at least from his point of view. I received a note from him indicating that he would look into the matter "in good time" and then contact me. Weeks passed, but still no clarification from Aldrich. In as diplomatic a manner as I knew, I dropped him a note reminding him of our proposed get-together. He replied rather bruskly that he had told me he would consider the matter "in good time" and that this still pertained.

In the absence of the clarification, I wrote Fleschner on August 12 explaining the autonomous status of the laboratory as I understood it, what the alternatives (associated with or complete separation from or absorption by the Department of Biological Control) were, and that I hoped that Aldrich would call a conference to clarify the matter. In late September, I asked Fleschner, whom I knew had an upcoming appointment with Aldrich, to try his hand in getting the latter's decision on the status of the laboratory. Nothing came of Fleschner's attempts either.

Finally, in mid-February, 1960, university Dean (meaning statewide) Aldrich met with Dean (meaning on a particular campus) Ryerson and me for a briefing on the matter. This was my first face-to-face meeting with Aldrich, about whom I shall have more to say in a subsequent section. I found him to be energetic and direct in his questions relating to matters in the Department of Biological Control and the Laboratory of Insect Pathology. It was clear that he had little understanding or knowledge of the undercurrents involved; he explained this lack by saying that he had been depending on his campus deans to keep him informed. He quickly decided that the matter was of sufficient import to call a meeting of interested parties on both campuses (Berkeley and Riverside). Then he suddenly turned to me asking why I had published a question-and-answer leaflet on insect pathology, which had made reference to the administrative organization of the laboratory, without it having had his approval. Fortunately, I was able to abate his apparent indignation by telling him, and later documenting, that the leaflet that he had referred to was prepared in cooperation with his own office of Agricultural Publications in response to questions continuously received from growers, agriculturalists, and scientists, as well as news services requesting the information from Agricultural Publications. The form of its presentation was decided by the latter office and not by me or by anyone in our laboratory. I recount this incident because it gave me an insight (later to come in very handy) into at least one way of working with Aldrich—a man easily excited and made indignant, but quick to forget and forgive once a matter was properly and adequately explained. I was relieved at the outcome of this particular exchange, else our request for separate departmental status would have probably died right there.

To make a long story shorter and less tedious, Aldrich called a meeting of Experiment Station Director Paul Sharp, Citrus Experiment Station (Riverside) Director A. M. Boyce, Dean Ryerson, Fleschner, and myself. Perhaps the date of the meeting was prophetic for it was 1 April 1960—April Fools' Day. With the aid of charts, diagrams, and mimeographed material I presented the laboratory's case as forceably as I knew how. Dean Aldrich complimented me on the presentation I had made; however, although Ryerson supported me in principle he was not in a position to do much about it when it was obvious that Sharp (who was a strong supporter of our work but not of our organizational aspirations), Boyce, Fleschner, and possibly Aldrich were against my position. In spite of the pressure against me from Boyce, Fleschner, and Sharp, Aldrich refused to make a decision on the spot. He said he would think it over and let us know. I am sure that everyone in the room felt that although I had made a good case, and had won the battle, I had nevertheless lost the war. Flechner and Boyce expressed their condolences, but I did not need them because, surprisingly, I was filled with a feeling of euphoria—partially, I imagine, from a sense of relief that it was over, partially because I knew in my heart that I had made a good fight for our cause regardless of the ominous outcome. We all shook hands and parted in good company and fine spirits. As I left I turned to thank Aldrich for at long last giving me my day in court; to which he responded, "I haven't rendered a verdict yet." I replied that it was clear to me what it would be and that I would now be sending some telegrams in response to offers from other institutions.

As I returned to my office I recalled that during the discussion

Aldrich had called me a "zealot," and one of the men from Riverside commented that I was a "promoter." I had been called a "promoter" before while trying to build the laboratory so this didn't bother me. Upon reaching the office I went to Webster's International Dictionary to survey the variety of meanings these words might have. I found that a zealot was someone filled with zeal, which in turn meant "ardent and active interest; enthusiasm; fervor." Fine. Among the definitions of promoter is "one who encourages progress." Regardless of what the speakers intended the words to mean, these definitions were good enough for me. I was content to be a "zealot" and a "promoter"—building and responding to a challenge were my "racket."

Four days later, on 5 April, at about 8:10 in the morning, Dean Ryerson called saying that Dean Aldrich wanted to see us in his office as soon as possible. Within fifteen minutes we were down the slope to University Hall and hearing Aldrich say: "We have decided that as of 1 July next, there shall be a separate Department of Insect Pathology and that you shall chair it." I was flabbergasted! What had happened? I asked about changing the name to the Department of Invertebrate Pathology. "No," said the dean, "this might cause some concern over in Zoology; at least for the time being let's keep it 'Insect Pathology' even though you may work on other invertebrates." This was a disappointment, but the fact that the laboratory was suddenly to become a department was completely unexpected after the conference of April 1. I clumsily expressed my gratitude, and said I would do my best.

Later I was to learn that actually the arguments I had put forward had sounded convincing to Aldrich. After giving the matter further thought and knowing that I was planning action of my own, Aldrich held a telephone conference with Directors Sharp and Boyce and Dean Ryerson. The decision to have a Department of Insect Pathology was made. Either the men concerned felt charitable, or, perhaps, they saw merit in my argument to build a strong department devoted to the basic study of disease in invertebrate animals—as in the case of vertebrate pathology and plant pathology. The Department of Biological Control would retain their principal interest in the matter —i.e., microbial control—with Irv Hall being a member of their department at Riverside and with one of our men transferring to Biological Control at Albany. Surprisingly, this latter man turned out to be Mauro Martignoni, whose primary interest lay in basic research. However, Tanada expressed a strong desire to remain with the new department, and Mauro was willing to help out with microbial control work on a part-time basis (roughly sixty-five percent) "for a year or two," after which he wished to rejoin the new department.

The actual announcement of the formation of the new Department of Insect Pathology included the following passages:

> The new department will be responsible for conducting basic research pertaining to the pathology and microbiology of insects and other invertebrates. It will also be concerned with all applications of insect (invertebrate) pathology and microbiology other than those relating directly to the use of microorganisms in the biological control of insect pests. These biological control applications will remain in the Department of Biological Control.
>
> The new Department of Insect Pathology will be located on the Berkeley campus of the University. During the coming years it will attempt to augment its present research program with additional projects and staff. Some of this planned expansion must await the availability of more space. By means of the new department it is intended to enhance the emphasis and activity relating to fundamental research on the diseases of insects and other invertebrates. This is being done with the conviction that before we can reap the full benefits of the applications of insect pathology and insect microbiology, the amount of basic research in this area of biological science must be greatly intensified. . . . The diagnostic service formerly maintained by the Laboratory has been transferred to the new department.

In connection with the release of this announcement, the entomologists on the Davis campus engaged in the study of disease in the honey bee were assured that we had no intention of interfering or usurping their work on the research, diagnosis, and control of the maladies of this insect; we would become involved only in areas where they did not wish to work.

If the reader has felt that my account of the evolution of insect pathology at the University of California from a one-man operation (1944–45), to a distinct discipline, to a laboratory within the Department of Biological Control, to an autonomous laboratory associated with the Department of Biological Control, to a separate Department of Insect Pathology as of 1 July, 1960, has been unnecessarily long and verbose, permit me to assure him that he has been spared reams of detail. I have six-inch thick files of documents with such headings as "Justification of Insect Pathology as a Distinct Unit," "Administrative Organization of Insect Pathology within the University," "Advantages of Insect (Invertebrate) Pathology as a separate Institute or Department," "Proposal for the Establishment of a Department or Institute of Invertebrate Pathology," etc., etc. These documents are accompanied by charts, graphs, descriptive data, and expository arguments of such a nature that I thought would convince the most pertinacious and obstinate administrator imaginable. Even now, as I peruse them while regretting that space will not permit me to share their noble contents with the reader, they sound convincing and, I believe, retain their basic validity. I write this paragraph also to assure all who were involved in these matters that there has been much which I omitted for fear of boring the reader; but that which I have elected to recount I have done so as objectively, as fairly, and as succinctly as I know how.

In the midst of this, two personal complications arose. I had been planning to take a year's sabbatical leave during 1960–61. Part of this time I had planned to spend at the University of Wisconsin, where I had been invited to be the Visiting Knapp Professor. Obviously, however, it was my responsibility to change my plans at least to the extent that the beginning of my leave did not correspond with the initiation of our new department, 1 July 1960. Accordingly, I decided to delay my leave until 1 September 1960, at which time, with the aid of a Guggenheim Fellowship, I would take a six-month leave in Japan and other parts of the Far East. This arrangement met with the approval of Dean Aldrich. Joe Tanada was appointed acting chairman during my absence.

Dean Ryerson retired 1 July 1960, but he continued his responsibilities as Senior U.S. Commissioner of the South Pacific Commission, and in this capacity he and I enjoyed further association with regard to the rhinoceros beetle (*Oryctes*) in the South Pacific. E. Gorton Linsley, then chairman of the Department of Entomology and Parasitology, was appointed the new dean of Agriculture for the Berkeley campus. Gort had been named to the position after G. B. Bodman had filled in for a while as acting dean and had been acting dean 1 July, when our new department came into existence. Gort and I were good friends, and I regarded him highly both as a man and as a scientist—but oh what anguish we were to go through together within three short years!

IV

Upon my return from a sabbatical leave which took me to flourishing laboratories and institutes of insect pathology at a dozen or more points around the world, the task of obtaining more building space and additional funds (largely federal) for our new department was foremost. As in most universities so it was at the University of California at that time; we really had only two sets of problems: one was space and money, the other money and space. With regard to space, I can honestly say that some of my best friends are space analysts and planners, but their profession and their activities within the university (in concert with those of academic space committees) have always constituted the principal bane of my existence—the thorn in my side. When, as they frequently have, they come up with space assignments which are not adequate even to hold the scientist's equipment, let alone a place for him to work, just because their formulae indicate that that is the amount of space a biologist should have, my "heat of mind" is difficult to subdue. I am sure that hordes of fellow academicians share my feelings in this matter. Here, again, it is interesting to go back to old files to read our plaintive, but fruitless, pleas for *Lebensraum*. It would be difficult to overstate the depressing effect this impasse for additional space had on our entire staff. An example of the unwillingness of the Berkeley administration to cooperate regarding the space situation for Insect Pathology occurred when we had the prospect of receiving a five-year $1,000,000 federal grant. The university absolutely refused to offer any available space for the project when the grant application was made. I had to go off the campus to find rental space to accommodate the expected three professionals and their technicians which the grant would provide. But extra space can be found when the university deems it necessary! About the same time that I told Aldrich that I would come to Irvine, I received word from the U.S. Public Health Service that we had received the above mentioned grant. After I had already made my commitment to Aldrich, more adequate quarters were offered for Invertebrate Pathology to accommodate the grant and as an incentive for me to stay at Berkeley, but I had already committed myself. Fortunately, Invertebrate Pathology was still given the larger space.

Given the fact that with our new department status we were forced to retain the name "Insect Pathology" rather than "Invertebrate Pathology," it made little sense for us to remain outside the Department of Entomology and Parasitology. What's in a name?—in this case a great deal—at least psychologically, and at that point in time. Indeed, my failure to bring about this change in name (for which I had now obtained the support of the Department of Zoology) so as to more accurately represent the scope and integrity of our discipline was one of the reasons (along with inadequate space and staff) that, on 7 March 1962, I informed the Berkeley dean—my friend Gort Linsley—that I was convinced that he should replace me with someone having greater powers of persuasion or influence. Linsley assured me that my case was sound, my demands reasonable, and that I had done all any department chairman could do to bring about the changes and improvements we felt necessary, but that we were at the mercy of the establishment, and that he, too, felt disappointed and frustrated in not being able to be of greater assistance to us. With kindness he admonished me, "Have patience, Ed." I said nothing aloud, but to myself I commented, "After years of fighting with space committees, academic senate committees, and a sympathetic, but what seemed to me to be an overly conservative, administration, my patience had run out!" Moreover, about this time an idea came from somewhere in the higher administration or from a committee in the Academic Senate—I could never learn precisely from where it came—that those departments (such as Cell Physiology, Biological Control, and Insect Pathology) which were essentially research units and did their teaching in another department could not be called the prestigious name "Department." Since there were no units known as "Institutes" in the Division of Agricultural Sciences (although they abounded throughout the remainder of the university), what were we to become? Were we to revert to a laboratory status again? Be absorbed by another department? Linsley, who had been informed of these rumors and said he believed them, professed not to know the answer. Could we, I wondered, retain our integrity and identity—and of particular importance, our own direct budget line—by becoming a division of an enlarged, over-all entomology department? Dan Arnon, chairman of what was then essentially a one-man Department of Cell Physiology, assured me that he was confident we could "beat this crazy idea." Nonetheless, I was left to ponder our future.

If it were to be our fate to limit our name and most of our activity to insect pathology rather than to invertebrate pathology as a whole, then it made sense for us to be part of an entomology complex, or at least so it seemed to me. Not being secure from Shakespeare's "worldly chances and mishaps," after the main part of the banquet at the June 1962 Pacific Branch meetings of the Entomological Society of America being held in Santa Cruz, California, Linsley and I found an isolated table at one corner of the room and held what amounted to a confidential discussion of the administrative problems he was having with the northern branch of the Department of Biological Control. During the discussion I somehow momentarily forgot our own aspirations as a Department of Invertebrate Pathology, and developed an intense empathy with Gort concerning his problems as they related to the areas of entomology and biological control as they were developing on the Berkeley campus. Mentally placing myself in his position, I suggested to him the possibility of having a sort of superdepartment in which there could be four autonomous divisions: Entomology, Parasitology, Biological Control, and Insect Pathology. Perhaps the super-department could handle all of the teaching curricula, while the four divisions administered the Experiment Station research in their respective domains. Also, Insect Pathology could be changed to Invertebrate Pathology and the name of the super-department be some term which would be all-encompassing. Linsley, in his characteristic quiet way, gave the idea an immediate positive response. At the same time I liked what I had extemporaneously proposed because, although we would surrender our departmental status, it would give us the name Invertebrate Pathology (implying that we could thus expand the scope of our work accordingly), and would permit us to retain our administrative integrity and autonomy. Or so I thought.

On 5 July 1962, I sent Linsley a memo in which I suggested that an appropriate name for the department made up of the four divisions might be Department of Invertebrate Sciences inasmuch as all four divisions were primarily concerned with invertebrate animals. (Later, I regretted I had not suggested that the administrative umbrella be called a "division" which would be made up of the departments.) I again outlined what I thought the advantages would be, especially from his standpoint as Dean of Agriculture. In a closing paragraph I somewhat heroically sounded the trumpets with "Biological Science is definitely going to be the next great wave in the march of science, filling the position which chemistry and physics have held in recent decades. And it is coming at a terrific speed! . . . we cannot afford to be caught with our thinking geared to the way things have been done in the past. Our Administration must be willing to adjust rapidly, to accept new names, new categories, new methods of organization, and new alignments of disciplines."

Shortly thereafter, Linsley met with Ray Smith, chairman of the Department of Entomology and Parasitology, Powers (Bud) S. Messenger, vice-chairman of the universitywide Department of Biological Control, and me as chairman of the Department of Insect Pathology. Gort and I had agreed that we would make no mention of our previous conversations so as not to prejudice my role in the discussions; however, he agreed to support the change in the name of our unit and the change in the name of the overall department to be all-inclusive of our several disciplines. After giving us a clear and general outline of the idea, Linsley left the three of us to discuss the matter and to come up with a recommendation for him, one way or another.

Needless to say, during the meeting I felt somewhat as though I were playing the role of a pretender or an affecter. At first Smith (Ray, that is; there seem to be *so* many Smiths in entomology and zoology!) did not warm up to the idea. Knowing Ray to be one who considered such matters with great care and deliberation, I presumed that he was worrying lest the new arrangement would leave him with a meaningless department and that it would allow the entomologists on the Davis campus (which were administratively a part of the Berkeley department) to break away to form their own department. Accordingly, I argued that the Davis people were on their way to autonomy in any case, and that as head of the overall department he would be in charge of all the teaching, and have the most prestigious administrative position. But in addition to these benefits, I believe Ray saw, as I had when I first talked with Linsley, that such an arrangement would solve many administrative problems for the dean, would be a means of aiding Biological Control to free itself administratively from Riverside control, and would make an entomological complex second to none in the United States. As far as Messenger was concerned, he knew that his staff would support the plan because it would give them their long-sought autonomy. I mentioned that I had long thought that the parasitology part of the Department of Entomology and Parasitology should be unitized. This was agreed, as was the logic of asking Deane F. Furman to head up the unit. All of this was duly reported to Dean Linsley

After appropriate consultation with key members of the respective staffs, and after processing the proposal through the appropriate administrative channels and committees of the Academic Senate, Linsley informed us that the idea was now a fait accompli. The official announcement read, in part, as follows:

> The formation of a new department of entomological and related sciences has been authorized on the Berkeley campus of the University of California as of March 1 [1963]. The name of the new department will be the Department of Entomology and Parasitology, and will be comprised of four divisions: Biological Control, Entomology and Acarology, Invertebrate Pathology, and Parasitology. Each of these divisions will be responsible for conducting its particular area of research activities in the Agricultural Experiment Station. The teaching responsibilities, in which all divisions will participate, will be centered administratively in the central office of the new department. . . . Essentially, each division will operate autonomously within the departmental framework. . . . The chairman of the new department will be Ray F. Smith, and the divisional chairmen will be P. S. Messenger for Biological Control, R. L. Usinger for Entomology and Acarology, E. A. Steinhaus for Invertebrate Pathology, and D. F. Furman for Parasitology.

In keeping with his previous promise to me, Linsley told the chairmen (who acted as an executive committee) that in order to facilitate the approval of the plan by the administration and by the various committees through which it had to pass, no change in the name of the Department of Entomology and Parasitology had been made when the formal proposal was forwarded. However, now that our organizational structure had been approved, we could decide on and propose a new overall or umbrella name. This, and the name Invertebrate Pathology for our own division, were the concessions I had been willing to make to give up our own departmental status granted us not quite three years before. I anticipated no difficulty in this; in fact, in the written announcement I circulated to the staff in Pathology, I stated that the name of the overall department, i.e., Entomology and Parasitology, was "for the time being" only. As the youngsters say, "I couldn't have been wronger."

Without going into the pros and cons of the argument that ensued when I brought this matter up at one of our first executive committee meetings, I can only report that what I had expected to be a readily accepted proposal was voted down. Ray, as chairman of the committee, remained noncommittal; only one other member of the group saw things as I did. While I am sure that the outcome of this vote was of no great consequence to most of the committee, and certainly they did not know that, as a result of my previous discussions with Linsley, we would have had his backing for virtually any name we proposed, to me it was a most depressing turn of events bordering on betrayal. Yet everyone had acted in good faith; I had sacrificed our departmental status and the wide scope (invertebrate animals) it stood for, to be swallowed up as a division of what could have been a new, challenging, and broadly drawn department. Although most of our work would still be with the diseases of insects, still invertebrate pathology could not be said to fit properly into entomology. And, although one can argue that much of pathology (i.e., disease) is a part of parasitology (as are the entomophagous insects involved in biological control), by no means is this always the case. Rather, parasitology is a part of pathology. Moreover, all of the noninfectious diseases of invertebrates, the number of which are legion, are outside the province of parasitology. To be sure, only a name was involved, but I had dedicated myself to a principle that I thought worth fighting for, and I could not do this under a banner I felt was restrictive and a misnomer. But I realized that it would be imprudent to reveal my feelings after having lost the decision; it was a time to be cooperative and to be a good sport for the greater good of all. Nonetheless, as I left the meeting, I am afraid I censoriously remembered the famous remark of Leo Durocher "Nice guys finish last."

VI

As the confusion raged in my mind as to what was happening to the unit my associates and I had worked so long and against such odds to build, my thoughts turned to another matter that I had been cogitating on during the past several years. It disturbed me that "what goes wrong" with life and living systems was almost always taught as a discipline completely apart from the rest of biology. Even general biology courses usually omitted any significant reference to pathology (disease in the broad sense) or to parasitology and pathogenic microbiology. Instead of the biologist always telling students what happens when the Krebs cycle works right, could he not give examples of what happens when a part of it—or any system of enzymes or hormones—does not work right? Is there really no place in biology for explaining briefly (granted, in depth treatment requires advanced or specialized courses) malfunctions of photosynthesis, transpiration, glycogen metabolism, schizophrenia—yes, even polyhedroses?

But my job was to teach and conduct research in insect or invertebrate pathology. I would hardly be welcomed by my colleagues over in Letters and Science spreading such doctrine when, to begin with, I felt that much of Berkeley's biology at that time was not being presented in a manner suitable for the times. Obviously I was not alone in thinking this about the *new biology*, because as I was licking my psychological wounds the annual All-University Faculty Conference was coming up.

I was a delegate to this conference held on the Davis campus. This year the topic was "The Student and the Quality of his Intellectual Environment in the University," and the dates were 7–10 April 1963. During one of the sessions I was sitting in one of the back row seats. Coming in and taking an empty chair beside me was university Dean of Agriculture Daniel G. Aldrich, Jr., who had recently been appointed chancellor at the new Irvine campus near Newport Beach in southern California. It was, of course, the same Dan Aldrich who had played the decisive role in permitting Insect Pathology to become an autonomous department, and now a division within a superdepartment. We greeted each other. In a purely casual manner I then asked Aldrich what his plans were for the biological sciences at Irvine. He started to answer when suddenly his eyes "caught fire" and he asked me to come over to a corner of the meeting room so that we could talk without disturbing others. He asked me what my ideas about the biological sciences were, how they should be organized, and how they should be taught. I told him that certainly biology was in a state of ferment, needed a great deal of reorganization, that much of the way biology was being taught throughout the country was, in my opinion, outmoded, and that he had a tremendous opportunity at Irvine to do something about it. Not knowing what his opinion was on the subject, I added that I thought that the quarrel going on between the molecular biologists and the so-called traditional biologists was ridiculous—I saw no reason why biochemistry and molecular biology could not be embraced without minimizing or harming traditional biology, although there was considerable deadwood in the latter that had to be cleaned out, for there was subject matter that could better be handled at the high school level.

We talked a few more minutes after which I was surprised to hear him ask, "Would you be interested in considering the position of dean of the Division of Biological Sciences at Irvine?" (In this case the term *division* was equivalent to *college* or *school* rather than a segment of a department.) He explained that by now he was well aware of my willingness to fight against great odds and my "ability to organize and administer new endeavors." Needless to say, I was somewhat taken aback at this assessment; apparently what I had considered a moment of weakness—i.e., offering to become a division of a large department complex—he had interpreted as a sign of strength through cooperating. I responded that since I had been giving some thought to new directions that biology could take, I might be willing to consider such an offer. However, I warned him that I would have to think it over carefully and for several weeks. What I did not tell him was that I was becoming quite disillusioned with the situation pertaining to invertebrate pathology at Berkeley, that I had thought it might be worth asking for a transfer to the new Santa Cruz campus, but that geographically southern California had no attraction to me whatever. We parted by my agreeing to put on paper for him some ideas I might have about how the biological sciences might be presented at a new university campus, whether or not I finally agreed to come to Irvine. Later during the conference he introduced me to Ivan H. Hinderaker, then assisting him at Irvine and soon to become vice-chancellor for academic affairs. Ivan had been chairman of the Department of Political Science on the Los Angeles campus of the university; he remained vice-chancellor at Irvine until 1 September 1964, when he was named chancellor at the University of California, Riverside.

It so happened that also attending the conference from Berkeley was my trusted friend Robert L. Usinger, famous for his classical research on Hemiptera and his writing of textbooks on zoology and other subjects. Later in the day, I sought Bob out, explained the offer Aldrich had made me, and asked him what he thought about it. As I remember Bob's counsel, it was that while he hated to see me leave Berkeley, he could understand my responding to the challenge of building an entirely new program in the biological sciences. In his gentle, deliberate way, Bob enunciated the pros and cons of the matter, then ended by saying, "And, you know, Newport Beach is a pretty nice place to live."

Upon returning home from the conference I, of course, discussed the possibility of transferring to the Irvine campus with my family. Typically, my wife was all for the idea if it was something I wanted. We decided that before I could say "no" we should at least go down to look over the area and to attempt to ascertain what it would be like to live in southern California. As we drove down the freeway from the Los Angeles airport toward Newport Beach, we noticed the heavy smog prevailing in the Los Angeles basin. We agreed that if the smog extended beyond Disneyland (Anaheim), we would let air pollution be the deciding factor. As if by Disney's magic, and as though the curtain of a great stage were rising, as we reached Disneyland the smog dissipated, and it was not long before we were approaching the rolling green hills of the great Irvine Ranch. I stopped off at the interim office building that had been constructed at one edge of the campus, the rest of which was barren except for cattle grazing; Mabry took the car in to look over the neighboring communities of Corona del Mar, Newport Beach, Laguna, and Costa Mesa.

While waiting to see Aldrich, I chatted briefly with Hinderaker and Jack W. Peltason, who was being recruited for dean of the College of Arts, Letters, and Science, but who was destined to replace Hinderaker as vice-chancellor for academic affairs when the latter was picked to be chancellor of the Riverside campus. (Peltason, after a two-year sojourn at Irvine was invited back to Illinois as chancellor of the Urbana campus.) Enthusiasm about what could be accomplished in a new academe, starting from scratch, flowed from both gentlemen. Shortly, Aldrich was free to see me. He was in his usual hurry, bounding with energy and exuding enthusiasm about everything. We talked in his office awhile,[10] went to lunch together, then drove over to an old buffalo barn on the Irvine Ranch that had been remodeled into offices and drawing rooms by the well-known architect William L. Pereira, who had recently completed the preparation of a master plan for the development of the 88,000-acre Irvine Ranch, especially that third of it nearest the Pacific Ocean. Here Aldrich and one of Pereira's assistants showed me models and drawings of the planned university and surrounding community that gave me a feeling that it was all an Alice-in-Wonderland world and that I must guard myself against the oversell common to land-developing situations such as this. I looked out the window at the grassland where, they were telling me, there would soon be not only a 1,500-acre campus but model communities, and even new cities, one of which (next to the campus) would be called Irvine. An original 1,000 acres had been donated by the Irvine Company to the university, which then purchased an additional 510 acres to serve as a buffer zone between the campus and the urbanization that one day would surround the campus. Some of this land would also serve as land for faculty housing when needed. They explained to me how they were preparing to develop some 10,000 acres surrounding the campus into a university-focused community of approximately 150,000 persons by 1990. Unbelievable, but such was the charisma of the moment that I, too, could see what the models and drawings showed as I looked out over the treeless hills with cattle grazing on them.

In the meantime, my wife was being caught up in the possibility of living in a brand new house in a brand new community near a brand new university campus, near the good old Pacific Ocean. A week or two later, my wife drove the children down—after all, this was going to have to be a family decision. The children, were quite against moving to a new and strange area. We were unable to transmit to them by word of mouth the sense of pioneering we were beginning to feel. However, as they approached the beaches of Corona del Mar and Newport Beach, their aversion turned to reluctance, then to curiosity, until by the time they returned home they had assumed at least a neutral position. With great magnanimity they left the final decision up to Mom and Dad.

In the meantime, immediately after the All-University Conference, I had gone to Dean Linsley and explained to him the offer Aldrich had made. I similarly informed Ray Smith, chairman of the Department of Entomology and Parasitology. According to university procedure, Aldrich should have approached me through the chancellor of the Berkeley campus and through Linsley, my dean. Since he did not, Linsley seemed to appreciate the fact that I had let him know of Aldrich's contact with me even though technically this did not free him to do any negotiating with me even if he wanted to; and Gort was one who adhered strictly to protocol and directives and rules and regulations. I must admit that I was pleased that Linsley's reaction was not one of joy at my possible departure and that he expressed sincere disappointment that I would consider leaving. He made it clear to me, however, that he and the administration had done all they felt they could for invertebrate pathology. I responded immediately that I had not intended to embarrass him with any bargaining, that I appreciated all he had done for us, and that the repeated requests I had made for more space and staff were matters on which I had decided the university would not act in any case.

This meeting with Linsley was perhaps the crucial one and probably determined to a great extent my ultimate decision. If Gort *had* indicated in any way that our grievances could be negotiated, I am sure that I would have seriously considered remaining at Berkeley, especially if he would have agreed to the re-establishment of our previous status as a department or as an institute. As I have indicated, however, and on Linsley's behalf, it must be said that a directive had been sent out from President Kerr's office indicating that no interference should be made with anyone's desire to transfer to one of the developing campuses. Gort, I am sure, felt that anything he might say to induce me to stay would be violating this directive. And certainly it would have been highly presumptuous of me to assume that there was any reason for him to want to encourage me to remain at Berkeley. I learned, however, that Gort did communicate with both university dean of Agriculture, M. L. Peterson, and Berkeley chancellor E. W. Strong, both of whom were displeased over the fact that Aldrich had approached me directly.

At this point the usual bureaucratic red tape and strangling chain of command inherent in the university set in. While the Berkeley officials waited for Aldrich to contact them (apparently so that they could "talk turkey" with me), I naturally became more and more involved in discussions with Aldrich and with the Irvine campus. About a month later, while I was attending a special meeting between the university's editorial committee and some biologists, I received a telephone call from Aldrich relating to some routine matters. I asked Aldrich if he had yet talked with Linsley as he had promised he would do when I had visited on the Irvine campus. He said he had not as yet done so; accordingly, when I told him that Linsley was in the next room, he agreed to discuss the matter right then with Gort. So far as I know this was the only official contact Aldrich ever made with the officials on the Berkeley campus, except that after I had finally agreed to transfer to Irvine, I understand that Aldrich and Strong had a verbal exchange about the matter at one of the regent's meetings.

I would not prolong this account of why I left Berkeley to accept the challenge at Irvine except that it illustrates the tangled web we can weave when we practice *not* to deceive! As Gort said after I had made my final decision, "Ed, the problem is we have both leaned over backward to be strictly honest with each other while we became entangled in university red tape. We have both been too ethical, if such is possible." As the time approached when I had to give Aldrich my final decision, Gort and I had prolonged discussions on the matter, but his hands were apparently tied. On the day I had promised Aldrich I would give my answer, I told him that I would indeed come to Irvine as of 1 July 1963. Ironically, that evening I received a telephone call from Linsley, who told me that he had just finished talking to Strong and that now he could arrange a meeting for the three of us to discuss space and other matters. I told Gort that I appreciated greatly all that he and Strong may have attempted to do on our behalf, but I had now committed myself to go to Irvine. I had told Gort previously that I had promised to give my decision on that day.

The next day, I conferred with Ray Smith and told him of my decision. I also explained that the matter of space was only part of the problem, that I had long desired to return to full-time teaching, research, and public service, and that I had had administrative duties long enough to warrant being freed of them. However, as it seemed fated for me to continue in administrative work, I might as well accept the challenge of being dean of Biological Sciences at Irvine, and see what I could do about presenting the new biology and about bringing into biology some appreciation of the phenomenon of disease and of what can "go wrong" with living systems. Inasmuch as it did not appear that I was going to be permitted to develop invertebrate pathology at Berkeley in the manner I felt was its due, I felt that greener pastures were my only hope. Ray and I both indulged in further rationalization, but agreed it was over for better or worse for I had made a decision. In order to show good faith, I agreed not to take (as was my right) any equipment obtained on federal grants I had held, and to leave at Berkeley a new Public Health Service grant for over a million dollars to expand work into the noninfectious diseases of invertebrate animals (the same grant the university refused to find space for when the application was made). Later the Public Health Service insisted that since Berkeley had not found someone to take my place, the grant had to be moved to Irvine or be canceled. Since it would be Irvine's first federal research grant, Aldrich urged me to have it transferred, although, in all honesty, I felt I had my hands full without it—and I did.

I attempted to explain the reasons for my action to the professional staff (Falcon, Martignoni, Tanada, and Brooks), to the technicians and assistants, and especially to my graduate students. I expressed my sincere regrets in leaving them and discussed with each student his continued graduate study, making arrangements for each one to select another major professor. The remaining two months or so at Berkeley were sad ones for me even though I looked forward with enthusiasm to my new venture at Irvine. Some of my colleagues thought me a traitor, others thought I was crazy to go to that "God-forsaken outpost of Irvine in Orange County," and others expressed sincere good wishes and success for my new venture.

VII

As I prepared to depart from the Berkeley campus—and to disrupt the love affair with her that has never ended—the search for my successor had already begun. I had assured Linsley and Smith, and truly believed it to be the case, that they would have no trouble finding a replacement for me. However, for reasons best known to the parties concerned, those pathologists interviewed for the job during the subsequent months either did not measure up to what the administration wanted, or, after looking into the matter, the applicants did not find the position and its responsibilities of the type that they cared to assume. I made a valiant, and I believe a successful, effort to stay out of the matter in spite of the number of communications I received from likely candidates as to "why did you leave Berkeley." My diplomatic reply was that I did not leave Berkeley—rather, I came to Irvine. I adopted the principle that for the good of an organization one should not have a significant hand in determining his successor. However, I was desperately anxious to have invertebrate pathology continue to grow and develop at Berkeley, as I was sure it could if those in power really wanted it to do so. I urged every acceptable candidate who conferred with me to take the job. I did not agree with one noncandidate who wrote, "The Executive Committee made up of divisional chairmen and headed by a powerful department chairman has reduced the chairmanship [of the division of Invertebrate Pathology] to a completely subordinate administrative figure. Thus there is no enthusiasm for this administrative post, and there cannot be." This would not be so if the divisional chairman were a strong chairman, especially since Ray Smith, as department chairman, operated in a strictly democratic fashion—at least in my experience. Perhaps the noncandidate was partially right, for another colleague wrote in 1967 that in his opinion "they [the university] failed, [to find a replacement] . . . because no one relished the battling with other groups in Entomology necessary to ensure the growth of Insect Pathology and the facilities it required . . . many of its [insect pathology's] most enlightened ideas for progress appeared to be curbed by opinions in other groups of Entomology." I have had to admit that the decision of the executive committee not to change the name of the new four-division department to an all-inclusive one indicated to me that perhaps there might be some lack of foresight. Invertebrate Pathology should have remained a completely separate department, or if it were to be a unit in an entomological complex, then the name Insect Pathology should have been retained. Most ironic of all, the events that led up to this sad state of affairs, the rumors about abolishing nonteaching departments in the College of Agriculture, never materialized.

Let it be understood at this point that I have no quarrel with entomology or with entomologists, for it was they and their discipline that gave us a home for insect pathology when it was a mere infant. We were nurtured and supported, and without their encouragement, we would have never matured into a full-blown discipline. For this I shall forever be grateful.

No new chairman had been found, and neither Martignoni nor Tanada wanted the burden of directing the division. Joe finally agreed to fill in until the matter could be resolved. Not having been on the scene, I am in no position to analyze what took place over the following months. A variety of versions reached me. Admittedly, I agonized over the situation, at times feeling that the administration was derelict in not recruiting in a more aggressive and persuasive manner, at other times feeling that there were forces or influences at work that militated against the welfare of the division. Obviously the morale of the unit itself dropped to a very low level. My sadness was profound; what was to happen to the endeavor into which I had poured almost twenty years of Churchillian sweat and tears? Then came the news that Martignoni was leaving to accept a position with the U.S. Forest Service in Corvallis, Oregon. I spoke with Mauro several times on the phone trying to persuade him not to leave the University of California. Joe, who was in Hawaii at the time conducting field studies, also tried and implored him to reconsider, but to no avail. Mauro confided the reasons for his frustrations and why he found it impossible to come to a logical and tolerable compromise with certain individuals as well as the establishment. I found that I had little ammunition with which I could refute his arguments or change his decision, and it was impossible to heed his and Joe's generous plea to "come back." At any rate, Martignoni's move to Corvallis was one that made him a happier man, and it kept him and his genius within the field of invertebrate pathology.

Eventually, discouragement over not having recruited a new chairman and expediency entered the picture. After consulting with the staff (now consisting of Falcon, Tanada, and recently appointed George O. Poinar, Jr.), the administration decided to combine the division of Invertebrate Pathology with the division of Entomology, of which Bob Usinger was chairman. I received this news with considerable happiness and relief; it meant that the cause and all we had worked for was not lost. It meant this because, of all the men in the entomology complex, Bob was one of those with the clearest understanding of the aspirations and needs of invertebrate pathology. Joining his division as insect pathologists gave the men and the discipline the security they needed. True, the opportunity for Berkeley to have a large, productive, independent department devoted to the pathology and microbiology of all invertebrate animals was probably lost. On the other hand, the past work of the Laboratory of Insect Pathology, then the Department of Insect Pathology, then the Division of Invertebrate Pathology had spawned a viable discipline now recognized throughout the world, with numerous other laboratories, institutes, and departments having followed its lead. This had been accomplished. Insect pathology could continue at Berkeley under Usinger's leadership of his division—and it could continue its programs in teaching, research, and public service.

[According to the outline of EAS and the materials that he had gathered together, the next section would have been on teaching insect pathology. Unfortunately, there was so little written, it was not worth including. MCS.]

4. IRVINE: PROMISE AND FRUSTRATION

> Soaring imagination generates soaring
> new ideas, and such ideas are the
> mutation of our intellect.
>
> *Nicolas H. Charney*

I

Sometime early in the 1950s the Board of Regents of the University of California, after considering enrollment and population projections, concluded that a new campus would be required in the area of Orange County. In January 1961, a deed was recorded consummating a gift to the university of 1,000 acres of the famed Irvine Ranch land. The regents purchased an additional 510 acres from the Irvine Company in January 1964 "to provide for future campus housing and ancillary services."

As I write these words, in 1969, and view the development of not only the University of California, Irvine (UCI), but that of the surrounding area as well, it is difficult to believe that when I arrived to assume my new duties in July of 1963 the territory involved consisted of barren rolling hills and empty grassland. In six short years there has been assembled a distinguished faculty and a student body of over 4,000, and an organized cluster of attractive buildings has been built.

Unfortunately for the university, by 1969 the socio-political climate of California had become conservative, and at times and places reactionary, in character. The state had instituted what it considered to be economies, the people had voted down bonds for further construction, and a period of austerity for higher education had set in. Orange County was the heartland of ultraconservatism, and, as stated in *Fortune* magazine (1968), a county where "little old MEN in tennis shoes fulminate against the Supreme Court, the U.N and the disappearance of God from the public school."

For some years, prior to coming to Irvine, I had been concerned about new approaches to biology, feeling that the "straight jackets" which the traditional departments of zoology, botany, microbiology, etc., had made for themselves prevented them from facing up to the fact that the biological sciences are truly the ultimate sciences because they follow naturally mathematics, physics, and chemistry, but no other branch of science follows biological sciences. What was needed were much broader concepts than the tradition-bound departments were able to deliver. The biological sciences are not ordinary sciences; they touch every activity of man and are destined during the next several decades to have more profound impact upon man's activities and public policy than any other science. The sheer necessity of survival demands that biological studies be a major concern of man.

I was thrilled to have an opportunity to help mold a new Division (later School) of Biological Sciences into what we all hoped would be a more meaningful approach. In my first proposal on biological sciences to Aldrich, I had felt that the middle-of-the-road approach would be acceptable to him, and once the camel's head was in the tent I could go ahead with the "new biology" concept I had been thinking about. Even though there are more differences between physiology and morphology and between ecology and molecular biology than, say, between several areas of chemistry or physics, there are still bridges for connecting these disciplines. We elected to use the "levels" approach (molecular, cellular, tissue, organ, organism, population, ecosystem, etc.) in place of the broad groupings, such as plants, animals, and microorganisms. These levels have been departmentalized accordingly at the upper division and graduate levels. Suffice it to say, since this is not a book on the development of Biological Sciences at Irvine, that by developing my own ideas, and borrowing the ideas of others as to how the new biology could best be approached from a pedagogical standpoint, I held to the hope that here was a once-in-a-lifetime chance to also incorporate into a new biology curriculum what "goes wrong" with the life and life system, as well as the usual presentation of life as it is in a healthy state. I was sure that there was a place for invertebrate pathology in our new approach to the natural sciences. However, preoccupation with a thousand and one obligations of my job as dean had to delay the realization of this hope. When I came to Irvine, I did not realize that attempting to build a School of Biological Sciences from scratch was going to be an eighteen-hours-a-day, seven-days-a-week job.

Before discussing the development of invertebrate pathology at Irvine, I hope that I will be forgiven for diverging from the subject at hand momentarily. The first few years at UCI were challenging, stimulating, and rewarding years indeed. The privilege of helping to plan and initiate a new university campus from absolute scratch is indeed a rare experience—at least it was at that time. And though the work was arduous, it was nevertheless a genuine pleasure. To work and operate—at least for the first two or three years—with a minimum of the usual academic and administrative red tape and obfuscations was, in itself, a glorious experience.

It must be said that the leadership of Daniel G. Aldrich, Jr., as chancellor was largely responsible for the vigorous manner in which UCI began. Not only did he provide the drive and energy with which the campus was built up from virgin ground, but by being able to give a speech at the drop of a hat and by having an outgoing, extrovert personality, he soon developed strong moral support from the surrounding communities in Orange County. At the same time two of the chancellor's right-hand men deserve more credit than they are generally given for initiating programs. These were Ivan Hinderaker, vice-chancellor for academic affairs and later, in 1964, appointed chancellor of the Riverside campus, and his successor, Jack Peltason. Their vigorous recruiting policies together with academic know-how were invaluable assets to the beginning campus. Vice-chancellor of business and finance L. E. Cox implemented the physical planning of the campus and organized the nonacademic departments so as to serve the academic needs.

I well remember the day in July 1963 when I arrived at the temporary building (or, as the architects called it, the Surge Building). I was met with good-natured teasing because the boxes containing my personal professional library had preceded me and completely filled one of the offices. This shipment of books apparently heralded the vanguard of academic flavor which up to then, of course, had only been a sort of anticipation and not a reality. Another, but empty, office was immediately assigned to me. Within minutes of my arrival for official duty, my desk and other office furniture were being delivered and put in place by a truly remarkable person, Tony Ercegovich. Tony, a former marine top sergeant, feigned gruffness and an intolerance of women, but perhaps took care of more tasks, helped and served more people, during those beginning days than anyone else. As UCI grew, and formal job assignments were made, Tony was forced to confine his activities and was assigned the important task of building and running the campus storehouse and receiving operation. Tony and I were to become good friends exchanging many choice items of UCI gossip and opinions as the years went by. Tony at the time of my arrival was also mailman, and he said that my coming increased the mail three times over. Those were the days!

In an earlier chapter I explained how Aldrich, while university dean of agriculture, had acted to preserve and further our progress in departmentalizing insect pathology at Berkeley. He was still to further our efforts in invertebrate pathology and pathobiology at Irvine. His support came enthusiastically and wholeheartedly even though I doubt whether he really ever fully understood the nature of what I was trying to accomplish. I have enjoyed the fact that this articulate man, who was never at a loss for words, never once, on the several occasions when he introduced me to groups, was able to explain precisely what my field of science was.

Chancellor Daniel G. Aldrich, Jr., was truly plain Dan Aldrich to most who knew him. As he gathered us together at Irvine to help launch the new campus, we became aware of his presence in all things. A tall man of six feet four inches, 210 pounds, well muscled, erect posture, and broad shouldered, he dominated virtually every meeting or every conversation in which he was engaged. He and his capable and likeable wife, Jean, were as unpretentious and down-to-earth as a couple could be. In their roles of leadership, they mingled easily with the rich and powerful of the community as well as with members of academe.

Clark Kerr, one of the university's better and more remarkable presidents, was quoted in the press (February, 1969) saying that Aldrich was "probably the only man I know of who could do the job that had to be done at Irvine. . . . We were acutely aware of the special problems that would be presented by the existence of a university campus in so conservative an area as Orange County. We knew the people there would be hostile to much of what went on in the University. . . . That's why I wanted Dan for the job. I knew him to be a dedicated defender of academic freedom. . . . More importantly, I knew Dan could communicate with the outside community and be trusted by them." Events proved Kerr to be right. Aldrich, a moderate Republican, did a superb job in walking the precarious tightrope between a generally ultraconservative community on the one hand and a generally liberal but dedicated faculty and anti-Establishment student body on the other.

Outwardly, Aldrich appeared to be a simple, straightforward, uncomplicated man. He was obviously ambitious, in the best meaning of the word, and in academe, as on the athletic fields he loved, he played to win. Possessed of a quick, volatile temper, he never, in my experience, bore a grudge. As one news reporter put it, however, Dan "can be a highly complex man, a whirlwind of conflicting forces—pride and humility, fury and placidity, strength and tenderness, tenacity and flexibility, maturity and puerility, culture and commonality," which was a fairly accurate, generalized profile of Dan Aldrich.

While dean of Biological Sciences I found my associations with Aldrich to be generally stimulating and rewarding. To be sure we had stormy encounters, but, in a sense, Aldrich admired a fighter so that locking horns with him did not involve personal animosity but engendered mutual respect. I knew I could not win an argument unless I had my facts right and could make a strong case. When I could confront him so armed, more often that not I was successful—whether it was to gain his authorization for more funds, faculty, and space, or to have study and eating places more equitably placed on campus. I definitely was not one of his favorite deans or administrators, but I believe I had his confidence; and, considering the demands I had to make of him on behalf of my faculty, he was tolerant, fair, and considerate.

While Aldrich received adulation and praise from much of the public, as well as from the university community, I observed attributes of his that have rarely been mentioned. For example, he was a man who found it difficult to give or accept compliments—his New England background may account for this. However, he had great compassion even though he often tried to present a tough or hard-boiled appearance. Although some felt him to be insensitive to the feelings of others, when he felt he may have hurt someone he went out of his way to make amends and to apologize. He is no saint, but he is a sincere, actively involved Christian for whom I have great respect and admiration. His decisions certainly have brought about several turning points in my life.

Being in on the beginning of any endeavor naturally enables one to accrue a number of firsts. Among those in which I take pride—but not credit—are: the first academician on the Irvine campus;[11] the first biologist, the first dean, the first pathobiologist, author of the first book, author of the first paper from UCI to be published in a scientific journal, recipient of the first federal grant, and so on. One in which I take special pride—my oldest daughter, Margaret Ann, graduated with honors in the 1969 charter class—being the first faculty member with a graduating offspring.

II

With the U.S. Public Health Service grant that had been transferred from Berkeley to Irvine I could initiate a program in pathology at UCI and help get things started in research in the biological sciences. In deference to the programs in invertebrate pathology at Berkeley and in insect pathology at Riverside, and in order to avoid any semblance of competitiveness with my former colleagues, I decided to use another term for my project at Irvine. Accordingly, I borrowed the word *pathobiology*, which had, apparently, first been used by a veterinarian, Frank P. Billings, in 1888 at the University of Nebraska.

The Public Health Service grant supporting the project provided for the employment of three professional workers and their technicians. While I would be able to do some research of my own on the project and give it general administrative supervision, clearly I would have to find individuals who were self-reliant and could work more or less independently. Inasmuch as the scope of the grant-supported project was a broad one (as indicated by its title, "The Diseases of Invertebrate Animals"), each of the scientists could fairly well select the area of study of his or her principal interest. However, the terms of the grant were such that emphasis should be on noninfectious diseases.

Because the salaries, as well as all other support, for the pathobiology project were on "soft money," as federal grant funds were frequently called, and hence could offer no more tenure than the five years for which the grant was given, it was difficult to recruit professionals whose interests fell within the parameters of the grant's subject matter. We were fortunate, however, by the end of the first year of the grant's duration to have obtained three qualified professionals and their technicians: Phyllis T. Johnson, already well known in medical entomology and one whom I had known since she was a graduate student at Berkeley, where she obtained her Ph.D. under M. A. Stewart; John C. Harshbarger, who did his graduate work in entomology at Rutgers under Forgash and who had just finished a postdoctorate fellowship under A. M. Heimpel, director of the Laboratory of Insect Pathology, U.S. Department of Agriculture at Beltsville, Maryland; and Ronald L. Taylor, who had just completed his Ph.D. at the University of Minnesota and had been recommended to us by Marion A. Brooks and A. Glenn Richards of that institution. Later, in November of 1967, when Harshbarger accepted the position of director of the Registry of Tumors in Lower Animals at the U.S. National Museum in Washington, D.C., we were joined by William T. Wilson, who had just received his Ph.D. at Ohio State under W. C. Rothenbuhler.

The grant expired on 31 January 1969. Because her work was yielding fine results, but not yet finished, and because she therefore needed an additional year to complete her work, Phyllis Johnson applied for, and received, a year's extended funding with an application of her own to NIH.

The results of the project were both intangible and tangible. It gave UCI's newly assembled purchasing and accounting departments practice in handling federal grant purchases and budgets. It paved the way for oncoming scientists at UCI to obtain grants, because before approving this one, the NIH made a site visit to be sure that there really was going to be an Irvine campus of the University of California. Once assured of this, the word was passed among the granting agencies in Washington, and subsequent applicants did not have to convince them that UCI was a viable part of the University of California. From the standpoint of invertebrate pathology, the project demonstrated a broader spectrum of pathology in lower animals and emphasized the need for the study of immunity as well as seemingly noninfectious disease in invertebrates.

The more tangible results, most of which were published and hence a matter of record, ranged over a wide spectrum. They included such contributions as an annotated bibliography of pathology in invertebrates other than insects, mortality rates in colonies of the clam *Donax gouldi*, comparative studies of the interaction of invertebrate coelomic cells and fluid and several species of bacteria, the defense mechanisms of certain marine invertebrates, a comprehensive literature review on neoplasms in insects, histopathological studies on the effect of carcinogenic chemicals, X-rays, and radioisotopes on certain insects, studies on tumorlike lesions in the cockroach *Leucophaea maderae*, and in several marine invertebrates, the physical structure of coalesced flagella in cultures of *Bacillus larvae* (a bacterial pathogen of the honey bee), cellular degeneration in the hydra, *Pelmatohydra pseudoliguctis*, teratologies in the beetle *Tenebrio molitor*, and a number of other observations and discoveries. [See Beeson and Johnson, 1967; Cantwell, Harshbarger, Taylor, Kenton, Slatick, and Dawe, 1968; Harshbarger, 1966, 1967; Harshbarger and Heimpel, 1968; Harshbarger and Taylor, 1968; Johnson, 1966*a*, *b*, 1968*a*, *b*, *c*, 1969*a*, b, c, 1970, 1971; Johnson and Beeson, 1966; Johnson and Chapman, 1969, 1970*a*, *b*, *c*, 1971; Preston and Taylor, 1970; Smith and Taylor, 1968*a*, b; Steinhaus and Zeikus, 1968*a*, *b*, 1969*a*, *b*; Taylor, 1965, 1966, 1967, 1968, 1969*a*, *b*, *c*, *d*; Taylor and Freckleton, 1968*a*, *b*; Taylor and Preston, 1969; Taylor and Smith, 1966; Wilson, 1970, 1971; Wilson and Combs, 1970; and Zeikus and Steinhaus, 1968. MCS.]

III

The pathobiology project had an additional and unexpected value. It paved the way for an effort to establish what we called a Center for Pathobiology. By this means I was able to shift from my role as dean of the School of Biological Sciences—which had by 1967 settled into the somewhat tedious routine characteristic of deanships—to that of resuming my activities and interest in pathology.

As of 1 July 1967, I was eligible for a sabbatical leave, which I applied for in order to begin the writing of the book you are now reading. The leave was graciously granted by the university administration after inducing James L. McGaugh, chairman of our Department of Psychobiology, to serve as acting dean. This inducement included, in addition to Jim's own magnanimity, the appointing of an associate dean—Patrick Healey—and a lightening of duties so that Jim could continue his research activities. In August I decided not to return to the deanship—why not, it occurred to me, quit while, apparently, I still had the good will of my colleagues? The administration, though considering other possibilities, had no difficulty in deciding that McGaugh should be asked to continue in the post. At first hesitant and dubious about being tied down with administrative duties, Jim finally decided to continue as dean beginning 1 July 1968; however, he resigned this administrative post on 30 June 1969, but it was understood that he would continue until 1 March 1970, when Howard A. Schneiderman would assume the deanship.

Soon after it began to be known that I had relinquished the deanship, I received several rather enticing offers from other institutions. Apparently they assumed that I wished to leave UCI. Of course, I did not; I merely wished to be a professor once again and to return to teaching and researching in invertebrate pathology. Except to close friends, I said nothing about these offers I had received, but Roger Russell, vice-chancellor of Academic Affairs, apparently got wind of them. He telephoned and asked that I stop by his office next time I was on campus. The next day, when I visited Roger, he inquired as to whether I was seriously considering leaving UCI. His question took me by surprise. I admitted that I had received the offers, but told him that only one of them really interested me and that I had told the parties concerned that I would think the matter over while spending the coming three weeks in New England, and give them my answer at the end of that time. I assured Russell, I very much doubted whether anything could induce me to leave UCI, providing I could continue with my interests in pathobiology. Russell asked if there was any particular aspect of pathobiology I desired to pursue and what I thought UCI should develop in this area. I responded that I had given some thought to an institute or center of some kind that would be dedicated to the advancement of the understanding of disease from the standpoint of the biological sciences; but with looming state and federal budget cuts for the university, this was probably not the time to propose such a venture. Russell, however, responded with a request that I prepare a formal proposal when I returned, and added that he believed the idea had merit and should be pushed.

Accordingly, in early December 1967, I submitted a proposal to establish on the Irvine campus a Center for Pathobiology to undertake research, teaching, and public service functions in the field of pathobiology. I took the position that "what goes wrong" with life is as properly a part of biology as what takes place normally and in a state of health. The value of studying disease types throughout the entire phylogenetic scale was clear. Moreover, disease should more thoroughly be studied on any and all levels of biology: molecular, cellular, organismic (i.e., the whole organism), population, or whatever. My proposal emphasized a point I had been making in scientific circles for some time—and, of course, I claimed no originality for it. It had become increasingly necessary to see disease problems from a broad biological viewpoint (as well as from clinical and other viewpoints) and that the study of disease encompasses all organisms, and that it be investigated as a natural phenomenon.

The proposal was approved and endorsed by Vice-Chancellor Russell and Chancellor Aldrich, and after the review by the usual committees of the Academic Senate it was forwarded to President Charles J. Hitch. Eventually, on 21 February 1969, the establishment of the center was approved by the regents. Of course, in the meantime—indeed, since 1 July 1968—the center had been a functioning unit operating, with local campus approval, with funds from the School of Biological Sciences.

Shortly after 1 January 1969, with the exceptionally capable help of Donna M. Krueger serving as administrative assistant, we released a small leaflet which succinctly stated some of the goals of the center:

1. To serve as a repository of typical and unique pathological specimens and related material. To gather and preserve such specimens and materials (including type and holotype specimens of microbial pathogens).
2. To collect and to preserve all books, reprints, literature, letters,

film, tapes, computerized programs, etc., pertaining to pathobiology.

3. To provide for the study of all pathological specimens, materials, disease systems, literature, tapes, films, and computerized information.
4. To prepare, program, and maintain computerized courses in invertebrate pathology and in pathobiology generally. At UCI these courses will be presented primarily in the curriculum of the Department of Organismic Biology.
5. To appropriately exhibit and display diseased animal and plant organisms of all kinds.
6. To publish serial accounts of the activities of the Center (including AV/TV and computerized programs). These accounts may be educational, scientific (i.e., investigational), or accessional in nature.
7. To conduct research into the biology of disease, especially that of lower animals and of plants. Included is mission-oriented research having applications in the solution of certain problems affecting urban areas, marine resources, agriculture, human and veterinary medicine, and other aspects of life affected by disease.

On the back of this first leaflet explaining the goals of the center was printed, almost hesitantly, an entreaty addressed primarily to comparative pathologists. It solicited diseased specimens (invertebrates, vertebrates, and plants) and type material, as well as all other informational, library, and museum articles relating to any aspect of disease. Also, rare works, notable manuscripts, private papers of pathologists, and other memorabilia were sought.

A little later the following announcement was sent out to pathologists:

AN INFORMATION SERVICE FOR
INVERTEBRATE PATHOLOGISTS

The Center for Pathobiology is now prepared to provide invertebrate pathologists, anywhere in the world, with photocopies of specific articles appearing in scientific journals. Copyright regulations limit the number of copies of any particular article to one copy. However, we shall be glad to attempt to provide single copies of any reasonable number of articles that may be needed. As yet we are not prepared to make literature searches according to subject areas; accordingly, please provide complete reference citations of the article you wish. Scientists residing in the United

States are requested to provide a stamped (estimate probable weight and indicate class of mail), self-addressed envelope of appropriate size. Requests from outside the United States may include an appropriate self-addressed envelope, and should indicate the class of return postage desired. Otherwise, at least for the time being, this service will be provided to anyone, anywhere, free of charge. Send your requests to:

Center for Pathobiology
University of California
Irvine, California 92664

[This section is obviously unfinished, since the center had just begun to function. On 16 October just four days before his death, EAS received word that the center had been granted its own funding—in other words, it had its own budget and was an independent unit within the University of California, Irvine. MCS.]

1. While admittedly I seek a certain degree of protection behind the phrase "as I remember it," I wish to assure the reader that I do not rely to any substantial degree on memory alone. As I write these words, I am surrounded with packing cases of documentation of all sorts. This statement is necessary here because at a few points in the following sections the reader may find my account at variance with several short "historical" passages by other authors. In taking exception to these passages or in handling "controversial" subjects, I have tried as much as possible to rely on valid documentation and verifiable historical research. Even so, the matter will unavoidably be flavored with my own interpretation of the data and documentation at hand.

2. The identity of the fungus distributed by Woodbridge is not clear. A letter dated 26 April 1915 from the well-known plant pathologist and mycologist H. S. Fawcett to State Commissioner A. J. Cook contains this statement: "I examined two different lots of cultures which Dr. Woodbridge is using for black scale, which came in through Mr. Beers. In neither of these could I find any trace of the *Isaria* fungus, which is present to some extent on black scale in Santa Barbara County. The only fungi that I found present in these cultures were a species of *Mucor* and a species of *Penicillium*. Either of these two fungi might, it seems to me, have gotten as chance contamination in the process of making culture test." In a letter, dated 17 August 1914 to Woodbridge, C. W. Beers, horticultural commissioner of Santa Barbara County, wrote, "For a number of years—at least three or four—I have observed a fungus disease at work in the county on the Black Scale. It seems to be very widely distributed, although in some particular sections its work is more complete than in others, and in some seasons it does more efficient killing of the Black Scale than at other times." There is no record as to whether or not this naturally occurring fungus was the same species as that marketed by Woodbridge. Inasmuch as he was making a commercial venture with his fungus, it is not unlikely that Woodbridge never did divulge its species, if in fact he had precise knowledge as to its identity.

3. At my request, my major professor joined me in publishing a part of this thesis (Steinhaus and Birkeland, 1939) in anticipation that additional research would yield a series under the imprecise heading "Studies on the Life and Death of Bacteria." This plan was, of course, thwarted by the emergence of what had been my latent interest in the microorganisms associated with insects. Nevertheless, the part of my dissertation that I did have published taught me the uselessness of attempting to evaluate one's own work. While studying aging cultures of bacteria, I had found that the number of cultivable bacteria remaining in cultures could be remarkably high even after two years of incubation. To me this appeared to be a significant observation inasmuch as most bacteriologists conceived of the growth curve of bacteria as declining fairly rapidly and steadily to its death after having reached its peak. To the long, slow period of numerical decline I gave the name *senescent phase*, certain that that this term would from then on be recorded in all textbooks of bacteriology as part of the terminology describing the growth curve of bacteria. It did not happen. Except for a rare reference to it, the paper and the term made little, if any, impression on any writer of texts. However, the fact that the logarithmic death phase tends to become asymptotic to the x-axis is generally recognized. Similarly, my description, in the same paper, of bacterial "cannibalism" failed to excite the bacteriological world. Could it have been that there really were other areas of general bacteriology that concerned bacteriologists—areas other than those in which a young graduate student had immersed himself for a year or so? Indeed, what scientist has not looked back on one or more of his published contributions in bemused consternation as to why his discovery either did or did not click or turn out to be the world-shaking discovery he envisioned it to be!

4. My credo or philosophy on this point is partially contained in a statement the famous American bacteriologist Hans Zinsser is supposed to have made: "As a rule, the scientist takes off from the manifold observations of his predecessors, and shows his intelligence, if any, by . . . selecting here and there the significant stepping stones that will lead across the difficulties to new understanding. The one who places the last stone and steps across to terra firma of accomplished discovery gets all the credit. Only the initiated know and honor those whose patient integrity and devotion to exact observation have made the last step possible."

5. This was the only one of my students with whom I ever coauthored a paper while he was my student or on work he did while he was my student. In this case, Thompson was my hired assistant and worked on a project I already had well underway. In a sense this was a bonus to him but one he certainly earned. It was at this point, however, that I made the decision with regard to the publication of my research that I have held to through the years.

Many colleagues feel they have every right (which indeed they do) to be a coauthor of papers published by students under their direction and supervision. I found myself unable to assume this privilege. Merely because I had been the student's major professor, no matter how much help or how many original ideas I gave them, I was their teacher; that was my job! Had I followed the usual practice, I undoubtedly would have had many more published papers. My feeling has been that, if the work were good enough to be considered the original work of the graduate student, it should be recognized as his alone. If acknowledgement of my help is felt by the student to be necessary, this can be placed in a footnote. Most of my students generously asked to add my name to the authorship of their published theses, but I have consistently declined. On the other side of the coin, I have been blessed with such talented and loyal assistants—be they student assistants or technicians—that I have always tried to include their names either in the authorship or, at least, in a footnote, because from the human standpoint without their hod-carrying I could not have laid the bricks. Occasionally, when I have felt that the assistant did exceptionally well, or generated the principal idea even though it was done under my supervision, I have been glad to have them as senior author or even as sole author. This was only just, and besides it served as a morale booster. Critics of this policy point out that footnote recognition is adequate for technicians hired to help, that technicians come and go and if their names are part of the authorship the literature becomes confusing, that especially women technicians are often temporary employees not remaining in science so it is wrong to burden the literature with transient authors. Perhaps the critics are right (see Howard, 1933, pp. 30–31). But the technicians are as human as the scientist, and in this world all too seldom is credit given where it is due. To my knowledge the policy I adopted back in 1949 has not hurt me. The really important thing is that the work is done, not who gets the credit. Yet, it must be admitted that a scientist lives on the recognition and respect of his fellows; and in academic circles much of an academician's promotion and welfare depends on the contributions he makes through the quality of his publications. But I am not convinced that a scientist suffers at the hands of a knowing judge if he shares the authorship of his papers. But I do not quarrel with those who disagree with me on this point. Greer Williams (1959) has rightly said, "The three things the researcher longs for, most of all, are satisfaction of his curiosity, recognition for his work, and the possibility that it will make a difference in the sum of knowledge and hence may benefit humanity. Persons trained in the profit motive and material gain may not appreciate how highly the scientist values these goals."

6. H. D. Burges (personal correspondence) says the following story was told to him by two different people: "Apparently you had a meeting with a number of representatives of commercial firms after your success with *Bacillus thuringiensis* against the alfalfa caterpillar and they misinterpreted your statement of dosage required, departing with the impression that only 1 g. per acre was required. This gave them great incentive to push ahead with production and may well have given very valuable impetus to the beginning of commercial production of this bacterium in the USA!" I wonder?

7. Prior to Bioferm's interest in microbial insecticides, at least two companies (or businesses) in the United States manufactured dust containing spores of *Bacillus popilliae* and *Bacillus lentimorbus* to aid in the control of the Japanese beetle. The Fairfax Biological Laboratory in Clinton Corners, New York, marketed a product known as *Doom*. Other preparations were sold under the names *Japonex* and *Japidemic*. As I have described in a previous chapter, the first microorganism ever mass-produced for the purpose of controlling insects was the fungus *Metarrhizium anisopliae*, but it was apparently not sold commercially. The fungus was mass-produced by Isaak Krassilstschik in 1884 in a small production plant at Smela. (For additional details concerning Krassilstschik's methods, as well as a splendid general review of the mass production of insect pathogens, the reader is urged to read the chapter written on this by Martignoni [1964]). Perhaps the first truly commercial manufacture and sale of a microbial-control product was the production of *Beauveria tenella* (=*Botrytis tenella* = *Isaria densa*) by Bribourg and Hesse, 26 Rue des Ecoles, Paris, in 1891. Tubes of cultures of this fungus were sold for the destruction of the common cockchafer, *Melolontha vulgaris*. Circulars issued by this firm gave the practical directions for using the fungus spores, which were placed on the market at fifty centimes for a trial tube, and six francs for "the commercial article." Prior to World War II, Laboratorie LIBEC in France manufactured a product known as *Sporeine*, which may, in fact, have been a preparation containing sporulated *Bacillus thuringiensis* or one of its varieties. As we indicated in a previous chapter, shortly after the turn of the century, the Florida Citrus Experiment Station distributed the so-called friendly fungi at cost ostensibly for the control of citrus pests. S. M. Woodbridge sold an unknown fungus to citrus growers in California to control the black scale. Apparently the distribution of cultures of *Beauveria globulifera* for chinch-bug (*Blissus*) control was made free of charge by a special experiment station at the University of Kansas in the 1890s.

8. In a paper, "Turbulence in Tolerances," presented before the Chicago Section of the Institute of Food Technologists, on 11 January 1965, Fisher recalled the incident as follows:

There is an interesting story in how an industrial fermentation company producing vitamin B_{12} yeast, and antibiotics, became an insecticide manufacturer. In 1956, we of course had no connection with entomology or insect pathology journals, and knew nothing about the insecticide field. Professor of Insect Pathology Ed Steinhaus at the University of California knew this all too well. For years he worked on the development of bacterial, fungal, and viral agents, but none of his work reached potential producers of these agents. So he pulled together a series of practical control tests and wrote an engaging and highly readable article on biological control for *Scientific American*. His idea worked. The day the August 1956 *Scientific American* arrived, our vice-president, Jerry Sudarsky, sent me to Berkeley to see Professor Steinhaus.

The Bioferm people, and especially Bob Fisher, were always overly generous in the credit they gave me in this enterprise. Moreover, the "engaging" quality of the article appearing in the *Scientific American* can be attributed to the editor of that periodical. And I think that Bob forgot that he, or someone at Bioferm, had also read my industrial-oriented article which appeared about the same time in *Agricultural and Food Chemistry*. In any case, the initiative was Bioferm's.

9. This virus was being tested at the time by Tanada of our laboratory on the cornworm on sweet corn. The virus appeared to be effective under certain field conditions (Tanada and Reiner, 1962).

10. Aldrich enjoyed telling a story which had two parts, but he knew only one of them. I never had the courage to tell him the other part. Paul Sharp, about to retire as director of the Agricultural Experiment Station at the time I was considering the opportunity to move to Irvine, warned Aldrich about hiring me because, in effect, I was "a fighter" and thus, in building a School of Biological Sciences at Irvine, was sure to be a thorn in his side. Dan liked to compliment me—and indeed it was complimentary—by telling me that he responded by telling Sharp that he knew what he was doing in offering the job to me, and that "a fighter" was precisely the kind of man he wanted. What Aldrich did not know was that Paul Sharp used exactly the same tactic on me. Sharp's motive—I was gratified to be told—was to keep me at Berkeley and in Agriculture by either scaring Aldrich regarding my alleged trait of "fighting and demanding for his people," or frightening me with "how hard it was to work for Aldrich, and why go to Irvine anyhow?" In any case, I was pleased that Sharp had not wanted me to leave Agriculture and Berkeley.

11. This first ignores the fact that my presence at UCI was preceded by Chancellor Daniel G. Aldrich, Jr., and Vice-chancellor for Academic Affairs Ivan Hinderaker. Yet, technically I preceded them because my employment papers moved with me (i.e., as of 1 July 1963), while those of the others were not moved from Berkeley and UCLA respectively until later. I became aware of this situation one day when President Clark Kerr visited the campus and asked me facetiously, "Well, Ed, where should we place the plaque indicating you were the first academician at Irvine." When I questioned his assertion and reminded him that Dan and Ivan had preceded me, he then explained how their transfers had followed mine even though physically they had come to Irvine during the previous year.

When the star of science becomes dim
in one country, sooner or later it
shines so much the brighter elsewhere.
Thus one nation after another becomes
the teacher of the world.

Rudolf Virchow

4

Insect Pathology Overseas

For sometime following World War II there were no general, broad-spectrum laboratories of insect pathology outside of North America. And none were associated with educational institutions or instructional programs. In mid-1947, one European entomologist, upon hearing that we had established our Laboratory of Insect Pathology in 1945, wrote disconsolately, "It is really time that people got the idea of the importance and necessity of a special laboratory for insect pathology. People in Europe will never get such ideas, especially under the present upset conditions. Well, in spite of this, I will do my best here as long as it is possible." Fortunately, his dire prediction was wrong; as we shall see, Europe developed excellent centers for research in insect pathology even though it remained reluctant to bring the subject—or its broader dimension, invertebrate pathology—into the classrooms of its universities. Accordingly, after the war, instead of Americans going to Europe to learn of the diseases of invertebrates, Europeans and others tended to come to America to study or train in this newly organized discipline.

Most of these seekers of knowledge were motivated by the possible applications insect pathology might have. Especially prominent in this regard was the hope that microbial control might eventually become a form of biological control having application in their homeland. At any rate, this was the justification for which many graduate students and postdoctorate fellows coming to this country from mid-1940s through the 1960s had their expenses paid. In many instances their hearts were set on doing basic research which, of course, was the type required for the Ph.D. degree. In the 1940s and early 1950s the University of California was the only educational institution offering a reasonably full program of courses and seminars in insect pathology, and as such we had many more applicants for advanced work than we could possibly accept. It was sad to have to turn away so many competent aspirants, especially those who were willing to make great sacrifices to come to us from overseas. In most instances the applicants sought scholarship aid from us. Although America had billions for foreign aid, the University of California had little in the way of financial resources it could offer the foreign student or visiting scientist.

So, although the principal intent of this chronicle is to trace the development of invertebrate—primarily insect—pathology in North America, I shall in this chapter digress to overseas developments to the extent that they affected the activities of our own laboratory, broadened our own horizons, and influenced our own attitudes and understanding of the diseases of invertebrate animals. In no sense should the reader interpret this chapter as an attempt to cover the history and biography of overseas invertebrate pathology or pathologists. Indeed, it will not even cover well the effect of overseas pathology on American invertebrate pathology that has taken place through the travels of others or through the published literature. Instead, I am afraid that what I write in these next few sections will represent only an abbreviated attempt to reflect, in a purely personal way, the impressions of one American scientist observing and sampling the activities of his overseas counterparts. However, what perhaps makes them worth recording is that they took place at a rather critical and formative period in the development of contemporary insect pathology.

1. GERMAN INSECT PATHOLOGY

I

In the first two chapters of this book, I explained how most of the early beginnings of interest in disease in insects in Europe and Asia arose through the investigations of disease in the silkworm and the honey bee. These activities continued after World War II had ended, but, as in North America, the diseases of other insects—particularly pest insects—were now gradually to come in for more attention.

With the end of the war in Europe in May 1945, American scientists proceeded immediately to avail themselves of whatever scientific literature had been published in Europe during the war years. I had heard of a group of Germans who had somehow, during the entire war, managed to continue working on the biochemistry of viruses at Kaiser Wilhelm Institut für Biochemie in Tübingen. Moreover, publications on the nature of insect viruses had appeared in German scientific journals in 1942 and 1943 by one Gernot H. Bergold and colleagues, but with the American involvement, it had been impossible to contact, or, indeed to obtain, copies of the publications in time to cite them in my *Insect Microbiology*, which, though published in 1946, had been submitted to the publisher in early 1944.

By late 1945, I had finally seen a fifty-five-page review titled "Über Polyederkrankheiten bei Insekten" published in a 1943 issue of *Biologisches Zentralblatt*. In retrospect, it is interesting to note that on the basis of the work Bergold and his colleagues did, they did not depart greatly from the concepts of Paillot, Gratia, Letje, Glaser, Stanley, and others, that the polyhedrosis virus, such as that responsible for jaundice in the silkworm, consisted of minute granules always less than 0.1 micron in diameter—in the case of those described by Glaser and Stanley (1943) the size of the particles was approximately 10 millimicrons. However, Bergold thought that the polyhedral bodies were made up of virus aggregates and that the polyhedral protein itself was the cause of the disease.

In 1947, it was my good fortune to meet Wendell F. Sellers, then stationed at the Citrus Experiment Station in Riverside, and later in Fontana, California, where he directed the California Investigations for the Commonwealth Bureau of Biological Control. I was delighted to make this contact because Bergold, who was actually an Austrian and had received his Ph.D. at the University of Vienna, had spent a good share of eight years as an assistant to Sellers while the latter was in Europe for the U.S. Department of Agriculture's Bureau of Entomology and Plant Quarantine. Bergold helped Sellers collect natural enemies (insect parasites and predators) of forest insect pests from throughout most of Europe. During the last four years of this time he "doubled in brass" by similarly working for W. R. Thompson of the then Imperial Institute of Entomology. Here permit me to quote from a letter Bergold wrote me in January, 1948:

> In 1939 the European Parasite Introduction work of the Americans and the English was stopped, and I was sent through the Reichsforschungsrat in Berlin to Tanganyika Territory, East Africa, to work on the control of coffee insects and various plant diseases, as well as to do cooperative research with a veterinarian on certain diseases of cattle. I was stationed at Mr. Bueb's coffee plantation, Kifumbu, near Moshi. Two months after arrival war broke out and interrupted everything. After being confined for five months to an internment camp in Dar es Salaam, I was repatriated to Germany in February, 1940. Then I had a chance to be a leader of a branch station of the Kaiser Wilhelm Institute of Biology and Biochemistry as a guest in the I. G. Farbenindustrie in Ludwigshafen. Starting there in 1940, I had to build up an entirely new laboratory for virus research, and I began with the polyhedral diseases of caterpillars. In 1943, I went to Tübingen where I am still working in an excellently equipped laboratory and in charge of the Virus Department for Zoology of the Kaiser Wilhelm Institute of Biochemistry, directed by Prof. Dr. A. Butenandt. I was married in 1941, and we have a daughter of 6 years. We are Austrian subjects.

Sellers had had recent contact with Bergold and told me what he knew of Bergold's work and aspirations. Encouraged by Sellers, I wrote to Bergold telling him that we were working on the polyhedrosis of the alfalfa caterpillar and the California oakworm and asked him for reprints and any comments he might wish to make about his recent work, especially since I was expecting soon to submit to the publisher the manuscript for *Principles of Insect Pathology.* An interesting part of his reply was that "It does not make too much difference that my publications are not available to you. It is true that we came to similar conclusions, a little earlier, than Glaser and Stanley, but according to my latest researches, we have both been going in the wrong direction. Through the help of Dr. [Gordon] Bucher I am sending you the proofs of my latest paper; it will probably be two months before I shall have reprints of it to send to you." The proofs he so kindly sent were those of his paper "Die Isolierung des Polyeder-Virus und die Natur der Polyeder" that was published in *Zeitschrift für Naturforschungen* later the same year (1947). From this, and subsequent papers, it was clear what Bergold had meant in his letter when he indicated that he, as well as Glaser and Stanley, had been "going in the wrong direction." Instead of the small, presumably spherical particles these workers had thought might be the virus, Bergold had found large rod-shaped particles embedded in the polyhedron; as was eventually well confirmed, these rods represented the "real virus."

Meanwhile, after making it known (Steinhaus, 1945) that the wilt disease of the alfalfa caterpillar, *Colias eurytheme,* was caused by a polyhedrosis virus and not by a bacterium as had been reported, I had decided to make a complete study of the disease in the hope that the virus could be used as a biological-control agent. If such an application were successful, I was confident that more funds would be made available for basic research on the diseases of insects, and thus an expanded program in insect pathology justified. By the time of my 1947 communications with Bergold, we had completed preliminary studies on the symptomatology, histopathology, and epizootiology of the disease in *Colias,* and had just begun an investigation into the nature of the polyhedron and associated virus. Whether or not we would have gone the erroneous early route of Bergold, or of Glaser and Stanley, is difficult to judge. Because our centrifuges were of relatively low speed, we might have been lucky and not have thrown down—and out—the large virus particles (40 by 300 millimicrons), as had our predecessors. In any event, the receipt of Bergold's paper determined the direction we should take, and without difficulty we isolated from the polyhedra virus rods similar to those Bergold was finding in the silkworm and the gypsy moth (Bergold, 1947, 1948; Steinhaus, 1948).

As mentioned in chapter 3 Bergold and I also had an extended correspondence regarding granulosis—he working with one from fir-shoot roller, *Cacoecia murinana*, and we with the variegated cutworm, *Peridroma margaritosa*. Bergold beat us to the publication of a new type of virus disease in insects just because we were "taking our time." Bergold's interest in and knowledge of diseases of insects, especially the viruses, made him a logical candidate for an addition to our staff. However, our efforts to hire him were thwarted by the U.S. government red tape. The Canadians, more adroit in being able to hire aliens—particularly biologists—succeeded.

With Bergold's departure from Germany, I had no clear image of where in that country one might find others interested in the diseases of insects. In November 1947, I had received a hint of someone who might be destined to provide the leadership for such efforts; it was contained in a letter from Jost M. Franz, which began with the words, "As collaborator of Dr. Bergold (Tübingen). . . . " It was a polite request for a copy of *Insect Microbiology* and an expression of his interests in the diseases of *Cacoecia murinana*. Franz had been employed by the Imperial Bureau of Biological Control, working on entomophagous insects. Like Bergold, one of his principal tasks was to collect predators and parasites for the bureau, mainly for shipment to, and study by, scientists in Canada.

It was sometime in 1946 that de Gryse, disguised in an Army uniform as was necessary for North American visitors in those immediate postwar days, visited the bomb-damaged Institut für Angewandte Zoologie in Munich, headed by the well-known zoologist W. Zwölfer. De Gryse was attempting to contact everyone he could who was familiar with *C. murinana* in order to arrange for importation into Canada of as many as possible of its natural enemies, including microbial pathogens, to aid in the North American battle against the closely related spruce budworm, *Choristoneura fumiferana.* Jost Franz had written his thesis on *C. murinana*, but de Gryse's visit highlighted for him the fact that no one knew the cause of the death of the blackened pupae he had observed. Indeed, de Gryse's inclusion of microbial pathogens among the natural enemies he sought caused Franz to appreciate more fully the significance of disease as a control factor.

About a year after de Gryse's visit, Gordon E. Bucher, an energetic young Canadian working for the Imperial Bureau of Biological Control on the population dynamics of the fir budworm in the Black Forest and the Vosges, visited Franz in Munich. Bucher was headquartered, in 1947, in Bergold's laboratory in Tübingen and had come to ask Franz's help in locating outbreaks of *C. murinana* and advice in otherwise handling this tortricid. This Franz did; at the same time, in his contacts with Tübingen, he became familiar with Bergold's work on insect viruses and learned to appreciate the ecological significance of disease in insects generally.

As Franz told me later, in 1951 and 1952 during his study of a predator, *Laricobius erichsoni*, of the balsam woolly aphid, *Chermes picea*, he was confronted with an unexplained mortality of the beetle. In the autumn of 1951, during the Deutsche Planzenschutz-Tangung at Würzburg, he for the first time contacted Erwin Müller-Kögler, who as early as 1941 had published on fungus diseases of insects. Hopeful of obtaining a diagnosis of the diseased *Laricobius*, Franz had sections of diseased material to show Müller-Kögler but they could not arrive at any definite conclusions concerning what they thought might be a bacterial disease of the insect. This, and other examples, convinced Franz that "we in Germany were not in a position to identify either insect pathogens or entomophagous insects of the more difficult groups." At the next year's plant protection meeting, he made a point of this, and the necessity of developing insect pathology within the framework of biological control. In his published paper (1953), he asked, "But is it not grotesque that we have to ship insects dead from diseases to North America for a diagnosis?"

When in 1953 the Biologische Bundesanstalt für Land- und

Forstwirtschaft established a laboratory for biological control research in the former Colorado potato beetle laboratory in Darmstadt, Franz was made its director. He carried his ideas concerning the importance of insect pathology with him. One reason it was decided to emphasize work on the microbial diseases of insects was that the equipment available to the new laboratory was more appropriate for such studies than for the handling of entomophagous insects, and it was felt that close cooperation could be had with plant pathologists of the Biologische Bundesanstalt already dealing with infectious agents. The latter did not prove to be as easy as contemplated, so Franz decided to staff his laboratory with co-workers interested in the principal types of microorganisms known to attack insects. The first to join him, toward the end of 1953, was Müller-Kögler, to be followed in April 1954, by Aloysius Krieg. Langenbuch, who had been on the staff of the former Colorado potato beetle institute was asked to turn his attention to problems involving the histopathology of diseased insects. Then followed O. F. Niklas, with a background in entomophagus insects and ecology. Perhaps largely because of Franz's contacts with the Commonwealth Institute of Biological Control and the Laboratory of Insect Pathology at Sault Sainte Marie, Canada, the first project of the new Institut für Biologische Schädlingsbekampfung at Darmstadt was the nuclear polyhedrosis of the European pine sawfly, *Neodiprion sertifer*, somewhat along the lines being pursued by F. T. Bird in Canada. The success of field applications of the virus to control the sawfly in Europe contributed greatly to the acceptance by the German administration of insect pathology as a worthwhile field of endeavor.

This was the situation at the Darmstadt Institute when I dropped in for a visit on 23 August 1954, except that diversification of projects and interests had begun to develop. Müller-Kögler, whom I visited in the hospital where he was recovering from a relapse of what was believed to have been a case of Q fever, was concerned with entomogenous fungi and the diseases they cause. Niklas was then interested in the epizootiology of both a rickettsial disease in the cockchafer *Melolontha melolontha*, and the nuclear polyhedrosis of the European pine sawfly. Krieg was working on the *Melolontha* rickettsia and had in preparation the paper in which he named it *Rickettsia melolonthae*. Knowing that C. B. Philip was in the process of revising the Rickettsiales, intending to place most of the insect-pathogenic rickettsiae in a separate genus (*Rickettsiella*), I suggested to Krieg that he consult Philip before publishing the name of his new species. Although circumstances of timing did not give Krieg an opportunity then to place his species in a separate genus (Philip did so later), I believe that his contacts with Philip, plus his own dynamic curiosity, caused Krieg for a while at least, to be interested in rickettsiae generally (Krieg, 1963). My two days in Darmstadt were all too short, but it was clear that Franz was sailing a tight ship and that he was laying the foundation for what was to be the principal center for insect pathology in Germany.

On a subsequent visit to the Institute in Darmstadt in late November 1960, I found Franz and his staff to be in somewhat of a discouraged mood. The federal government's support budget, as well as salaries, were considerably lower than what one would expect in view of the amazing economic recovery made by West Germany following the war. One of the professional staff members inquired of me as to the availability of positions in insect pathology in the United States—his salary was virtually the same as that of his technician's and below that which he believed he could survive decently. Franz was fully aware of the prevailing low morale, but there was little that he could do about it. Even their long-promised new building (they were occupying a large residence, with Franz and his family living on the third floor) had been delayed, and its realization was nowhere in sight.

However, Franz saw to it that research continued apace. The staff was about the same as at the time of my first visit except that Langenbuch had retired in 1958 and his position had been filled by Alois M. Huger. W. Herfs, a young visiting scientist, was working with Krieg largely on *Bacillus thuringien-*

sis. Franz himself had just finished two comprehensive reviews on biological control, which included the use of microbial pathogens (Franz, 1961*a*, *b*). Krieg (1961) was awaiting the publication of his book *Grundlagen der Insektenpathologie.* Müller-Kögler was deep into his monograph *Pilzkrankheiten bei Insekten*; this classic work was finally finished in 1965. Niklas was studying and preparing publications on the natural enemies of populations of *Melolontha* larvae, as well as being concerned with the ecology of certain sawfly diseases. And Huger was enthusiastically pursuing his research on histopathology, and on microsporidia, rickettsiae, and on the granulosis viruses about which he wrote an excellent review of the knowledge of these agents up to that time (Huger, 1963).

During this 1960 visit I gained a deeper insight into not only Franz's institute but how his laboratories cooperated with others in Europe, particularly those of the French. Franz was reaching out to make academic contracts with a local college at which he eventually was to hold a professorship concomitantly with his federal position. Shortly after we arrived in Darmstadt, I was asked to give an informal talk to about sixty people associated with the zoology department at the college, Die Technische Hochschule Darmstadt. The head of the department, who made the introductory remarks, was a descendent of Martin Luther, and still carried the surname of Wolfgang Luther. The audience enjoyed a difference of scientific opinion between Franz and myself during the discussion period, which ended up in a local wine cellar in the best spirit of conviviality.

This 1960 visit to Darmstadt was a delightful one in many ways. Six years (since my last visit) had accomplished much in the rebuilding of the city which had been badly damaged by the war. Although it was almost a month before Christmas, the season was in the air. Decorations of great beauty—without the gaudy signs of commercialism all too often blatantly in evidence in American streets—were everywhere. Probably because as a child I had attended a German Lutheran Sunday school, Christmas that year nostalgically seemed to belong with the German setting.

II

It was during my 1954 trip to Europe that I first met Hans Johannes Blunck, professor emeritus in the Institut für Pflanzenkrankheiten at the University of Bonn. I found him at his home in Pech bei Bad Godesberg, a city nestled in the green and rolling hills and situated right on the edge of the Rhine River. He had converted the elevated basement of his home into a laboratory complete with an adjacent glasshouse and, through the courtesy of the university in nearby Bonn, was being assisted by three technicians and two secretaries. Although a general plant protectionist and an applied entomologist, in recent years Professor Blunck had become interested in the microbial pathogens and the insect parasites known to attack the European cabbageworm, *Pieris brassicae.* In so far as the diseases of this insect were concerned, he was especially interested in those caused by Microsporida, although he had also reported some observations on entomogenous fungi on other insects.

In the fields and cabbage patches in the area about Bad Godesberg, Blunck found that approximately ten percent of all populations of *Pieris* were more or less regularly infected with Microsporida, especially *Nosema polyvora.* Because of the constant low-level incidence of infection he did not consider the protozoan to be very pathogenic and felt that it did not have great promise as a microbial control agent. He had published briefly on this microsporidan (Blunck, 1952, 1954) by the time of my visit. At the Tenth International Congress of Entomology in Montreal in 1956, he presented a review of his findings with regard to several species of Microsporida found in Pieridae (see Blunck, 1958). He reiterated his conviction that "until now there is little chance of using Microsporidia for biological control of injurious Pieridae." Following the congress, Blunck traveled about Canada and the United States examining the populations of the imported cabbageworm, *Pieris rapae,* to determine the incidence of its natural enemies (Blunck, 1957). Interestingly, unlike the situation in Europe, he found virtually no Microsporida in the North American Pieridae examined. During the time he was using the facilities of our department to examine his samplings of larvae, he displayed an enthusiasm and zest that was truly an inspiration to all of us.

When I had visited Blunck's laboratory in 1954, we had to converse in German. Frau Blunck could speak English, however. When he decided to come to North America, he decided to first master the English language in order to "better get along" while in the United States and Canada. Undoubtedly Frau Blunck helped him in his resolve, but he accomplished his goal without the aid of Berlitz or any of the present-day crash methods of learning foreign languages. Such an undertaking as his would not be unusual, or worthy of comment, except that he was almost seventy-one years old at the time, and learned the language for the purpose of this single trip. This act typified the spunky diligence and determination of the man. He was a pleasant, congenial, cultured, thoroughly delightful, sensitive man.

Unfortunately, not too long after Professor Blunck's return to Germany he became ill, and died of cancer of the pancreas on 12 January 1958, at the age of seventy-two (see Rademacher, 1958). What an impact this man would have made on insect pathology had he found interest in the diseases of insects at a time earlier in his career!

III

If not from the standpoint of insect pathology as such, then certainly from that of insect microbiology, the great German school of symbiotology established by Paul Buchner should be mentioned here. In particular, I refer to those advances having to do with the mutualistic and commensal relations between insects and microorganisms. It is difficult to find any European biologist with a more loyal following of students and associates than that possessed by Professor Buchner. This is clear from the accolades accorded him on the occasions of his sixty-fifth and eightieth birthdays (e.g., see Koch, 1951,

1966; Schwartz, 1966) and the accounts of his life and work, especially by Koch. Buchner was born on 12 April 1886. Except for a footnote or two, it would be superfluous for me to attempt to add to the well-documented life of this outstanding man and scientist.

As my interest in the biological relationships between insects and microorganisms gestated during my years as an undergraduate and graduate student, it is difficult to describe the genuine thrills I experienced each time I found in the literature an account of some new dimension of this subject. Such was certainly the case that day in 1937 when I first came upon Buchner's 1930 classic *Tier und Pflanze in Symbiose* (later to appear in revised editions as *Endosymbiose der Tiere mit Pflanzlichen Mikroorganismen* [1953] and still later, translated into an English edition, *Endosymbiosis of Animals with Plant Microorganisms* [1965]. That insects could regularly contain within certain of their cells living microorganisms which were transmitted by special mechanisms from generation to generation of the insects fascinated and entranced me beyond description. Finally, in 1939, I summoned enough courage to write Buchner requesting reprints of some of his publications, but the onset of the World War II prevented a reply.

In October 1946, I wrote him again, after seeing a note in an American journal that he was living on the island of Ischia in the Bay of Naples and was without food, clothing, and income. In response to my inquiry as to how I might be of help to him, I received a poignant letter explaining his need for food, and especially for warm clothing because they had no means of heating their home during the approaching winter. Also, "A great desire of my wife's would be some washing-soap, which costs an awful lot here." Although America was still suffering shortages of its own, there was no difficulty in sending him the items he and Frau Buchner needed. Thus began a long and pleasant exchange of correspondence and reprints over the next quarter of a century.

One of his letters, dated 21 December 1946, is of particular historical significance and interest because it indicates the state of the Buchner school of symbiosis in Germany at the time as well as what his immediate postwar intentions had been. Among other things, he said:

> In the meantime I have been summoned to Munich as a successor to von Frisch, who apparently has preferred to exchange the destroyed city and the equally greatly destroyed Institute (which was originally founded by the Rockefeller Foundation) for the city of Graz [Austria]. His move was no doubt motivated by the fact that he had lost his house and all his belongings in Munich. It is still uncertain as to when I shall actually begin my new job. . . . I have not as yet been able to remove from Leipzig (in the Russian Zone) any of my belongings which remain. Thus I would have to begin life in Munich without any of my scientific equipment, and without my furniture or household utensils, and with the few clothes I have here [Ischia] with me. I am certain that life will be even more difficult in Munich than it is here. It will also be very difficult to surround myself with a group of co-workers which would be properly trained in my field of interest inasmuch as most of the younger workers--since they had in bygone days been forced to belong to the [Nazi] Party--do not, at present, qualify for such positions. Of course, many capable workers died during the war; among them, for example, is E. Ries to whom we are greatly indebted for his interesting study of the symbiosis of Pediculidae. Professor A. Koch had to leave his institute in Danzig, and for the time being lives with his mother in Stockdorf-Planegg (near Munich) at 41 Forstkastenstrasse (American Zone). I am sure that he will be very glad to hear from you. I certainly would be more than happy to offer him a position at the institute with which I shall be associated. Professor H.-J. Stammer, another student of the field of symbiosis, is professor at the Zoological Institute of the University at Erlangen (American Zone). Dr. H. J. Müller, the author of the excellent monograph on the symbiosis of Fulgoridae, is unemployed and lives in Halle Saale (Russian Zone) at 3 Weinbergweg.

One can judge from this excerpt why Buchner eventually decided not to accept the chair at Munich (which was later filled by Koch), and to remain on the beautiful Italian island of Ischia.

Professor Buchner, originally a cytologist and a student of Theodor Boveri, Richard Hertwig, and Richard Goldschmidt (later to join the faculty at Berkeley), credits much of his inspiration and interest in symbiosis to reports on this subject by U. Pierantoni (Naples) and K. Šulc (Brno). Buchner presented his first paper on symbiosis at a 1911 meeting of the Gesellschaft für Morphologie und Physiologie. He worked and taught at the University of Munich, Greifswald University, Breslau University, and Leipzig University. After the end of the war he considered returning to Munich, as we have seen, but eventually decided to continue his research from his home at Porto d'Ischia under the aegis of the university. It was here that I was privileged to spend a never-to-be forgotten day with Paul Buchner.

At least parenthetically, it should be said that Buchner's residence in Italy since World War II did little to diminish his influence in Germany. He was not retired as professor emeritus from the University of Munich until 1959. Prior to that time, however, he had received numerous honors, awards, and honorary degrees. In 1951, the Institut für Experimentelle Symbioseforschung founded in 1947 by Buchner's student, Anton Koch, in Munich was renamed the Paul Buchner-Institut für Experimentelle Symbioseforschung. The Federal Republic of Germany honored him with the Grand Cross of the Order of Merit in 1961 on his seventy-fifth birthday.

After passing his doctorate examination in 1909, Buchner received an academic grant to study for a year at the famed Zoological Station in Naples. This was the beginning of his fascination with life in Naples, for here he met his wife, Miliana, an accomplished painter. During a return visit to the station in 1927, he decided to establish a vacation residence on the island of Ischia, and accordingly bought a vineyard on the island. During vacation periods he supervised the construction of his house and workshop. Together with his son, Giorgio Buchner, who became an archeologist, he studied as hobbies, the archeology, geology, and natural history of the island. His archeological and geological findings were significant enough that they are now housed in a small museum, Museo dell' Isola d'Ischia, founded by the Buchners. For his work on symbiosis, he had built himself a roomy, glassed-in portico off the living room which served as a laboratory well equipped with microscopes, camera lucida, drawing board, microtome, an impressive slide collection, and all the other furnishings necessary for his painstaking research.

Not only did Professor Buchner not allow his departure from Germany to interfere with his scientific work, but neither did he permit advancing age to serve as an excuse for the cessation, or even a diminution of his productivity. Indeed, when I first met him, on 23 September 1954, I could not believe that this white-haired man, gentle, small of stature, and in his late sixties was the scientific giant still leading the Buchner school of symbiotology. Even as I write these lines in the spring of 1968, I have received a letter from the professor, along with a reprint of his latest work. And at this time he is in his eighty-third year!

Inasmuch as at the time of my 1954 visit to Italy I did not know precisely when I would be in Naples, I failed to make an appointment to see Buchner on a specific date. Therefore, upon my arrival in Naples, I sent him a telegram (he could not be reached by telephone) requesting an appointment. After a day or two, having received no reply, I decided to take the chance of visiting him by simply going to Ischia with the hope he would be able to see me for few minutes, in any case.

Buchner's professional address in Porto d'Ischia was the small museum I mentioned, and I erroneously assumed that he probably worked there most of the time. I did not know he had a home laboratory. The taxies on the isle were one-horse carriages. My family and I engaged one and set off for the museum. When we arrived, we found all the doors locked (I have never been able to understand the rules, if such there are, by which Italians open and close their public institutions). Returning to the carriage I was expressing my disappointment about missing Professor Buchner. Since my Italian was limited to the usual tourist-used phrases and sentences, I couldn't discuss my predicament with my non-English-speaking driver. Nevertheless, from our conversation, he picked out the word "Buchner," and with accommodating enthusiasm he said in the form of a question, "Professor Buchner?" I replied, "Si!" with an affirmative shake of my head, and we were off immediately at a trot toward the opposite end of the island. Long before we arrived, we could see the Buchner residence, high above the harbor of Porto d'ischia, identifiable from a distance by a huge eucalyptus tree. The driver stopped at iron gates and went inside, soon returning with a smiling, gentle-faced, grandfatherly man whom he announced to be Professor Buchner.

Our reception by Professor Buchner was accompanied by some bewilderment because he had not received the telegram nor had he received a letter which I had written earlier. However, he welcomed us enthusiastically and literally with open arms. Buchner did not speak English but he was kind enough to converse in simple German phrases, so that we managed to understand each other. He explained that Mrs. Buchner had gone on an errand, but would return in an hour. He conducted us on a tour of his beautiful garden enclosed with a high stone wall, showing us the several vistas of the magnificent view he had from his house. (No wonder he loved this place and found it hard to go back to Germany.) He told us of fifteenth-century castles, churches, and prisons, and of ancient Greek civilizations that had one time inhabited this isle which he had come to know so well.

While in his garden he explained how, when the allies were driving the German armies back up the boot of Italy, he had taken the precaution of hiding his valuable scientific equipment in the soil beneath his flower bushes. Whether he feared the retreating Germans or the advancing English and Americans was not clear to me, but he seemed satisfied that he had done the wise thing. He was taken to an English war camp for a while. Beginning with the fall of 1944, Buchner never again made his usual shuttle trip to Leipzig and back. Since the University of Leipzig was in the USSR Zone he was invited back to the University of Munich, but he did not feel that he had the time or energy to rebuild his former institute. Instead he gratefully accepted a professorship in absentia, but with frequent contacts with Munich.

It was during this visit with Buchner that I importuned him to revise his famous *Endosymbiose der Tiere mit Pflanzlichen Mikroorganismen* and attempt to have it published in an English edition. Strange as it may seem, at that time, although the symbiotic (especially mutualistic) associations between microorganisms and insects were well known among European biologists, except for the entomophilic protozoa in termites they were only slightly appreciated in the English-speaking world. This was especially true of the intracellular symbiotes. Indeed, until about twelve years after the end of World War II, my own publications contained the only reviews of the subject—representing especially the work of the Buchner school—appearing in the English language. Fortunately, Walter Carter and Bertha Mueller of the University of Hawaii—with the aid of a federal grant that I was able to help secure—were to provide the opportunity so badly needed. The English edition appeared in 1965, and there is no question but that this new *Endosymbiosis of Animals with Plant Microorganisms* had, and continues to have, an emphatic impact on the development of interest in endosymbiosis in North America, as elsewhere.

What Paul Buchner once said of a report by Šulc on intracellular symbiosis can also be said of Buchner's life work: it "takes a bandage away from our eyes." To become familiar with only a small portion of Buchner's findings, largely morphological and cytological in nature, is to realize the ocean of research yet to do in cellular biology and physiology generally before the true basic significance—surely great in scientific import—of the relationships between microbial symbiotes and their invertebrate hosts is known.

2. FRENCH INSECT PATHOLOGY

I

Post-World War II invertebrate pathology in France was a bit slow in getting started, but it soon made up for this delayed beginning by the enthusiastic activity of those who chose to study the diseases of insects and of certain other invertebrates. Investigators of insect diseases were located in the Institut National de la Recherche Agronomique of the Ministère de l'Agriculture with stations at Alès in southern France (where Louis

Pasteur did his monumental work on pébrine and flacherie in the silkworm), at La Minière par Guyancourt (later moved to La Minière par Versailles), and at the Institut Pasteur in Paris. Also individual workers, such as Nadine Plus of the Laboratoire de Génétique Formelle, Gif-sur-Yvette, Odette Tuzet of the Université de Montpellier, and J. Théodorides of the Université de Paris, as in most major countries, were making contributions from their own laboratories at educational institutions.

In addition to Constantin Toumanoff, who had made outstanding contributions to French apiculture and medical entomology as well as general insect pathology, two giants were soon visible on the postwar scene: Constantin Vago, initially concerned with the diseases of the silkworm and edible snails, and Pierre Grison, interested in promoting the use of microbial pathogens, as well as entomophagous insects, in the control of insect pests. Today, in 1968, as I write this paragraph, these two men are still leaders of world-known laboratories concerned with both basic and applied aspects of insect pathology.

However, before saying anything more about these men and their colleagues, I must explain that one of the principal reasons I visited France in 1954 was to complete pilgrimages to the laboratory (now a domicile and barn) in Alès where Pasteur had worked, and espeically to the laboratory in Saint-Genis-Laval where André Paillot did so much to further general insect pathology between World Wars I and II.

The appropriate place for an insect pathologist to pay homage to Louis Pasteur is at what was once a lonely house and barn at the foot of the Mount of Hermitage on the outskirts of Alès (Alais). Pasteur set up his laboratory here when he began his work on the diseases of the silkworm in mid-June 1865, and after he left about three years later the buildings returned to being a domicile and barn. And, except for the attachment of a commemorative plaque high on the side of the house, it was still so when I first saw the famous site on 10 October 1954. It was on this date, a Sunday, that my family and I arrived in Alès. We decided to drive about Alès to see what we could see for ourselves. By coincidence, as we drove toward one end of town, my eye caught a familiar sight: in the distance I could see virtually the same scene (*Habitation du Pont-Gisquet, près d'Alais*) that serves as the frontispiece to Pasteur's 1870 famed *Études sur la maladie des vers à soie*. We had been led to the location as if guided by some patron saint. As we arrived in front of the house, we were surprised to find that the buildings and grounds were not open to the public. France, surprisingly, had not made this one of their national shrines or monuments.

As we approached the house, flashes of the great man's life, and some of his words, went through my mind. I remembered how, as I had read and re-read his life and works, he had displayed the human emotions of anger and impatience as well as compassion and consideration; how he was at times overly confident and boastful, at others times uncertain, humble, and modest; how he began the preface of his famous report on the diseases of the silkworm by saying, "I should begin this work by excusing myself for having undertaken it." I remembered that it was to Paul de Kruif's *Microbe Hunters* I owed the revelation of Pasteur's stirring words to students:

> Do not let yourselves be tainted by a deprecating and barren skepticism, do not let yourselves be discouraged by the sadness of certain hours which pass over nations. Live in the serene peace of laboratories and libraries. Say to yourselves first: What have I done for my country? until the time comes when you may have the immense happiness of thinking that you have contributed in some way to the progress and good of humanity.

I was also reminded of the biographer René Vallery-Radot's (1937) philosophical thoughts of Pasteur, for example:

> There are two men in each one of us: the scientist, he who starts with a clear field and desires to rise to the knowledge of Nature through observation, experimentation and reasoning; and the man of sentiment, the man of belief, the man who mourns his dead children, and who cannot, alas prove that he will see them again, but who believes that he will, and lives in that hope, the man who will not die like a vibrio, but who feels that the force that is within him cannot die. The two domains are distinct, and woe to him who tries to let them trespass on each other in the so imperfect state of human knowledge.

I had always liked, and was in complete sympathy with, Vallery-Radot's comment that Pasteur never accepted an invitation to those large, impersonal, social gatherings "which are a tax on those who have nothing to do, on the time of those who are busy."

My principal contact in southern France was Constantin Vago, then *chef* (or *ingénieur*) de la Section Pathologie at the Station de Recherches Séricicoles in Alès. I had an appointment with him the next morning at 9:30. However, I was also scheduled to meet Mauro Martignoni at the railway station at 8:30. It had been arranged for Mauro to meet us in Alès and to accompany us through most of France, both as an interested companion and as an interpreter.

The train arrived precisely on time, and Mauro, bubbling with his customary enthusiasm, was all set to go. As we all climbed into the car, I sensed that the group of us must have made a sight—Mauro and I, both tall men, my wife, and two children, plus baggage, certainly strained the dimensions of the diminutive Hilman-Minx. After checking Mauro in at the hotel, he and I set off for the sericulture station, which I had located on our drive the day before. As we approached the station, I realized that something had been added. We were both surprised and pleased to see—in our honor—the flags of Switzerland and the United States flying above the main entrance and below and on either side of the French banner. The portent of a friendly welcome was clear.

Vago was waiting for us. Our greetings were mutually warm and enthusiastic. My previous contacts with Vago, other than reading each other's scientific publications, had been only by correspondence, beginning in March 1951. At the time of our visit in Alès, Vago appeared to be about forty years of age. His disposition was pleasant and outgoing, making it easy to discuss all aspects of invertebrate pathology. He was born and raised in Hungary and had come to France in 1946 to accept the position he held at the station. He came by his interest in the phenomenon of disease naturally inasmuch as his father, in Hungary, had been a veterinarian-pathologist. By the time he was fifteen years old, he had developed marked expertise with microscopes; seeing this, his father encouraged him to pursue a career in medicine or in the natural sciences. Upon entering the University of Debrecen, Vago registered to major in both medicine and the natural sciences. However, his interest in biology, and especially invertebrate zoology, caused him to abandon medicine. Upon graduating in the sciences, he took a position working in the laboratory of St. Georgy in Hungary. In 1939, he came to Austria as director of a pathology laboratory of the national health service. His original assignment at the sericulture station in Alès was to work on the diseases of the silkworm. This he did with great competency, but it was easy to discern, while talking with him, that his interests and aspirations in the diseases of insects were general and not limited to those of the silkworm.

Parenthetically, I should explain here that La Station de Recherches Séricicoles d'Alès, the only such station in France, was established 1 April 1897, in response to a request from members of the parliament coming from the Department of Gard, from the village of Alès, and other Departments in France in which sericulture was an important industry. It functioned under the administration of the Institut National de la Recherche Agronomique of the Ministère de l'Agriculture of France. Its first director was Mozziconacci, who served until his death in 1923. At first located in a domicile, in 1927 it was moved to specially constructed buildings, at which time a greater emphasis on laboratory research was instituted. Following World War II, André Schenk was made director, and six specialized units were formed, all of which were in operation at the time of my 1954 visit—the biology unit, hatchery unit, breeding unit, mulberry unit, the technology of silk unit, and, of course, the pathology unit headed by Vago. The station was also serving as the site of the International Sericulture Commission and as the editorial offices of the publication *La revue internationale du ver à soie.* It also created and coordinated several regional and national organizations and committees to promote the production and use of silk, as well as conducting extensive services or "seasonal schools" in apiculture. At the time of my visit, I was told that the production of silkworm cocoons in France amounted to about 400 tons; in addition it produced and exported a large quantity of eggs of *Bombyx mori.*

The pathology unit was, at the time, concentrating on bacteriological and virological studies. However, I suspect largely because of Vago's interests and orientation, and also through its contact with other agronomic research stations, this unit performed diagnostic services pertaining to all diseased insects submitted to it; furthermore, as I have indicated, its research included work on the diseases of some insects other than the silkworm. A disease of edible snails was also being investigated. Interestingly, if I understood correctly, the pathology unit was also concerned with problems relating to the soundness and "integrity" of cork, and the "health" of chestnuts—indicating that the station was alert to the needs of the area (especially Cévennes) in matters other than silk production.

At no time during my visit to Alès, or during meetings in subsequent years, did I hear anyone call Vago by his given name, Constantin. Breaching the European protocol Martignoni was constantly trying to have me maintain, I asked Vago to drop the formalities and call me Ed—the insistent "Professor Steinhaus" was becoming too much for me. While doing so, I took the liberty of asking him if his friends called him Constantin, or some variation of it. "No," he replied, "I prefer to be called 'Szilard'." However, I have had difficulty in doing so because it was such a departure from his real given name, and because "Vago" itself had the ring of a first name.

Poor Mauro Martignoni! What a source of trial, trouble, and embarrassment I must have been to him then. Although he had spent half a year in my laboratory in the United States as a graduate student, my "typical American attitude of informality," as Mauro described it, was obviously a strain on him whenever he saw me initiating an overtly friendly rapport with one of my European hosts. One of his most embarrassing times with me was at a reception given toward the end of our visit at the station—a delightful and memorable occasion.

It was about five o'clock in the afternoon, and we had just returned from a drive through the countryside where silkworm rearing was being done in the homes of the farmers, and from a visit to the building in which Pasteur had lived and conducted his famous research on pébrine and flacherie. (Vago's effort to persuade the housewife to allow us to see the interior was in vain.) In showing us the family-type rearing operations maintained by the families (usually the wife and daughters) of farmers, Vago very objectively had us see examples of rearing done under very sanitary conditions—and thus producing silkworms relatively free of disease—as well as examples of rearing carried out under careless and insanitary conditions and thus highly subject to disease.

When we returned to the station, I sensed that something special was about to take place. Director Schenk had called together the staff of the station, as well as some local dignitaries and newspaper reporters. Monsieur Sernoux, president of the local Chamber of Commerce was there to greet Mauro and me on behalf of the city of Alès.

The proceedings began quite formally with Dr. Schenk reading an embarrassingly laudatory statement, speaking clearly and slowly so that I could understand. All through the reading I kept wondering how and if I should respond. I am sure that Mauro was even more worried than I about what I might say! At the end of the speech the applause and the nodding heads indicated that a verbal response from me was expected. I rose to speak, proceeding immediately to commit a faux paux which my French colleagues accepted in good grace. Presumably because I had been traveling in Germany, where I had gotten by reasonably well with my fractured German, I had unintentionally begun my response speaking in German. Suddenly realizing that this would be offensive to my French hosts, I attempted to switch to my almost nonexistent French, finally ending up apologizing in English and asking Mauro to translate my remarks into French. The reception ended with a lovely cake, decorated with Swiss and American flags, and Veuve Cliquot champagne. As it broke up, the newspapermen asked questions and took photographs. From their accounts in the next day's papers I gained an understanding of how Frenchmen—including French newsmen—could be diplomatic and even kind to a bewildered, surprised, and bumbling American.

The following day, Director Schenk and Dr. Vago, in an effort to give us a thorough understanding of their work in French sericulture, generously took Mauro and me on a trip to Vals-les-Bains, which took us through one of the principal silk-producing regions of France. We saw commercial silkworm-rearing rooms and a large filature to which sericulturists from throughout the countryside brought their cocoons to be processed. We had a wonderful dinner in Vals-les-Bains complete with local wines and champagne. Following our meal, we sat comfortably about the fireplace eating roasted chestnuts, and in a relaxed manner discussed at length the present state and the future of insect pathology. It was one of those conversations worth a host of scientific papers or a library of treatises.

Although Vago and I were to see each other at international congresses and were to exchange correspondence periodically, we were not able to get together in the same relaxed manner again until mid-August 1962 when he visited Berkeley. Again our dialogue was one that clarified for each of us the continuing development of insect pathology in Europe and North America. By this time, of course, Martignoni had joined our faculty at Berkeley and, together with Joe Tanada and our graduate students, we participated in the interchange of information and spirit as we had on other occasions with such foreign leaders as Jost Franz, Jaroslav Weiser, Keio Aisawa, Hisao Aruga, A. Krieg, Kenneth Smith, and others on their visits to our laboratory.

II

From Alès we (all five of us plus luggage in our Hilman-Minx) proceeded up the Rhone Valley to Lyon and Saint-Genis-Laval, the part of France in which André Paillot did most of his work.

Soon after arriving in France in 1954, as well as on subsequent visits, I was amazed to find that Paillot and his work were not held in the highest regard by a surprising number of his fellow Frenchmen. To this day I puzzle over this impression. To be sure, as is the case with all productive scientists, there are mistakes to be found in the published works of Paillot, and some of his speculations have not been borne out, but these seem to me to be insufficient cause for the almost uniform light regard for his accomplishments on the part of most of the French insect pathologists with whom I discussed him and his research. Or so it seemed; and there were some exceptions.

Apparently, Paillot was essentially a "loner" and somewhat of an irascible man with little time for social amenities. Some described him as stubborn, rude, and very discourteous, apparently reflecting unfortunate encounters. One who had worked with him for two years and considered him a good friend said that at times—possibly because of preoccupation with an experiment—Paillot would not speak to him for several days. On another occasion he discounted the accuracy of a colleague's findings because in the latter's report the name of the stain carmine was mistakenly spelled "carmen." But these represent frailties of personality common to all men, and because of them and some errors in his scientific observations and conclusions, we must not close our eyes to the great amount of good he did to aid our understanding of relationships—mutualistic, commensal, and parasitic—existing between insects and microorganisms. The honors he received during his lifetime bespeak considerable respect for him and his work. One can but express sympathetic compassion for Paillot during his last years when his children were all in the Resistance, one being deported to a concentration camp. This must have weighed heavily upon the mind of Paillot and undoubtedly affected his work as well as his health. As has been voiced by others, it is frequently the fate of a pioneer to be attacked by persons whose knowledge and authority have been disturbed. Usually there is some room for fair criticism, but in the balance, the discoveries and contributions he made weigh heavily in their benefit to mankind.

So it is because I still feel that Paillot played an important role in our understanding of disease in insects, and because I personally found encouragement in the published writings of this man at a time when so few saw value in such studies, that I must take time here to pay homage to this French scientist. Unfortunately, I never had the opportunity to meet Paillot (although we corresponded briefly), but I always found his work and publications intriguing, making appropriate allowances for the times and conditions under which he labored. His bibliography is an impressive one (nearly 200 papers), but it is not my intent to review the content of his many publications. To an extent, when feasible, I have done this in my *Principles of Insect Pathology*. Paillot's (1937) own earlier summary of his contributions as well as the recapitulation by Bounhiol, Bruneteau, and Couturier (1947) give a review of his scientific work.

I became familiar with many of Paillot's publications, especially his 1933 book *L'infection chez les insectes*, when I was a graduate student. In July 1939, I wrote him, rather presumptuously for a graduate student, for more information on the morphological and physiological characteristics of the bacteria he had described or reported from insects and for any reprints of his work he might care to send me. A month later I received a brief letter from Paillot explaining that he no longer had reprints of his work on the bacterial diseases of insects, but that I could find what interested me in his book which could be sent to me for 115 francs. He also sent reprints of several of his other papers which he thought might interest me. Somehow this contact must have left Paillot with a feeling that someone in the United States was interested in his work; in the spring of 1940, apparently anticipating the invasion of France, he sent me voluntarily and without any accompanying letter or message, a virtually complete set of reprints of his publications—in some cases with several duplicates. I was not to hear from him again.

Not knowing what fate the war had brought Paillot, I wrote him on 28 May 1945, expressing my best wishes and briefly explaining some of my activities during the intervening years. Finally, in August, I received in the customary black-bordered envelope a sad letter from Madame Paillot telling me that her husband had died suddenly of a cerebral hemorrhage "two days before Christmas, 1944." I felt a true sense of personal loss, as well as a loss for what by now I was dreaming could be an integrated field of endeavor called "insect pathology."

Madame Paillot's letter was the first of a series we exchanged about her husband and his work, but it was one of the most poignant, so I take the liberty of quoting from it, in rough translation:

> The kind letter you sent my husband unfortunately reached me some months after his death. It would have moved him deeply for all the sympathy you have manifested to him with regard to the great misfortune our country has gone through. He endured these sufferings with great courage and without ever losing faith in the country's liberation; and never, not for a single day, did he abandon his research which he pursued so passionately. He was fully engaged in his work when two days before Christmas, 1944, he suddenly died of a cerebral hemorrhage. Certainly the terrible trials of our country, the absence of our four children, all of them in the Resistance, and the youngest one arrested and deported to a horrible concentration camp, had broken his heart. After the Liberation our two oldest sons and our oldest daughter continued to fight in the army, so my husband did not have the joy of seeing them again, or to be present when our poor little daughter who had been deported returned to us three months ago.
>
> I was deeply moved, in reading your letter, by the feelings you expressed toward the work of my husband. My dearest wish is that all the scientists who have understood him and followed him in his research should keep alive his work to which he had given all his heart, in order that what he has sown in the Realm of Science should continue to germinate. It is a heritage which I entrust to all who work in this field.

Passages from subsequent letters give us insights into André Paillot, the man and his work:

> Alas, his laboratory here [at St.-Genis-Laval] is now only half alive. I would say that no one here is able to continue his work. In a similar manner, the laboratory for the study of the pathology of invertebrates established by him under the patronage of the Practical School of Higher Studies has been closed. He worked in great solitude conducting his scientific research without leaving any disciples. The man who is going to take his place in the laboratory will do research on the San Jose scale, but at the moment [letter dated 14 January 1946] nothing is yet well organized. It must be said that the Experiment Station was a very modest one. My husband had only a secretary for his staff, and sometimes a caretaker. But he preferred this arrangement because

he disliked everything that dealt with administrative problems or that diverted him from his work.

Thank you so very much for the packages of food, soap, and other items which have arrived safely. Thank Mrs. Steinhaus for the trouble of gathering the "layette." Would it be possible to send some baby powder for our latest grandchild? . . . My husband would have been so pleased to see the happiness of his four children, all married according to their hearts. Unfortunately, he was able to know only one of his grandchildren. I am rich with all these lovely little grandchildren [three grandsons and a granddaughter as of April 1947].

No one here has asked me for the research material of my husband; I would be very happy to send it to you. Unfortunately much of it is unclassified; he worked alone and had no real disciple to care for it. I can send you slides of insect blood and histopathology slides of *Euxoa segetum*, *Bombyx mori*, and *Pieris brassicae.*

I have asked my friend, Mdme. Grange, who was my husband's secretary and co-worker, to send you whatever slides of his work she has classified and can send. I cannot tell you how happy I am that my husband's work can be useful to you. I hope that despite her other commitments she will not delay too long. If your laboratory were not so far away, I assure you that I would have handed you all the research material of my dear husband because I am certain it would thus be in good hands and saved from oblivion. [Eventually a small assortment of slides were received from Mdme. Grange.]

Most of the pertinent facts relating to the life and works of André Paillot have been recorded in obituary statements (e.g., those by Lapicque, 1945; and Bounhiol, Bruneteau, and Couturier, 1947), and need not be recounted here. Suffice it to say that he was born 8 August 1885 in Jura; was for a while, like his father, a school teacher; began working for the ministry of agriculture under the entómologist E. L. Bouvier, and then Paul Marchal, director of the entomological station in Paris, who encouraged him to study the diseases of insects (which he did during the winters at the Pasteur Institute in Paris beginning in 1912); fought and was wounded in World War I, during which he won the Croix de Guerre; worked for a while as a bacteriologist in the military hospital at Lyons where he was convalescing; and then upon his recovery, was named to direct the work of the entomological station at Saint-Genis-Laval, near Lyons, in 1915. After receiving his doctor of science degree in 1923, he was given accommodations at the medical school at Lyons by the histologist A. Policard and was there able to pursue much of his notable work in the histopathology of insects. In 1936 the École des Hautes Études of the faculty of medicine of Lyon created for Paillot a Laboratory of Invertebrate Pathology, which he directed, along with the station at Saint-Genis-Laval until his death. In general, he conducted most of his basic research at Lyon during the winter and most of his applied research at the station during the summer. He did not like teaching and abandoned it to devote his entire time to research and its application.

Paillot authored three book-sized publications dealing with insect pathology: *Les maladies du ver à soie, grasserie et dysenteries* (1928), *Traité des maladies du ver à soie* (1930), and, as I have already mentioned, *L'infection chez les insectes* (1933). In the introduction of the latter we find summarized the main thrusts of his works and interests and his aspirations. There we find that he was concerned not only with the microbial diseases of insects, but with certain mutualistic relationships between microorganisms and insects (particularly aphids) and cellular and humoral (especially the latter) immunological phenomena in invertebrates. While rereading the introduction of his book in preparing to write these lines, I was impressed by his strong conviction that invertebrate pathology should be accorded adequate attention as a part of so-called comparative pathology. I was embarrassed that in the several diatribes I have spoken and written on this same matter, I have heretofore failed to credit adequately this Frenchman for the position he took, which he emphasizes by quoting from the inaugural address of G. E. H. Roger at the First International Congress of Comparative Pathology: "pour édifier une véritable pathologie générale, il faut envisager les troubles morbides dans toute la série des êtres vivants en commençant par les inférieurs, Protozoaires et Protophytes, pour s'élever progressivement jusqu'aux Mammifères et à l'Homme." How ironic, that I, ignorant of Roger's statement, should years later, at the Eighth International Congress of Comparative Pathology feel compelled to berate (Steinhaus, 1967) this same organization for its neglect of the pathology of invertebrates!

The day in mid-October 1954 when, in the company of M. Martignoni, I first visited the station in Saint-Genis-Laval, once directed by Paillot, was both fascinating and disappointing. Fascinating because here was the four-story house in which Paillot had lived and in which he had had his laboratory (mostly two rooms on the second floor) from which most of his scientific productivity flowed; disappointing because now as a laboratory to study the San Jose scale and other insects, most of the evidence of Paillot's one-time presence had been removed. Indeed, in his laboratory all the new director, C. Benassy, could show me of Paillot's scientific equipment was an antiquated centrifuge. One of the assistants did remember that, dumped in a large box in a closet, there were thousands of coded, but unidentifiable, slides of sections of diseased insects; these together with some left-over reprints of a few of Paillot's papers, to which the present occupants kindly told us to help ourselves, were the only evidence that Paillot had ever worked here. It was useless to attempt to rescue the slides without having the key to the numbered code. What a waste of precious and historically important material that might have been preserved.

III

Leaving Saint-Genis-Laval, we proceeded on to Paris with the L'Institut Pasteur and the Institut National de la Recherche Agronomique our destinations. Paris and its environs, this October, were radiantly beautiful with foliage everywhere exhibiting those fall colors which artists have been painting for centuries. Although subsequent trips to this magnificent city, which to Balzac was a "veritable ocean," were to be as scientifically interesting, this 1954 visit, being my first, could scarcely be surpassed when one included the beauty of Versailles and the countryside.

I had been warned that my first meeting with Constantin

Toumanoff at the Pasteur Institute would be in the nature of a confrontation, so I was anxious to get it over with. I called and made an appointment; I was assured that he would be pleased to see me at ten o'clock on the morning of 15 October. Although Toumanoff spoke English, as we drove to the Pasteur Institute, I was feeling that I would have special need of Martignoni's interpreting services as well as his moral support. I had learned that the Russian-French scientist had been offended by a comment I had made in a paper on *Bacillus thuringiensis* published that year. Certainly I had no intention of insulting him or his laboratory and I was anxious to make amends. However, I had to know directly from Toumanoff just what he had been offended by.

After extricating ourselves from one of those indescribable French traffic snarls in which the gendarmes wait on the side doing nothing to get the traffic moving until after all participants in the impasse have had their say, we arrived at the appointed hour. Toumanoff, a large man in his early fifties accorded us a warm and cordial welcome. He was outgoing and talkative, giving no hint of having suffered any insult. He smoked heavily, which gave him a persistent cough and, he told us, caused him to worry about developing throat cancer; he also assured us that he loved wines of all kinds. After exchanging a few more pleasantries, I decided to face the matter of the offending published statement head-on. Toumanoff responded most graciously. He told me that his anger had now abated, although at the time he first read the paper he was so upset that he "did not sleep for two nights."

Briefly, the unintended offence was this: Following the publication of Toumanoff and Vago's description of *Bacillus cereus* var. *alesti* in 1951 and 1952, I wrote to the former requesting a transfer of their culture because I suspected it to be closely allied to strains of *Bacillus thuringiensis* that I had been studying, and I was hopeful of including their strain in our comparative study. Receiving no reply, I wrote to Vago making the same request. In a letter received from Vago, he explained that the strain was in the collection at the Pasteur Institute. In the meantime, the two French workers had apparently discussed the matter, for on 17 November I received a letter from Toumanoff in which he stated, "Our studies, with M. Vago, on *Bacillus cereus* var. *alesti* are not finished now and some publications concerning this strain are in preparation. We prefer to send you the culture when these works are completed, and also after testing it on silkworms in the spring." Early in 1954 we did receive two transfers from Paris.

Immediately an examination of the transfers for the presence of crystalline inclusions in the sporangia of cells, which had characterized the strains of *Bacillus thuringiensis*, was made. These examinations, plus plate cultures on nutrient agar, caused us to suspect that the cultures we received had more than one strain of spore-former present; three colony types were finally distinguished, one of which produced the crystalline inclusions (Steinhaus and Jerrel, 1954) similar to *Bacillus thuringiensis*. We were convinced that this particular organism was a strain of *Bacillus thuringiensis*, and if it was the one which the French workers considered to be variety *alesti*, their organism should be designated as *Bacillus thuringiensis* var. *alesti*. I wrote Toumanoff to find out which of the three organisms they considered to be *alesti*, and received no reply. I wrote again, and still no reply. By March the paper was scheduled to go to press, so in the absence of any response from Paris, I assumed that the isolate that we made from the French culture possessing the crystals was the one Vago and Toumanoff had referred to as *alesti*. We apparently unintentionally left the impression in our paper that the French workers had been using impure cultures. When I learned that this was Toumanoff's feeling, I immediately offered to publish an explanatory note in any journal of his and Vago's choice removing all implications regarding the mixed culture we had received from Toumanoff's laboratory. Both Toumanoff and Vago emphatically assured me that they did not desire such a statement, and their comments convinced me that the matter had now indeed been settled amicably. I mention this instance only because, to my knowledge, this is the only time in any publication that I have caused a fellow scientist personal displeasure (except until this present tome appears!).

Toumanoff arranged for a guided tour of Pasteur's living quarters and original laboratory rooms in the institute building, as well as a visit to Pasteur's tomb on the ground floor of the same building. Both experiences were deeply impressive, and probably do as much as anything can to bring substance to one's vision of Pasteur and his accomplishments. What one sees has been described in booklets issued by the institute. In the tomb there is one wall with a beautiful mosaic portraying the work Pasteur did on the diseases of the silkworm. Which reminds me there is a statue in a public square in Alès, and also one side of a monument to Pasteur in Paris, depicting the debt owed to Pasteur for saving the silk industry in France.

Since it is not the purpose of this account to be giving a history of European insect pathology, I shall not deal further with the many contributions of Constantin Toumanoff to this subject, as well as to apiculture, medical entomology, and tropical medicine; summaries of these may be found in obituary accounts, such as that on page 149 of volume 11 of the *Journal of Invertebrate Pathology*, 1968. It was a sad day when I received from Mme François of Professor Toumanoff's laboratory a brief note saying that he had died at his home during the night 26–27 April 1967. With approximately 235 published contributions to science, this friendly, emotional, and talented man had served invertebrate pathology, as well as other fields, well and long. In addition to getting to know Toumanoff personally, I had enjoyed intermittent correspondence with him since 1946. In one of his letters, in April 1947, he had asked my help in finding him a position in the United States; it was during a period of great scientific depression and discouragement in parts of Europe. In this same letter he mentioned that since the passing of Serge Metalnikov and André Paillot, there appeared to be no interest in France in the diseases of insects, but that he himself was still interested in the subject and was wondering "if the Pasteur Institute of Paris will organize something in this way." Another letter regarding Metalnikov seems worthy of quotation:

> Professor Serge Metalnikov was born in Simbirsk [Ulyanovsk], USSR. He obtained his secondary school and high school training

in Saint Petersburg [Leningrad]. He was the assistant and good friend of Professor A. O. Kowalewsky in the University of St. Petersburg, and after that in the Academy of Science Laboratory when Prof. Kowalewsky became the President of the Russian Academy of Science.

In 1895 Metalnikov was sent to Heidelberg, where he studied under the direction of Professor Butschli on the physiology of Infusoria. He spent 1897 in Naples, in Prof. Dohrn's laboratory, studying phagocytosis in marine animals.

He came, for the first time, to the Pasteur Institute in Paris early in 1898; after that he came back twice (1900 and 1908) to work on immunity and phagocytosis under the direction of Professor Metschnikoff. In Russia, Metalnikov had been the Director of the Lesschaft Institute in St. Petersburg until the Revolution. He left Russia definitely in 1919.

Born April 23, 1870, he died September 15, 1945. During his career, Metalnikov produced many scientists, and his goodness for everyone who worked with him cannot be forgotten.

Toumanoff himself was one of those guided and inspired by Metalnikov. He learned much concerning immunity in insects from the latter, with whom he published several papers. They also worked together some on entomogenous fungi. The last manuscript written (in 1967) by Constantin Toumanoff involved these fungi.

IV

During my 1954 excursion to Europe, the renaissance of insect pathology in France was further revealed when, with Mauro Martignoni still the staunch guide, interpreter, and guardian of protocol, we visited the Laboratoire (later Station de Recherches) de Biocenotique et de Lutte Biologique at the Station Centrale de Zoologie Agricole of the Institut National de la Recherche Agronomique on the outskirts of Versailles. Here Pierre Grison was the Maître de Recherche and a strong proponent of microbial control as well as other forms of biological control. Grison's responsibilities were largely those of administration and organization. Most of the actual experimental work on the use of microbial insect pathogens was being conducted by E. Biliotti, who joined our all-too-brief conference.

Although our time here had to be short, there was a charisma about Grison that helped us to become good friends. He displayed the typical and friendly attributes by which most Americans characterize most Frenchmen. He was talkative (including arms and hands), outgoing, extroversive, yet courteous and kind. He bubbled with enthusiasm. Most surprising, however, was the fact that at that time, he was the first then non-English-speaking Frenchman I had met who spoke clearly, slowly, and simply enough for me to understand. Our visit coincided with a time when Grison was excited over architectural drawings he had on his desk—drawings of a new institute the French government was to build for him and his colleagues at La Minière, about four miles from Versailles. With great enthusiasm he described for me how his new institute would consist of three main sections: one for the identification of insect parasites, predators, and pathogens; one for specific experimental infection and parasitism; and one for general research on insect pathogens, entomophagous insects, and biological control. Seven years later, in 1962, I was to visit France again, and it was with great admiration that I witnessed the realization of what, on my first visit, had been great expectations.

During my 1954 visit, it was clear that Grison's approach to microbial control, as to biological control generally, was essentially an ecological one. Or, I should say, one based on what he and his associates believed to be sound ecological principles. Inasmuch as this approach agreed with my own ideas on the matter, I was naturally gratified and heartily assured him of my concurrence. I had become convinced that before true and reliable success from microbial-control methods could become a reality we needed to know a great deal more about the epizootiology of insect disease, as well as about the nature of the various pathogens and host reactions. Also clear, in 1954, was the fact that both rivalry and cooperation existed between the several centers involved with insect pathology. For example, Grison's group depended upon the Pasteur Institute in Paris for its biostatistics and the mass production of microbial pathogens. It was understood that, at least for the time being, most of the basic research relating to the microbial control of pest insects would be done by Vago in Alès and by Toumanoff at the Pasteur Institute. Most of the actual field trials were under Grison's jurisdiction and performed by Biliotti.

By the time I visited Grison's station in 1962, his staff had been augmented by several researchers concerned with the diseases of pest insects and with the nature and properties of the causative agents. Among those it was my privilege to meet then were B. Hurpin, who was working on rickettsial and other diseases of the cockchafer and who later became director of research, and A. Burgerjon, who at the time was concerned with methods of standardizing microbial-control preparations. The leadership of Grison and the work of his staff have had a very strong impact on insect pathology, not only in France but internationally as well.

Grison and I next met at the Scientific Group meeting in the fall of 1962 called at Geneva by the World Health Organization. (This meeting is briefly described in chapter 5.) Throughout most of the discussion during the meeting, which was concerned with methods of controlling medically important insects (especially vectors) without the use of chemical insecticides, we found ourselves in essential agreement. On this occasion I was very much aware of his political savoir faire, as well as his direct, but diplomatic, approach to the problems being discussed. Indeed, this talent appeared to be well-developed in all the Europeans (e.g., Franz, Weiser, and Rubtzov) present, and I envied them their cryptic skill in this art.

In 1967, while going through the United States on a trip to and from Asia and South America, Grison was kind enough to divert his course so as to come by Irvine. This turned out to be a very important meeting because talking directly, he could explain much of the behind-the-scenes politics and balances of power involved in European science, especially as these affected invertebrate pathology and biological control. It was important for me to have a better understanding of these matters at that particular time because we were in the process of establishing the Society of Invertebrate Pathology, and we were hoping that in spite of the apparent reluctance of many Western Europeans, they would nonetheless help us establish the society on an international basis.

Back to 1954. After finishing our visit with Grison, Martignoni and I returned to Paris. On the way back we stopped and had a short visit with Harry L. Parker, whose field station at Malmaison was well known for its work in collecting parasites and predators in Europe for the U.S. Department of Agriculture. The few diseased insects (mostly codling moth larvae at that particular time) he did find were being routinely sent to the insect pathologists in Beltsville, Maryland. In Paris we had time for a few minutes with Lucie Arvy at the Sorbonne. In the course of other work on insects, she had observed in the blood cells of the tent caterpillar *Malacosoma neustria* what she considered to be the refringent polymorphic inclusions that Paillot (1924) had reported as characterizing an unusual type of virus disease in larvae of the cabbage butterfly of Europe, *Pieris brassicae*. Because no one else had, nor to this day has, confirmed Paillot's findings, I was anxious to see any slides or other material she might have. Unfortunately, the only material she had possessed had been sent to Kenneth Smith in England for examination. He had returned to her electron micrographs of the small particulate "virus," but the photographs were quite unconvincing as representing portrayals of virus particles. Apparently, Smith must have thought so too; he refers neither to them nor to Arvy's work in his 1967 treatise *Insect Virology*. Miss Arvy promised to send me some infected larvae the next year when the insect was again abundant; although she was sure that diseased specimens would be available, alas, the next spring no diseased specimens were to be found.

3. SWISS INSECT PATHOLOGY

I

Zurich, Switzerland, was about a halfway stop during the 1954 trip. Mauro Martignoni was holding a full-time position at the Schwizerische Anstalt für das Forstliche Versuchswesen (Swiss Forest Research Institute). At the same time, he was pursuing his graduate studies in the nearby Entomologisches

Institut at the Eidgenössische Technische Hochschule (ETH) in Zurich under the direction of Professor P. Bovey.

We arrived in Zurich from Germany on 31 August and were immediately caught up in the hospitality of the Martignonis. The days with Mauro were spent most profitably. We made a field trip to Otelfingen, where Mauro knew of a population of scale insects (*Lecanium corni*) on young walnuts heavily infected with a fungus which, upon my return to Berkeley, was identified as *Cordyceps clavulata* (immature stage: *Hymenostilbe lecaniicola*). Mauro introduced me to a number of his scientific associates. Included among these was H. P. Wille, who was working on certain diseases of *Melolontha* and was one of a group (others were Bovey, Wikén, and Martignoni) concerned with the possible use of microorganisms in the control of this destructive cockchafer. Wille had found an interesting spore-forming bacillus pathogenic for the insect, but his chief, T. Wikén, the professor of bacteriology at the Hochschule, refused to give me a transfer of their culture "until the work is finished." I was particularly impressed to meet K. Mühlethaler, the well-known head of the Institute of Electron Microscopy. It was the first, and at that time the only, entire department devoted to this now commonplace technique as far as I know.

Mauro was doing masterful research for his doctorate on a granulosis, and its epizootiology, of the larch budworm, *Eucosma griseana*, which abounded in the Engadine in southeast Switzerland. The damage (represented by reddish brown areas in otherwise green forests) caused by this tortricid could be seen dramatically from the pass as we drove over the ridge going down into the beautiful Engadine valley to the Larchenwickler-Station in the village of Zuos. The intensity and scope of the current outbreak of the budworm appeared to have been restricted by the outbreak of a disease. In 1953, Mauro had discovered that the dying larvae were being killed by a granulosis virus. His research on this problem was the basis for his doctorate. While his published thesis (Martignoni, 1957) is an outstanding publication, it does not do justice to the skill and precision he brought to the problem.

As an undergraduate, Mauro took numerous courses in microbiology, plant and vertebrate pathology, as well as in other sciences. Soon after graduating he became a research and teaching assistant of the well-known and well-loved Otto Schneider-Orelli, then head of the Department of Entomology of the Swiss Federal Institute of Technology (see *Journal of Invertebrate Pathology*, *8*, 135–6), who apparently was responsible for initiating Martignoni's interest in insect microbiology and pathology. When a scholarship opportunity came for Mauro to come to Berkeley to study, Schneider-Orelli generously encouraged it even though he did not have similar interests. When Mauro had to return to Switzerland, he worked for the Food and Agricultural Organization (FAO) of the United Nations for a while, then decided to finish his graduate work under P. Bovey.

Mauro had been extremely helpful in Switzerland, Italy, and France, acting both as an interpreter and one who explained European attitudes. This being my first European trip I am afraid that my informality and naïveté were a constant source of embarrassment. When we were in Switzerland, I had unintentionally kept one hand in a coat pocket as I talked to Mauro's associates. This was entirely too informal a posture to take when one was meeting the members of a faculty of a European university. Also, though I had managed to get through Germany with my poor German—and even thought that I was doing fairly well—a male secretary at Mauro's institute told me that, although he did not know English, he could understand my English better than he could my German!

Professor Bovey and his colleagues were, understandably, none too happy about our "pirating" this promising scientist. After Mauro had joined us, at a congress later, Bovey told me that they expected, and were still expecting, Mauro to return to Zurich. Fortunately, their hopes were in vain.

By 1959, Bovey apparently had accepted the fact that Mauro was not coming back. Accordingly, he engaged the services of another young and capable man, Georg Benz, who had received his doctorate under the well-known geneticist, E. Hadorn. Before taking up his duties as an insect pathologist in Bovey's institute, it was arranged for Benz to spend a year at different laboratories in North America to familiarize himself with methods of research and instruction in insect pathology on this continent. His two principal sojourns were at the Insect Pathology Laboratory at Sault Sainte Marie, Ontario, and at our laboratory in Berkeley. In Canada, George learned methods of studying the diseases of forest insects; in Berkeley he did some postdoctorate research and took our course in general insect pathology. He was in Berkeley from January through June 1960. Upon the completion of his time in North America, he returned to Zurich, and has been leading the institute's work in insect pathology ever since. By his penetrating investigations, he has well repaid insect pathology in North America for whatever help or insights it may have given him.

In mentioning Benz's mentor, Hadorn, I must acknowledge the great contributions the Swiss scientist and his students and associates have made, not only to genetics generally, but to insect pathology by virtue of his observations on the genetic diseases and abnormalities in insects. As with so many Western geneticists, most of Hadorn's work with insects was accomplished using *Drosophila*; however, some Europeans and Asians have reported teratologies and physiological abnormalities in other insects such as the silkworm and in different species of beetles. So far as Europe is concerned, notable recorders of teratologies (nongenetic as well as genetic) of insects of different orders were W. Bateson, S. A. Arendsen Hein, H. Singh-Pruthi, P. Cappe de Baillon, and J. Balazuc. The writings of these biologists did much to inspire my own brief concern with certain teratologies in *Tenebrio*.

While in Switzerland, I had hoped to call on the well-known honey-bee pathologist W. Fyg. Martignoni and I drove down to the Liebefeld station (near Bern), at which Fyg worked, but alas he was away on vacation. There is no question but that Fyg's contributions influenced man's knowledge of the diseases of the honey bee all over the world. On a visit to another experiment station in Wäldenswil, near Zurich, I enjoyed meeting the very competent authority on insect dispersal and migration Dr. Fritz Schneider (a son of Professor O. Schneider-Orelli).

In discussing Swiss insect pathologists, one should not fail to mention Ruth Lotmar, one of the first observers of a cytoplasmic polyhedrosis. In 1951, I wrote Lotmar concerning her then startling 1941 paper in which she had reported the presence of polyhedra in the cytoplasm of the midgut cells of the webbing clothes moth, *Tineola bisselliella.* (The only previous report of a cytoplasmic polyhedrosis was that by Ishimori [1934] in the silkworm.) Miss Lotmar very generously sent me microscope slides of a nuclear polyhedrosis affecting primarily the adipose tissue and hypodermal cells, and of the cytoplasmic polyhedrosis with polyhedra confined to the cytoplasm of the epithelial cells of the midgut. At the time, we had been finding polyhedralike bodies in the cytoplasm of the midgut cells of the variegated cutworm (*Peridroma*) but could find no virus particles within them before running out of infected larvae. I had, therefore, hoped we could obtain infected larvae of *Tineola* from Switzerland to repeat Lotmar's work and in addition demonstrate the virus particles. Of course, now that so much more is known about cytoplasmic polyhedroses, perhaps it is difficult to understand the eagerness I had then regarding the virtually unknown type of virus disease.

Going south we stopped in Lausanne long enough to inspect the microbial culture collection (Centre de Collections de Types Microbiens de l'Institut d'Hygiène) maintained by P. Hauduroy, who served as director. Even then I had a growing concern about the preservation of type and holotype cultures of entomogenous microorganisms. Inasmuch as Hauduroy had requested, and we had sent, all strains of entomogenous bacteria, fungi, and yeasts that we had in our laboratory, I was interested in learning to what extent Hauduroy might be willing to make a special effort to obtain entomogenous species from all parts of the world. Unfortunately, Hauduroy was vacationing in Spain, but an assistant of his talked briefly with me. However, I soon found it expedient to leave because upon learning I was an American she began berating me in stinging terms because the American Type Culture Collection charged for the transfers it distributed, whereas their collection—and, she implied, all others—distributed transfers free of charge. I attempted to explain that I had no connection with the American Type Culture Collection, but I was sure that since it was not a subsidized operation, it had to charge for its services in order to maintain itself. In any case, it was reassuring to know that they were maintaining at least some of the strains we had sent them, but even more gratifying to learn somewhat later that, so far as entomogenous bacteria were concerned, as complete a collection as possible was being maintained by Oleg Lysenko in the Laboratory of Insect Pathology headed by Jaroslav Weiser in Prague.

4. ITALIAN INSECT PATHOLOGY

I

When we left Switzerland, our ultimate destination was Naples, where as I explained earlier, we visited with Paul Buchner on the island of Ischia. Enroute we stopped in Lodi to pay homage to Agostino Bassi, the inadequately recognized and appreciated pioneer of insect pathology and the "founder of the doctrine of pathogenic microbes" (see chapter 1). We had to hurry on for an appointment with Professor Enrico Masera in Padova. Mauro and his wife were conveniently vacationing at the same time in Padova, so that he accompanied me on my visits with Masera.

Masera is a most remarkable man in several respects, and at that time was, through his work and numerous publications on the diseases of the silkworm and the yellow mealworm, undoubtedly the leading insect pathologist in Italy. At the time of my visit with him at the Sericultural Experiment Station in Padova, he had but recently returned—after having resigned in protest over certain policies of a former director—to take charge of all of the station's research on the diseases of the silkworm. I found Masera, then in his early fifties to be a large, friendly, man, anxious in every way possible to accommodate and please his guests. Years later, in 1968, it happened that this Italian scientist was unanimously elected the first Honorary Member of the Society for Invertebrate Pathology. In proposing him for nomination, Martignoni circulated a brief statement concerning Masera; because of its succinctness and accuracy, I take the liberty here of quoting:

> Enrico Masera was born on December 2, 1902, in Fondo, in the Province of Trento, in northern Italy. He attended Trento's Technical Institute (a college preparatory high school), where he graduated in physics-mathematics, and then studied biology at the University of Padova. In 1928 the University of Padova awarded him his doctoral degree in the natural sciences, with the highest possible grades (110 out of 110 points). . . . [After some teaching and postdoctoral research] in 1933 he entered and won the competition for a staff research scientist position at the Sericultural Experiment Station (Stazione Bacologica Sperimentale "E.Verson") in Padova, where, except for the war years, he remained throughout his career. . . . As Professor at the University of Padova, he taught sericulture and apiculture. Having reached the mandatory age for retirement, on the last day of 1967 he retired as Director of the Experiment Station.
>
> About half of his over 200 papers are scientific publications covering the following fields of research: pathology of the silkworm and of other insects, with special emphasis on mycoses, pebrine and flacheries, and silkworm physiology, with emphasis on the study of the dorsal vessel, digestive system and cocoon formation. He also investigated the economic aspects of sericulture as well as the history of the evolution of sericulture in Italy and throughout the world. Among the major contributions of Professor Masera is the thorough study of the microbial flora of the intestinal tract of the silkworm published in 1954 ("Sul contenuto microbico intestinale del baco da seta e sull'etiologia della flaccidezza," Agric. Venezie *8*, 714–735). On the basis of Masera's careful analysis (the result of three years' work), the flacherie complex can now be subdivided into a number of distinct morbid entities with characteristic syndromes. Professor Masera is also the author of what still is the only Italian treatise in insect pathology (*Le malattie infecttive degli insetti*; Licinio Cappelli, Publisher; Bologna, 1936). This treatise contributed significantly to the development of insect pathology, particularly in Europe. Furthermore, this book contains the most complete, monumental bibliography of insect pathology up to 1936.

My first contact with Masera was in the summer of 1947 when I wrote to him inquiring as to how our laboratory might obtain a copy of the treatise mentioned in Martignoni's statement, as well as any available reprints of his scientific papers. Then, and subsequently, he generously sent copies of most of his papers pertaining to diseases in the silkworm or in other insects. With Masera's work, as with much of the Japanese work on silkworm diseases, the published accounts are filled with much valuable data and reveal basic principles; yet rarely are they recognized by other insect pathologists, most of whom have not appreciated the "guinea pig" role the silkworm has played in elucidating many of the principles of disease as it occurs in insects, and in other forms of life.

During my visit with him in 1954, we continued verbally a most amiable discussion that had started by correspondence: the significance of infection caused by the bacterium *Serratia marcescens* and the true identity of Pasteur's "vibrion à noyau," which most authors designated by the name *Bacillus bombycis.* (I have discussed both of these subjects in detail elsewhere; the relevancy of the speculations regarding *Bacillus bombycis* has become somewhat diminished since the now generally believed notion that the flacherie studied by Pasteur was, in fact, caused by *Bacillus thuringiensis* var. *alesti.* The only disquieting aspects of this conclusion are that Pasteur's illustrations of the sporulated bacillus he considered to be the cause of the disease did not show the characteristic toxic crystals associated with *B. thuringiensis* and Paillot's evidence that *B. bombycis* is a secondary invader to a virus has not been entirely disproved, especially in the light of the Japanese findings of a virus they consider to be the sole cause of flacherie or a form of flacherie.) In any case, Masera's and my conversation and correspondence on these and other matters were interesting and intellectually titillating; and in retrospect it is fascinating how frequently the French and Italian silkworm pathologists disagreed in their interpretations and conclusions of essentially the same phenomena. At this writing I still am unable to make a clear synthesis of the conflicting evidence and conclusions, but I have the feeling that the present-day sophistication of some of the Japanese work on the diseases of the silkworm will aid in bringing the best from the knowledge compiled by both the French and the Italians. In the meantime, insect pathology generally would be well served by a treatise comprehensively extracting, reviewing, and synthesizing the great storehouse of information and thought existing in the world's literature pertaining to the diseases of the silkworm.

Masera proudly showed us the restoration and remodeling taking place at the station that would provide him with adequate and modern facilities. And he was also very proud of the fact that his station was the second one in all of Europe to have been established for the scientific study of sericulture including the diseases of the silkworm. The first such institute was that established at Gorizia, near Trieste, in 1869. Masera made a point of the fact that the French laboratories came later, just as the officials at Lodi emphasized that Bassi's discoveries preceded Pasteur's, even though no one had ever given Bassi the credit he deserved. Later in the day, Masera took Mauro and me to a meeting of the National Italian Zoological Society taking place that day at Padova University. Pierantoni, of symbiosis fame, was to present a paper. Alas, for reasons no one knew, he failed to appear. Nevertheless, it was a delightful way in which to end a busy and informative day. But best of all I had met and visited with Enrico Masera.

Driving south through Italy, we stopped at the University of Portici near Rome. Even though I had heard of no insect pathology work being done there, I was hopeful that some might be going on. Professor F. Silvestri of the Laboratory of Zoology was famed for his work on entomophagous insects, and it seemed possible that his laboratory might be interested in this phase of control. However, I learned from G. Russo that he knew of no work in insect pathology being done at the university or in any place in that part of Italy, nor was any contemplated. This was disappointing information.

Some days later, in Florence, I had a pleasant visit with G. Florenzano at the university of that city. Florenzano (1949) had written a review paper on bacteria in insects, and I was anxious to learn whether this paper was a harbinger of what we might expect from him in the future. Unfortunately, he wrote the paper because of his general interest in the subject, not because he had done or planned to do any actual research in the field. This, of course, I was sorry to hear, but it gave me an opportunity to learn of some of the problems facing academicians in Italy at that time. Florenzano exemplified the situation: Most university teachers received little salary, and virtually no funds for research, and had to work at other enterprises to earn enough to support themselves and their families. Few could afford to work solely at the university. In spite of this Florenzano was of a cheerful and optimistic disposition. Unfortunately, there is still little insect-pathology work being done in Italy. As of 1967, there were only five or six persons working in the field.

It should be mentioned that across the Adriatic, in Yugoslavia, L. A. Vasiljević and others were engaged in the study of the diseases, primarily viral and bacterial, of insects. However, it cannot be said that this activity had much direct impact or influence on the development of insect pathology in North America. I have had the pleasure of meeting Vasiljević at international meetings, and found him friendly and enthusiastic. At these occasions, it was a pleasure to recall that the University of Zagreb was one of the collaborating institutions involved in the American-supported International Corn Borer Investigations (chapter 5) and that B. Hergula and V. Vouk included their observations on entomogenous fungi among their contributions to the *Scientific Reports* of the investigations. In addition to Vasiljević other present Yugoslav insect pathologists include Ž. Kovačević, L. Schmidt, K. Sidor, and I. Tomasec.

5. INSECT PATHOLOGY IN THE NETHERLANDS

I

The development of insect pathology in northern Europe, particularly in the Low Countries and England, has been intriguing to watch since the early 1950s. And, although quantitatively not as significant to the history of insect pathology in North America as some other countries, they have had a "twist" and vigor peculiarly their own, this is especially true of the British insect pathology.

While driving about Europe in 1954 attempting to obtain an overall view of the status of insect pathology, I made it a point to drive to Delft, Netherlands. Not only did I wish to talk with the yeast experts at Centraalbureau voor Schimmelcultures of the Laboratory of Microbiology there, but what person, trained in microbiology, could fail to visit the birthplace of the great Dutch microscopist and discoverer of bacteria, Anton van Leeuwenhoek! His biographer, Dobell (1932) says, "he was the first bacteriologist and the first protozoologist." Strolling about Delft, we tried to locate the home of Leeuwenhoek, thinking that surely the home of a man so famous would have been preserved. After some inquiries we were directed to the spot, supposedly marked with an appropriate plaque, where his home *had* been. When we arrived at the spot, the plaque was nowhere to be seen. Then, as I examined the area of the high fence where the plaque was supposed to be, I discovered that it was there alright—but covered by a traffic sign.

The indignity accorded the plaque commemorating Leeuwenhoek provided me an unexpectedly interested welcome the next day by A. J. Kluyver, bacterial physiologist and director of the Laboratory of Microbiology. My comment regarding the covering of the Leeuwenhoek plaque was made purely in passing, but this information very much annoyed Dr. Kluyver. It seems that he had had something to do with the installation of the plaque in the first place; he wasted no time in contacting the city officials to see that matters were put aright!

Kluyver, a very busy man, was most gracious and generous with his time. After explaining to him my current interest in yeasts associated with insects, including *Pullularia* (or *Aureobasidium*, the "black yeasts") that I had been finding associated with certain scale insects, he arranged for me to confer with J. Lodder and W. Slooff. Dr. Lodder kindly came over from a nearby yeast factory where she was employed, and Miss Slooff served under Kluyver as curator of the Centraalbureau voor Schimmelcultures. What impressed me was that, as with the yeast experts in the United States, these world-famous (certainly this can be said of Lodder) authorities on yeasts knew only vaguely that insects harbored a great number of species of yeasts as intracellular symbiotes. However, as Miss Lodder patiently explained to me, until someone with certainty cultures the yeasts and yeastlike forms from insects, there was little that could be done concerning their taxonomy because so much of yeast classification is based upon the physiological properties of these microorganisms. Of course, I had already realized this, for my colleagues E. Mrak and H. Pfaff at Berkeley had expressed a similar opinion, but my hope was that these Dutch experts would accept the challenge offered by the yeasts found intracellularly and in the mycetomes of many insects. My enthusiastic tale elicited interest, but the pressure for more knowledge about industrial yeasts was such as to exclude time for such exotic species as those living mutually with and in insects.

The experience I had with the yeast people at Delft was similar to that I experienced with their counterparts, including G. A. de Vries, of the Centraalbureau voor Schimmelcultures at Baarn. However, de Vries, who was a student of *Pullularia* as well as fungi pathogenic for human beings, gave me the benefit of his perspective of the so-called black yeasts. From him I was able to purchase two cultures of what typically represented *Pullularia pullulans*, the species we had found on, and possibly in, the coccids or scale insects we had been studying (Steinhaus, 1955).

On my way to Baarn, while driving through Leiden, I stopped to have a short visit with P. H. van Thiel, whose work on trematodes in mosquitoes had interested me. Van Thiel was professor at the Laboratorium voor Parasitologie der Rijks-Universiteit te Leiden. We had a fruitful conversation, but still I found no sign of anyone in the Netherlands overtly concerned about the diseases (or mutualisms) of insects as such. But that was in 1954.

In 1959, through an exchange of correspondence, I became aware of the general interest in insect pathology, and the application of microbial control, by P. A. van der Laan, an editor and leader in nematological circles and one involved generally in economic entomology and crop protection. Van der Laan was stationed in the Laboratorium voor Toegepaste Entomologie der Universiteit van Amsterdam. Although this Dutch scientist had done some work on parasitic fungi of nematodes, he admittedly was not trained or experienced in insect pathology; but in a letter dated 24 February 1959, he wrote, " . . . some work will soon be started in this laboratory in the field of insect pathology." Preliminary tests using preparations of *Bacillus thuringiensis* to control a tent caterpillar had been completed, and liaisons were being established with insect pathologists and biological control people in France and Germany.

Van der Laan's interest and activities in insect pathology and microbial control continued to increase, although the extent of this interest and activity was not reflected in publications from his laboratory. We became the direct beneficiary of his concern when, beginning in 1962, he enabled L.P.S. van der Geest to attend the University of California. This young man's doctoral research, conducted under the general supervision of R. Craig and, until 1965, M. Martignoni, was concerned with the biochemical changes in the variegated cutworm (*Peridroma*) infected with a nuclear-polyhedrosis virus. I was gratified to see returning to van der Laan's laboratory a promising young scientist oriented toward the basic and fundamental aspects of insect pathology.

I have referred elsewhere to the International Colloquium on Insect Pathology and Microbial Control held in Wageningen, 5–10, September 1966, and to the expert editing of the *Proceedings* rendered by van der Laan. However, it should also be mentioned that he played a key role in the organization of the colloquium, and this involved the delicate handling of numerous difficult nationalistic, political, and personality problems. He was one of the early members of the Society for Invertebrate Pathology when it was formed in 1967, and as a member of the nominating committee was instrumental in getting the French insect pathologists—who had been dragging their feet—to join the society in time for one of them, C. Vago, to be elected the third president of the organization. And, as I write these words in the spring of 1968, I know of at least twelve Dutch scientists concerned with one or another aspect of invertebrate pathology; most of the insect pathologists appear to have received, in one way or another, encouragement to pursue their interests from van der Laan and others at his laboratory.

The three months (August, September, and October) that I had spent in western Europe in 1954 visiting laboratories, some only vaguely connected at that time with what one would call insect pathology, convinced me that the field was indeed a viable and a distinct discipline. It was extremely important that there be a closer liaison between all foreign scientists and those in North America if the fields of insect and invertebrate pathologies were to come into full blossom.

[This chapter is only partially finished. EAS's contacts with insect and invertebrate pathologists in Great Britain, Czechoslovakia, Poland, and the Union of Soviet Socialist Republics have certainly had their influence on North Americans, and sections on each of these countries had been planned. Japan has one of the largest contingents of insect pathologists of any country in the world. A large section was to have been devoted to Japanese insect pathology and its impact. Those countries in which insect or invertebrate pathology still does not have a strong foothold but which have sent scientists to Berkeley either as students or visitors, and, in turn have been visited by EAS, were Australia, China (Taiwan), Egypt, India, Israel, Philippines, and Thailand. A section covering these countries was also planned. MCS.]

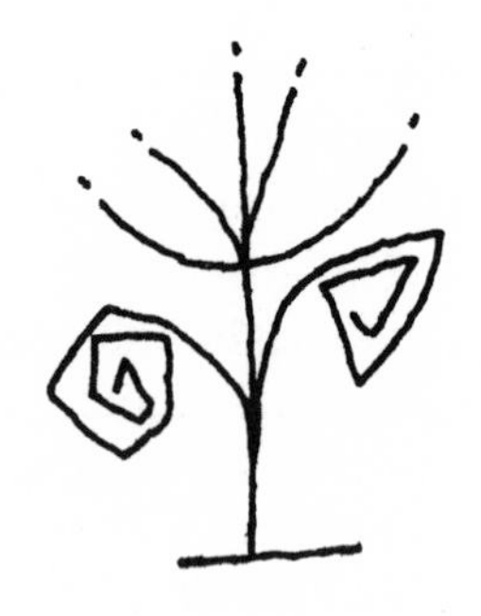

Conference [maketh] a ready man.

Francis Bacon

5

Come Now, Let Us Reason Together

1. SOME BEGINNINGS

I

There is little doubt but that small groups of apiculturists and sericulturists, in times past, assembled to discuss and to exchange information regarding the afflictions of the honey bee and the silkworm respectively. As shown by literature the early observers of disease in insects generally used the meetings of early academies and scientific societies for the purpose of announcing and describing their observations. Since in this book we are concerned primarily with the development of invertebrate pathology in North America, I shall leave to some more industrious fellow the pleasant, but very likely time-consuming, task of tracing the local or provincial meetings in other parts of the world.

In this chapter, I shall recount some of the ingredients and activities of the "curiosity cabinets," concerned, in one way or another, with the diseases of invertebrates. Curiously, however, primary emphasis will be placed on international gatherings. Purely American assemblages of insect (or invertebrate) pathologists did take place, but, to me, some of the most interesting meetings of insect pathologists were international in character.

Cursory examination of the transactions of the earliest scientific societies in the United States reveals little relating to the diseases of insects. Probably the first such society in the American colonies was that established under the leadership of Increase Mather in 1683 in Boston; it became the Boston Philosophical Society. Within a few years it died, apparently because it could not withstand the arguments between its members over politics and theology. Benjamin Franklin, in 1727, initiated a group, later called "the Junto," which met every Friday evening in Philadelphia to discuss philosophical subjects and natural history. The Junto eventually became the American Philosophical Society. Gradually, groups in the colonies tended to meet according to their specialized interest; thus a medical society was organized in Boston in 1735, and at New Brunswick in 1766. As indicated by Gardner (1965), the general effect of the early scientific societies was to encourage organized observations and experiments, to stimulate a new spirit of curiosity, to bring together those interested in science in order to discuss and report these interests, and to provide a means of publishing and circulating scientific treatises.

II

We have already noted that, setting aside the diseases of the honey bee and the silkworm, the first reported disease of an insect in North America was what we now know to be a polyhedrosis of the spring cankerworm, *Paleacrita vernata.* In 1795 Peck mentioned the disease, which he called "Deliquium," in *Massachusetts Magazine*, but the next year his account was published more prestigiously in the *Rules and Regulations of the Massachusetts Society for Promoting Agriculture.* To what extent he presented his observations verbally before amateur clubs and learned societies is not known. Inasmuch as he concerned himself with the cankerworm over a period of several years, there is a good likelihood that he did appear before interested groups.

The nineteenth century saw the formation of several societies and associations dedicated to apiculture and to sericulture.

References to the matter of disease of the honey bee and the silkworm respectively can be found in the records and transactions of these groups, e. g., those of the American Beekeeping Federation and the California Silk Culture Association. Undoubtedly, also, reports (regarding bee diseases especially) were read before meetings of local, state, and national clubs and societies of entomology and of general natural history. But it was before the twenty-second meeting of the American Association for the Advancement of Science, held at Portland, Maine, in August 1873, that the American entomologist J. L. LeConte (1874) advocated the use of infectious microorganisms to control pest insects. There have been at least fifteen entomology organizations of substantial consequence in Canada, Mexico, and the United States. Among the early papers and reports presented before these organizations were those that dealt with foulbrood in the honey bee, and the fungal diseases of the chinch bug, citrus scale insects, and mealybugs. Since World War II there has been an increasing number of papers pertaining to insect pathology and microbial control read before every national meeting, and most branch meetings, of the Entomological Society of America—presently the largest society of entomology on the continent.

In contrast with most microbiologists, entomologists were keenly interested in the subject of infectious disease in insects—primarily, one can assume, because of the potential microbial pathogens had as control agents. (To be sure, besides the apiculturists and sericulturists, there were those with opposite motives, who were interested in the prevention or suppression of disease in beneficial insects.) I well remember my own first efforts to handle the subject before groups of entomologists. First, there was the seminar Professor Harry Smith arranged for me to give at the Citrus Experiment Station in 1945. I discussed primarily the extracellular and intracellular microorganisms associated with insects, when my audience was expecting me to talk on the use of microorganisms in the biological control of insects! I attempted to correct this faux pas in June of that year when I was invited to speak on "Insect Pathology and Biological Control" before the Entomology Club of Southern California meeting in Alhambra. Then, at the 1947 national meetings of the American Association of Economic Entomology, held in New York, I found the courage to stand before a room of entomologists (not forgetting that my formal training was not in entomology) and describe the nuclear polyhedrosis of the alfalfa caterpillar and our experiments in the use of the viral agent of this disease to control harmful populations of this insect

The point here has nothing to do with me as a person reporting on the diseases of insects before meetings of entomologists—the point is that the subject matter was of sufficient interest to entomologists that they invited the subject to their programs. As far as entomological societies are concerned, this recognition of insect pathology as a valid and important branch of entomology reached one of its peaks when, in 1959 (30 November–3 December in Detroit, the site of the first national meeting of entomologists in America in 1875) the Entomological Society of America, in a joint meeting with the Entomological Society of Ontario and the Entomological Society of Canada, dedicated its annual meeting to the theme of biological control with special emphasis on insect pathology—one of the few times this major organization has chosen to follow a thematic program. Jaroslov Weiser from Czechoslovakia was the honored foreign guest speaker, appropriate special symposia were held, and insect pathology was the theme of the address I was honored to give in response to receiving the Founder's Memorial Award. The founder being honored was C. V. Riley, who, in 1875, was elected the first president of the Entomology Club, which was the "first national meeting of entomologists" and happened to assemble in Detroit that year. Of course, this coincidence had nothing to do with choosing Riley as the founder to be honored; his great contributions to entomology are well known to every American entomologist, or should be.

The Entomological Society of America, as well as other societies of entomology in North America, continues to include both basic and applied aspects of insect pathology in its programs. In 1966, Mauro E. Martignoni initiated a special insect pathology seminar that has provided a means for insect pathologists to get together for informal discussions of subjects of general interest during the society's meeting.

In May of 1967, the Society for Invertebrate Pathology was formed. Its organizational meeting was held in Seattle, Washington. But I shall have more to say of this society toward the end of this chapter. Let us now consider meetings of an international character.

2. INTERNATIONAL MEETINGS

I

Although not devoted exclusively to insect pathology and the use of microorganisms to control insects, the first international meetings that included a serious consideration of insect pathogens in its program were those sponsored by the International Live Stock Exposition and called "The International Corn Borer Investigations." The initiative for the investigations was taken by the Union Stock Yards of Chicago, Illinois, beginning in 1927. They were administered by a special committee. Three North Americans were members of this four-member administrative committee: Arthur G. Leonard of Chicago, Illinois, was chairman; C. F. Curtiss of Iowa State College at Ames, Iowa, and G. I. Christie of the Ontario College of Agriculture at Guelph, Ontario, Canada, were members, along with the fourth member, O. Ellinger of Copenhagen, Denmark.

The first actual International Corn Borer Conference took place at the Pasteur Institute in Paris, 25–27 April 1929, under the patronage of the Minister of Agriculture of France and with the institute's director, Emile Roux, serving as official host. (Two subsequent conferences were held in Budapest [1930] and Warsaw [1931].) At the Pasteur Institute, facilities were set aside and designated as a corn-borer laboratory. Similar laboratories also were set up, or already established laboratories were used, in other countries in Europe. At any rate, the initial volume of the *Scientific Reports* of the International Corn Borer Investigations listed twelve original (one added the second year) collaborating institutions—all in Europe (Denmark,

France, Germany, Hungary, Yugoslavia, Poland, Roumania, Sweden, and USSR)—with about seven out of some thirty individuals being concerned directly with the diseases and insect parasites of the corn borer.

The history and nature of the International Corn Borer Investigations and their conferences are cryptically interesting. Undoubtedly one of the reasons it is difficult to piece together the story is that all of the records of the International Live Stock Exposition were destroyed in the Chicago stock yard fire of 1934. No special laboratory, associated with these investigations, was established in North America even though the financial support (about $100 thousand) was coming from sources in the United States. Apparently the Canadian and United States departments of agriculture felt they were already doing all they could to combat the insect introduced into North America, through Canada, about 1910. (Some accounts have it that the borer entered this country early in the second decade of this century in broomcorn imported "probably from Italy or Hungary" for use in broom factories in Medford, Massachusetts.) Financial backing for the endeavor came from "leaders of industrial and financial corporations in Chicago."[1] Concerning this arrangement, Miklés Siegescu of Hungary, speaking at the first conference in Paris, made this comment: "In the case of the International Corn Borer Investigations, we witness a strange but fortunate procedure. Americans have organized, in Europe, a chain of laboratories which are achieving remarkable results not only from the farmer's but also from the scholar's point of view. The paramount significance of these laboratories lies in the bringing about of cooperation among the nations of Europe, torn asunder by the Great War, and in serving thereby the cause of scientific truth and the liberty of research."

Tage Ellinger, director of research for the investigations and editor of the *Scientific Reports* (volumes 1, 2, 3, 4; 1928, 1929, 1930, and 1931), explained further in the preface of volume 1, "In view of the vital importance of the Corn Borer situation to the agricultural and business interests in the United States, and because of the international character of the problem, the investigations were initiated, on the principal that the best scientific minds in each nation should be enlisted in an effort to discover and develop methods which would eliminate the Corn Borer menace."

Judging from the *Scientific Reports*, the participants did not lack enthusiasm regarding the use of microorganisms to control the corn borer, at least initially. Indeed, Ellinger's preface to volume 1 contained the statement "Much progress has been made [during the past year] in the study of the infectious diseases of the Corn Borer. . . . Spraying with bacterial cultures promises to become an effective method of eradicating the Borer." By the time the second volume was published he was able to testify that "striking practical results have been obtained from the use of bacteria to destroy Corn Borers in the field." He was referring here especially to the results obtained by Husz (1928, 1929) and by Metalnikov and Chorine (1929) with *Bacillus thuringiensis*. The possible effectiveness of this bacterium against the corn borer was confirmed in subsequent volumes of the reports (Husz, 1930, 1931; Chorine, 1930) and by other investigators. In the fourth, and last, volume of the *Scientific Reports* (and in a similar review in another journal [Vouk, 1932]), Vale Vouk of Yugoslavia made a summarizing statement for the investigations. In his statement he spoke glowingly of the results obtained with certain bacteria and with the fungus *Metarrhizium anisopliae*. "The scientific problem has been completely solved. The practical development of the method is now merely a question of technical and economic character."

Among those attending the first International Corn Borer Conference in Paris in 1929 was John Clay, a leading American livestock business man. He reserved his remarks for publication in *Live Stock Markets*, a tabloid published in Chicago by John Clay & Company. In the issue dated 6 June 1929, we find Mr. Clay's rather euphoric description of the objectives of the investigations. Realizing that Clay was seventy-eight at the time he wrote his speech, we find his idealistic appreciation of scientists refreshing, even though having a bit too much ardent zeal.

After eulogizing the great Pasteur and complimenting the

hosting Pasteur Institute, he spoke glowingly of Tage Ellinger's enthusiastic organization of the project. He referred to Professor S. Metalnikov, whose concern was primarily with entomopathogenic bacteria, as "a Russian, who before the Revolution was a rich man, devoting his life to science in a Russian University, refusing to accept remuneration. During the Revolution he left the country and joined the Pasteur staff. . . . " Other delegates at this first conference who were also scientists interested in microbial pathogens of the corn borer were Paillot, Toumanoff, and Chorine of France, Prell of Germany, Husz of Hungary, and Vouk of Yugoslavia. These were joined later by certain of their associates as well as by colleagues from other European countries.

Included in Clay's comments are these: "Among the Americans present were Mr. George Ranney and Mr. Cowan, of the International Harvester Company, who generously contributed to this Corn Borer Fund. . . . Corn raisers unfortunately are more or less careless, the renters especially. . . . We have to check [the corn borer] by destroying the corn stalks with fire and machinery, for the little devil secrets itself in the corn stalk during the winter months. There are also tiny wasps, bacteria and other microbes which kill large numbers and which are now being propagated by the scientists, but it takes time. . . . As a farmer, as a live stock producer, as a handler of vast numbers of live stock at eleven points in the States, as a banker loaning money against stock—mostly in the feedlot—and interested in the general prosperity of the country, I look upon the corn-borer as a great menace. . . . What we want is a parasite to counteract [the corn borer's] work."

One of America's earliest workers in the fight against the European corn borer was F. F. Dicke, then with the U.S. Department of Agriculture. Mr. Dicke has related to me some interesting aspects of the apparent lack of cooperation between the International Corn Borer Investigations and the state and federal efforts to control the insect. At the time of Dicke's first connections with the government's European corn borer project, from 1927 to 1930 at Monroe, Michigan, it was staffed with people who had worked on the gypsy moth project in New England, and apparently were not familiar with midwestern agriculture. At any rate, organizational politics began to set in throughout the corn-borer areas, and some believe that the New-England-in-charge factor had a bearing on the lack of confidence held by midwesterners, including the livestock people, in the government's efforts at quarantine and control. Also, there appears to have been strong political pressure on the lake states for legislation to implement and enforce "clean-up" programs such as were practiced in the eastern states. Since this program would involve new equipment, entomologists found some promotion for the "clean-up" program coming from manufacturers of farm machinery. C. F. Curtiss, dean of the College of Agriculture at Iowa State College at Ames, found himself aligned with elements which pushed the "clean-up" program. He had contacts with the International Live Stock Exposition and "other interested Chicago-based concerns." Thus, in the opinion of many, strained feelings about the "clean up" campaign and dissatisfaction as to the progress being made by federal entomologists led to the organization of the International Corn Borer Investigations, strongly promoted by Tage Ellinger, who was able to interest such highly reputable institutions as the Pasteur Institute in the endeavor. There was no active cooperation between the United States Department of Agriculture and the members of the investigating committee, although on occasion Thompson and Parker of the USDA Paris staff may have been consulted.

These investigations were unique in that for the first time an insect-control problem was attacked from all angles: life cycle of the insect; strains of corn resistant to the borer; alternate host plant possibilities; and investigation of bacteria, fungi, protozoa, and chemical substances, organic and inorganic gases, as possible control agents on an international basis. Also one finds a number of statements indicating that the researchers were well aware of the need of learning the answers to the basic questions before proceeding to try to solve the problem of eradication. Vouk (1931) summarized the four years of investigating the corn borer as follows: "The Corn Borer problem had hitherto been considered a specific problem of so-called applied entomology. . . . It . . . must be treated as a biological problem in the widest sense of the word . . . all branches of biological science must collaborate harmoniously for the solution of the Corn Borer problem."

I have been unable to learn precisely what factors brought about the cessation of the coordinated activities of the International Corn Borer Investigations, which published the last volume (4) of its *Scientific Reports* in 1931, for the preface of this volume promised a fifth volume that never appeared. In the United States legal battles ensued regarding the "cleanup" program recommended, and there was considerable controversy among farmers, the government, industry, and others. One obvious reason for the cessation might be that because the Great Depression of 1929 intervened, support for the European work could no longer be maintained. However, coincidentally, populations of the corn borer spontaneously declined during the years 1928–30. The pressure for control eased off about the same time that support for the European Corn Borer Investigations had to be withdrawn. And with its withdrawal an interesting chapter in the possible employment of microbial-control methods closed without adequate data on which to base a sound judgement as to the feasibilities of such methods.

From the content of the *Scientific Reports* and the activities of the scientists involved, it would appear that investigations were administered from the viewpoint that since apparently the corn borer was a less destructive insect in its native haunts and since this appeared to be because of the presence of natural enemies in Europe and other areas where it appeared to be stabilized, the first order of business was to find these natural enemies, study them, and whenever feasible, introduce them into North America. It would seem that it was this rationale that lent so much enthusiasm to the study of the corn borer's insect parasites, predators, and microbial pathogens. Probably also, in the meantime, the U.S. Department of Agriculture had undertaken an intensive importation of insect parasites of *Pyrausta nubilalis* into this country, the American sponsors of the "chain of laboratories" in Europe felt that the research being conducted by the United States and Canada made their support of similar work in Europe superfluous. (Writing in 1952, W. G. Bradley stated that twenty-nine species of insects are known to parasitize the corn borer in the United States. Millions of individuals were originally introduced from Europe and the Orient to initiate colonization—about ten species constitute those released and best established.) Unfortunately, however, the work with microbial pathogens, initiated by the European scientists engaged in the International Corn Borer Investigations, was not continued by our governmental agencies to any significant degree. I say unfortunately because some of the work, such as that by Husz and by Metalnikov and his group, with *Bacillus thuringiensis* should have been continued in light of what we know of this bacterium today. As reflected by the literature of that time, when the American support ended, the Europeans (this is especially clear in the case of Metalnikov and his co-workers), in spite of the encouraging results they were obtaining, turned to the study of the diseases of other insects.

II

At first glance it seems incredible that following the conferences of the International Corn Borer Investigations, no special meetings of consequence dealing with the diseases of insects in general, and involving North Americans, took place until the 1950s. Upon reflection, however, it is clear that part of the explanation of this quarter-century hiatus was the relatively few scientists working in the field, the Great Depression of the 1930s, and the occurrence of World War II. To be sure, occasional papers were presented at regular meetings, mostly of entomological societies, but no special sessions or symposia involving North Americans were devoted to insect pathology until 1956. (A meeting devoted to the subject of disease in insects was held by Soviet scientists and government officials in Leningrad in November 1954.)[2] During the period of 17–25 August 1956, the Tenth International Congress of Entomology was held in Montreal, Canada. Because of the extraordinary leadership Canada has always shown in insect pathology, I have a hunch that it was precisely because the congress was under the management of the Canadians that a full day of sessions was devoted to insect pathology and microbial control.

At any rate, on the morning of 24 August, under the chairmanship of Gordon E. Bucher, a session titled "Utilization of Microorganisms in Insect Control" was held. The afternoon session was titled simply "Insect Pathology." I was invited by the president of the congress, W. R. Thompson, to chair this particular session; it turned out to be a more interesting experience than I had anticipated. Also, concurrent with the morning session there was one on apiculture, which consisted of papers concerned with the foulbroods of the honey bee; apparently the competition of interest between these two sessions was not realized by the program committee. Interestingly, although the afternoon insect pathology symposium dealt exclusively with basic aspects of insect pathology, it was nonetheless placed under the sectional heading of the application "Biological Control." Insect pathology as a distinct branch of entomology in its own right had yet to be achieved. But there should be no complaint regarding this misplacement; had insect pathology not been allowed to flourish under the protective umbrella of biological control, it would not have revived as rapidly as it did or generally have been supported as well as it was. Among those scheduled to participate at the sessions were Grison, Weiser, Bird, Baird, C. G. Thompson, Bucher, Heimpel, Angus, Briggs, Yamafuji, Krieg, Vago, Plus, and Bergold—most of whom were or became well-known names in the annals of insect pathology and microbial control.

Professor Kazuo Yamafuji, a biochemist at Kyushu University, had spent several weeks as a visitor in our laboratory prior to attending the congress. His work at the time was quite controversial—discounted in some quarters—so we were pleased to have him as a guest in the hope that he would, in face-to-face conversation, clear up for us some of the questions raised by the avalanche of papers he published. Originally, he claimed that "On the basis of many experiments I believe that the [polyhedrosis] virus molecules can be formed experimentally in the silkworm body. Indeed, using hydroxylamine or barium peroxide I have succeeded in such artificial formation." By 1956, however, he was more inclined to speak of "inducing" polyhedrosis in the silkworm than of creating the virus "artificially." He agreed to demonstrate his methods to us during his visit. Unfortunately, his attempts to reproduce his results in our laboratory in the short time he was here did not succeed; Yamafuji felt that one of the reasons might have been the fact that we were using an Hungarian strain of the silkworm rather than one of the Japanese strains with which he had worked in his own laboratory.

While on the Berkeley campus, one of Professor Yamafuji's hopes was to meet and have an opportunity to talk with Nobel laureate Wendell Stanley of tobacco mosaic virus fame. I arranged for such a meeting, which took place in Stanley's office. Stanley met Yamafuji most graciously. (Out of the corner of my eye I could see on a stand in the far corner of the room a stack of the Japanese worker's reprints, so I imagine Stanley took the wise precaution of briefing himself before the visit.) After the exchange of proper amenities, the two got down to a serious discussion of the validity of Yamafuji's claims with regard to the de novo formation of viruses. Firmly, but politely, Stanley explained that since it was Yamafuji who had made the claims of producing viruses "artificially" it was up to him to prove and establish his claims and not the responsibilities of others to confirm or disprove his contentions. (Yamafuji had made some statement to the effect that he felt that others should now repeat and confirm his work—actually this was being attempted at two or three laboratories, but without success. Most scientists believed that the results obtained by Yamafuji could be explained by the theory that the chemicals which he fed to the experimental silkworms acted as stressors, and triggered occult virus or latent infections.) The visit with Stanley ended amicably, and the Nobel laureate then invited us to join members of his staff for coffee hour. Unfortunately, while circulating around the group one of the better-known prima donnas of virology rudely declined to talk with Yamafuji because he considered the latter's work of spurious value. How lacking in human kindness and modesty are some of those one occasionally meets who are under the impression that their own work is superior in importance and at the cutting edge of scientific discovery!

Toward the end of Yamafuji's visit with us it was clear to me that a large part of his problem in projecting his concepts and points of view was his difficulty in articulating the English language. I am not sure that he was aware of this because technically he knew English well, but all of us in my laboratory had difficulty understanding him, and hesitated to embarrass him by asking him to repeat himself. That is, all but one of us. Mariece Batey, my technician, somehow was able to understand him when the rest of us could not; accordingly, she frequently acted as an interpreter for us. No doubt his language difficulty contributed to the problem that he encountered on a trip to Tijuana, Mexico (across the border from San Diego). He had no trouble going into Mexico because, unless the border guards are suspicious, they do not require passports of people passing over the border. However, because there was a problem of Orientals passing illegally into the United States from Mexico, the American customs officials on his return asked to see his identification and passport. Alas, being used to going about without his passport while in California, he had forgotten to bring it with him on his trip into Mexico. He was detained by officials overnight, and was only cleared when he was able to contact officials of the Japanese consulate in Los Angeles.

Because of the language problem, I realized that the long paper he planned to read in English before the forthcoming entomological congress in Montreal would not be understood by most of the audience. Inasmuch as I was to be the moderator of that particular session, I suggested that he circulate a mimeographed version of his paper just before his presentation, that he prepare a second document which succinctly summarized his position, and that he list his reasons for feeling that he had proved his case in one, two, three order without any flourishes or additional verbiage. If there were a dicussion of his paper, he could perhaps clarify matters by reading this boiled-down summary. Or so I thought. At any rate, he presented his paper as planned—and as I had anticipated virtually no one understood him except what was gleaned from the mimeographed copy of his talk. He then returned to his seat, whipped out his pocketed summary, and asked me "Now?" Why not, I thought, so I nodded my head and explained to the audience that Dr. Yamafuji wanted to be certain that his concepts of virus origin were completely understood and had therefore prepared a summarizing statement he would like to read. He read the statement, which sounded fine to my ears, but inasmuch as I had helped him prepare it, I am not sure whether I was understanding it from foreknowledge or whether, indeed, it in any way helped the audience. At any rate, there was little discussion of Yamafuji's controversial work, either because most of the virologists present dismissed it out-of-hand or, more likely, because they hesitated to engage in a discussion with Yamafuji when they knew they would be unable to understand his responses. Enough was said, however, to make it clear that this congress had not provided Yamafuji with the opportunity to convince insect virologists that he could indeed chemically manipulate the DNA and other chemical elements of the cells in such a manner as to have the cell actually form particles of infectious virus.

Discretion—the plague of contemporary history—prevents me from relating other interesting and now amusing aspects of this afternoon session at the congress. I cannot forget, for example, how the director of one of the leading laboratories of insect pathology came up to me, after the session, to thank me for dashing cold water on a disagreement which two of his staff members began to air during a discussion period after one of the papers was presented. Shortly thereafter one of the men involved left the hall, I imagine because he did not wish to precipitate a more violent discussion at the end of the session; however, some participants concluded that he left because he did not want to be questioned about certain aspects having to do with median lethal dose values of the data he presented in his paper. Because I knew well all the personalities involved, the scene was not so mystifying to me as it apparently was to most members of the audience.

And there were delightful light touches. Jost Franz has reminded me how the charming Nadine Plus very articulately presented her paper in French as she wrote in English on the blackboard; and in response to a question, began answering in English as she wrote in French on the board.

III

It was at the 1956 Montreal International Congress of Entomology that I first met Jaroslav Weiser. I was to enjoy the friendship of this remarkable Czechoslovakian insect pathologist and parasitologist for years to come. Perhaps even then there was brewing in his mind the feasibility of calling an international conference on insect pathology in Prague. At any rate, this history-making event was held at Prague's International Hotel from 13 to 22 August 1958 and was called The First International Conference of Insect Pathology and Biological Control. The conference was organized under the auspices of the Czechoslovak Academy of Sciences and the Slovak Academy of Sciences by Weiser, head of the academy's Laboratory of Insect Pathology (in the Institute of Biology) in Prague, and by A. Huba, Laboratory of Plant Protection of the Slovak Academy in Ivanka Pri Dunaji. This important meeting, "the first time that in this field of scientific activity workers from the West and the Socialist countries have met together in order to inform each other of their work by means of lectures and discussions," deserves extended comment.

I journeyed to Prague via Stockholm where, coincidentally, the Seventh International Congress for Microbiology was in session from 4 to 9 August. There I joined Gernot Bergold, and together we continued on to Prague via Copenhagen and East Berlin. Except for basic microbiological information, especially in virology and production of microorganisms, the Stockholm gathering offered nothing of special interest for the insect pathologist as such. Bergold and I attended a meeting of the Virus Subcommittee of the International Nomenclature Committee held during the congress. At one session on 7 August, Bergold presented his resignation as chairman of the Insect Virus Group, and I agreed to succeed him in this capacity.

However, I did so with considerable apprehension because I had been warned that the chairman of the subcommittee, C. H. Andrewes, was running the subcommittee in somewhat less than a democratic manner. While I did detect a Sam Rayburnish type of conduct on his part, I could not help but conclude that in a situation where there were many divergent opinions, disregard and lack of understanding of basic principles of nomenclature, and plain stubbornness, it took someone with imperious or dogmatic qualities to lead and make progress in the very confused area of virus nomenclature. As a member of the Insect Virus Group during Bergold's chairmanship, I participated in protesting some of the decisions that we felt were being railroaded through, but in the end we more or less had to go along with the decisions being made by Andrewes and the Animal Virus Group although, in truth, at that time the insect viruses appeared to fall into much more logical groups than did most other animal viruses. Among the points we agreed on was that of ending the name of each genus of viruses with the suffix *-virus* (thus *Borrelina* became *Borrelinavirus*); however we objected to Andrewes's suggestion that "the species concept should not be applied to viruses at all" (Andrewes and Sneath, 1958). As this is written, general agreement as to virus nomenclature still has not been attained.[3]

As we boarded the plane for Prague, I began to worry about my companion, Bergold. When he left Canada, his visa for Czechoslovakia had not come through. While in Sweden he had gone to the Czech consulate and explained his situation, but the most that they could do for him was to give him a boarding pass onto the plane headed for Prague. What would happen when we landed in Prague? Would he be allowed to enter without a visa? Gernot seemed unworried, but nevertheless I assured him that if the authorities did not allow him to pass through customs, I would refuse to pass through also. My concern was needless. Upon landing I was surprised to be met not only by Weiser, but an official delegation from the Academy of Sciences. They had not been expecting Bergold on that flight, so after being officially welcomed, I explained Bergold's situation. Immediately, the head of the delegation turned to a custom's official, and within minutes Bergold had his visa. As we passed through customs with our hosts, our baggage was not opened or checked in any way. Thus the power and prestige of the Czechoslovak Academy of Sciences were evident from the outset.

At that time automobiles were possessed primarily by members of the Communist party. Weiser, not being a member of the party, did not have a car, but he had arranged for transportation through a member of his staff, Jiřina Vaňkova, who, together with her husband, was a member of the party. Dr. Vaňkova and her husband drove us from the airport to the International Hotel, where the congress was to be held. John Briggs, of the United States, had arrived the day before, so he was on hand to welcome us. He assured me that all was well and the Czechs were going all out to make us Westerners feel welcome. After arriving at the hotel, Jary Weiser handed me an envelope filled with Czech currency. Upon my expression of puzzled amazement, he explained that these were funds provided by the academy to take care of all of my official expenses during the time I was to be in Czechoslovakia.

As an indication of the great care with which the Czechs strived to demonstrate impartiality and equal courtesy to the West and the East, they selected I. A. Rubtzov of the USSR and myself as guests of honor to represent the obvious political polarization. We were also members of the Honorary Committee, which, in addition to ourselves, consisted of Liu Chung-lo of mainland China, H. Sachtleben of the German Democratic Republic, N. A. Telenga of Ukraine, USSR, F. J. Simmonds of the British Commonwealth (then stationed in Canada), and J. Franz of the German Federal Republic. I had not been warned that I was to have such a role of honor; as a consequence I was unprepared for the several extemporaneous talks I was called upon to give during the course of the congress, the interviews by press and radio, the address and official toasts at the grand banquet, and the final address at the last session of the congress. On behalf of the Eastern nations, Dr. Rubtzov performed similarly—whether or not he had been forewarned I do not know.

Parenthetically, let me say that the official banquet of the conference was held on the evening of 11 August at the Národní Klub in Prague. It was a gala affair. Only once before—at the Russian consulate in San Francisco right after the end of World War II—had I experienced the Slavic penchant for repeated toasts—those to your health, to your country's people (never their governments), and even to the evening's menu. I found that the great future of insect pathology was a safe subject to toast. Speeches were sometimes interpreted, sometimes not—but the conviviality did not appear to mind the language barrier. I was reminded of a story told of the well-known American musician Cab Calloway, who was once cited by a Hebrew organization for his efforts in promoting interfaith co-operation. During the banquet in his honor, a joke was told in Yiddish. When Calloway joined in the ensuing laughter, the toastmaster expressed surprise. "I didn't know you understood Yiddish," he said. "I don't," replied Calloway, smiling, "but I have *confidence* in you fellows!" And so it was throughout that memorable occasion.

The conference (which was really in the nature of a congress) was divided into two main sections: insect pathology, containing four symposia (insect bacteriology, insect mycology, insect virology, and insect protozoology and helminthology) and biological control, also containing four symposia (taxonomy of entomophagous insects, evaluation of results of introductions, rise of the effect of parasitic insects, and monophagous and polyphagous insects). Microbial control was included in the insect pathology section. In addition there was a symposium devoted to international cooperation, and a session at which resolutions of the conference and closing speeches were presented. The sessions were all well organized, the sound and projection equipment worked perfectly as did the multilingual (English, German, and Russian) earphone simultaneous interpretation system, and the entire meeting was recorded on tape. The delegates sat comfortably at felt-covered tables, each amply provided with bottled mineral water. The organizers had thought of everything, including stationery and lapel pins bearing a specially drawn insignia portraying a polyhedral body for the insect pathologists and a coccinellid for those concerned with entomophagous insects.

The conference opened on the morning of 13 August with an address of welcome by academician Ivan Málek, head of the biological sciences division of the Czechoslovak Academy of Sciences, as well as one by Jaroslav Weiser in his capacity as secretary of the conference. Weiser's speech was fascinating in the careful balance it contained with regard to paying respects to the USSR on the one hand and the Western nations on the other and in acknowledging the political factions making up the various segments of European science. His theme was to accomplish through this conference "an introduction of the workers of the East with those of the West." There was more of Weiser's heart and feelings in his address than is apparent to those who may now read his remarks in the *Transactions* of the conference. In addition, he wanted so much for us to be aware of the majesty and beauty that was Czechoslovakia's. Read the final sentences of his address keeping in mind the world situation as it was in 1958:

> Many of you might have come to our country with an uncertain feeling and perhaps apprehension as to how our people will behave towards you, the conditions you will find. I am asking you to go and see with your own eyes how our people live, find out what their troubles and their pleasures are. It is a pity that only very few of you have known pre-war Czechoslovakia. This would have been a great help in appraising our achievements. Permit me to welcome you once more and to wish our conference successful work and good results.

It so happened that the symposium for which I was moderator was the first one of the conference. Since it was the opening session I was anxious that it go off well. I believe it did. It was not without its interesting, even comic, aspects, however. I shall mention only one. The fourth speaker on the agenda was Dr. C. Toumanoff of the Pasteur Institute in Paris. Just before he took the rostrum he informed me that although French was not one of the official languages and he could speak as well in Russian or English, he was obliged to give his paper in French because "it would be an insult to my country, France, if I did not." I told him that under the circumstances he should go ahead. He indicated that at the end he would give a summary in Russian after he had finished his address in French. Much to my surprise, when he had finished, he did not give a summary in Russian, but proceeded to start to repeat his entire paper in Russian. Taking this amount of time would play havoc with the scheduled program, so I passed him a note reminding him that his remarks at this time were to consist of a summary only. He ignored the note. As he continued with his Russian address, I decided that in fairness to those who were to follow, something had to be done to get him to stop. Accordingly, I walked up to his side and audibly requested him to end his remarks. He expressed his indignation, then rapidly finished. I worried lest the delegates from the USSR would be insulted because, after all, Toumanoff had been using their language. During the intermission, a group of Russian delegates came up. I feared the worst. Much to my surprise, they thanked me and expressed indignation toward Toumanoff. The fact that he re-presented his paper in Russian after giving it in French implied that the Soviet delegates did not understand French. They were, therefore, mildly insulted by the speaker and appreciative of my action. This encounter seemed to break down the wall between us and the Russians and, as a consequence, served as an entrée for subsequent scientific discussions.

All of the sessions of the conference had their share of interesting points and were well attended. Their contents may be found in the excellently executed *Transactions*, which were published in 1959 in a fine hardback edition of 653 pages, well printed and well illustrated. Especially to be prized are the photographs throughout the volume of small groups of delegates. The frontispiece portraying most of the delegates is historic in the sense of being the first showing of such a relatively large group of insect pathologists, as well as those interested in entomophagous insects, representing both the East and West.

As is the case with most international congresses, the resolutions passed by the last plenary session of the conference were well intentioned but rather platitudinous. Most of them can yet, today, be expressed as valid aspirations of insect pathology and "entomophagology." The closing speeches contained the usual expressions of thanks—but certainly they were well deserved in this case. I had been especially impressed by the interpreters in their skill in handling difficult jargon. In the closing session I thanked them not only for the way in which they translated so skillfully but also for maintaining such happy dispositions even under adverse conditions. They had a small party after the closing of the session at which they said that they had interpreted for many conferences and meetings of all types at the International Hotel, but none had ever made an effort to cooperate with them or thanked them so warmly. They very much appreciated my remark about their "happy dispositions" because they knew that for me, as for most participants at a congress, they were essentially faceless voices. I was touched by their toast to my health, for my expression of thanks was no more than they deserved. I had found pleasure in talking back and forth to them during the conference, which the speaker on the podium could do using a separate line.

During the conference, I was allowed to roam about Prague as I wished, and on several occasions I was taken to a particular place that I wanted to see. Except for a fascinating and relaxing trip to Carlsbad and Marienbad, famous Czech spas, I did not go on the postconference excursions or the tour to Brno and eastern Czechoslovakia (Slovakia). On my leaving, I was courteously swept through customs, past armed guards, and up the ramp to the plane. As I fastened the seat belt and leaned back in the seat, I knew that no matter how many or what kind of congresses involving insect pathology the future might hold, there would be none which, in my opinion, would have the historic importance of this one. It was a success in every way, and the enhanced spirit among the countries represented was most gratifying. I was only sorry that more pathologists from America had not attended. (John Briggs and I were the only representatives from the United States; Hank Thompson had intended to come, but he was unable to obtain a visa in time.)

It would be a mistake to assume that, with the rapid development of insect pathology and its association with microbial control, conferences, meetings, and colloquia were not being held on the diseases of the silkworm and the honey bee. Indeed, the diseases of these two insects constituted part of the scientific programs of international congresses of sericulture and apiculture. However, since most of these did not relate prominently to North American insect pathology, they fall outside the scope of this book, except that American pathologists were on occasion invited to participate—especially in conferences on bee diseases. For example, the International Congress of Apiculture, meeting in Madrid, designated the days of 20 and 21 September 1961 as the "International Days of Bee Pathology." This occasion was supported by the Office International des Epizooties. The preliminary announcement was sounded in clarion terms: "These days will be reserved to all seekers and scientific workers in laboratories and research centers of apiculture, in all countries, without any distinction. The reports presented must be solely reports of pathology of a *scientific interest* (those on practical pathology must be presented at the [regular session of the] Congress of Apiculture.). "Further information could be obtained from the Commission Internationale de Pathologie Apicole in Nice, France, which I mention simply to remind insect pathologists of this commission dedicated to the study of the diseases of a single insect species. Moreover, the concern regarding the diseases of the honey bee is a continuing one, being a part of every international gathering of the Commission Internationale de Pathologie Apicole. It has been a source of real regret that conflicting commitments prevented me from accepting several invitations to attend the congresses sponsored by the commission. Especially was I sorry to have to decline the kind invitation extended by Maurice Rousseau to serve as moderator for a session on the virus diseases of the honey bee at the apiculture congress held in Prague in August 1963.

IV

International meetings of a general nature, such as the Pacific Science Congresses held under the auspices of the Pacific Science Association, gave a platform of sorts to insect pathology, particularly through their sessions on microbial and biological control. The first of these I had an opportunity to attend was the Ninth Pacific Science Congress held in Bangkok, 18 November to 9 December 1957. Approximately 700 scientists from forty-three countries attended. On 19 November a group of entomologists met on the campus of Chulalongkorn University, where several papers relating to insect pathology were presented. My own presentation, titled "New Horizons in Insect Pathology," was given with some apprehension because of its rather general platitudinous nature. Surprisingly, I received an unusual number of reprint requests for the published paper, because apparently it said something—I am not sure what—to entomologists in Asia and elsewhere who had been seeking approval for the inclusion of insect pathology and microbial control in their teaching or research programs.

At another session, Paul Surany then stationed in Kenya, gave a report on his work up to that time on the diseases of the rhinoceros beetle (*Oryctes*)—the significance of which I have mentioned elsewhere in this volume. [A portion of the book never written. MCS.] Paul's report was a detailed rendition of histopathological studies made of diseased beetle larvae and adults (especially those having what he called the "blue disease" and those having "histolytic disease") collected during his extensive surveys of the South Pacific. The possibility of using pathogens to aid in the control of this insect was discussed at some length by the participants at the meeting. C. H. Hofmann read a paper by C. G. Thompson on microbial control, which generated a question and answer period. Because of Thompson's absence the questions were directed to me; their number and variety indicated an intense interest in this application of insect pathology. At the end of that particular session I made a brief review of the use of microbial pathogens in the control of insect pests. A summary of these comments appeared in a mimeographed report on the congress to the U. S. agricultural officer to the Philippines, written by J. Alex Munro, then technical consultant of the International Cooperation Administration of the Philippines. A delegate, Dr. Lew, from China (Taiwan) made a similar report to his colleagues upon returning home. Later,

I realized that unintentionally I had generated hope and enthusiasm that microbial control might be the answer to some nagging control problems in the South Pacific and the Orient. I could only hope that my admonition that practical results depended upon a solid foundation of painstaking basic research would be heeded and not taken as a mere platitude.

Although a bit of a diversion, I am inclined to relate one or two interesting sidelights of this journey to Thailand and back. Harold J. Coolidge, executive director of the Pacific Science Board, National Academy of Sciences, had arranged with the Department of Defense to have a group of about fifteen American scientists along with military personnel and dependents, a total of about seventy flown out to Bangkok on MATS. We left Travis Air Force Base, near San Francisco, enthusiastically at 8:00 A.M. on 13 November, arriving in Bangkok at 11:30 P.M. on 16 November. My companion on the entire trip was the late Robert L. Usinger, the world-famous hemipterist. The plane was of the four-engine *Constellation* type. All went well until we landed at Hickam Field in Hawaii, where we had our first indications that the plane was not all it might be. It was discovered that the hydraulic system was not working, hence we had no wheel brakes; moreover, the nosewheel could not be steered properly, so we had run off the landing strip and had to be taken to the terminal by bus. From Hawaii we were to fly to Kwajalein, but were diverted to Iwo Jima because of a typhoon active near Guam, and then flew on to the Philippines, where we spent the night at Clark Field near Manila. The next day we enplaned for Bangkok. About thirty minutes out, the cabin began to fill with what appeared to be smoke and we began to lose altitude rapidly. The propeller on our number-one engine had stopped. The general conclusion was that the plane was on fire, that our engines were failing, and that we were headed for a ditching operation. The passengers remained remarkably calm, fastening their seat belts, looking for their May West safety jackets, and preparing for a landing on the ocean. But of all this, my most poignant memory was that of Usinger's concern—as we braced for our landing—that I had my seat belt properly fastened.

In fact, he bent over from his seat beside me and tugged on the fastener to be sure that it was secure. (All the time we were making these preparations, no announcements came over the loudspeaker system to explain what was happening; the crew was probably much too busy handling the crisis to talk to the passengers.) Flying at a few hundred feet elevation, we returned to Clark Field accompanied by two planes that had been sent out to "fly us in." We landed with an ambulance and fire engines driving along side in much the same manner they had at Hickam in Hawaii. Obviously, it was time to change planes.

The emergency had not been as serious as we had thought. In layman's terms what had happened was this: Engine number one had developed a defective piston. To keep from blowing a cylinder head, the pilot quickly shut off the engine, which shut off a supercharger, before closing the air vent. Thus cold air from the outside entered the cabin and condensed into clouds of moisture that we took for smoke. Since we were also losing cabin pressure, the pilot decended to the lower levels more rapidly than would ordinarily be the case, giving us the feeling that we were headed for the brink.

After a few hours we boarded another aircraft and proceeded to Bangkok. In addition to attending the congress, Bob and I took in the sights of this then very exotic city. We quickly learned the art of bargaining with the local merchants and taxi drivers, but I must admit Usinger had greater skill than I. We were of course tremendously impressed by the beauty of the palace and its surrounding buildings and grounds, the king and his truly beautiful queen, who hosted a reception for the congress (the first time this honor had been extended to an international group meeting in Bangkok) as did the American ambassador and his staff, the indescribable early-morning boat ride on the clongs or canals of the city, the "Thieves' Market," the extraction of snake venom at the Pasteur Institute outside of Bangkok, and the numerous other entrancing aspects of this land of the Siamese. But most impressive of all were the magnificent Thai people. Not only were our congress hosts most gracious and kind to us, but the people generally were quiet, gentle, kind, smiling, and altogether charming. They are a handsome people, always well groomed and cleanly dressed—the men usually in snow-white shirts—which seemed incongruous considering the not-so-clean river and canal waters. The same was true in the rural areas and at Lophuri, about 200 kilometers north of Bangkok, where we visited ruins of ancient temples, witnessed folk dancing, water buffalo and farmers at work in the rice paddies, and religious rites. To the extent that the West has in recent years had an impact on the culture of these "unspoiled" people, I can only mourn. I am grateful for having visited this magnificent land in 1957 before it adopted so many of the ways of the West.

I must mention that during a brief visit to the Faculty of Agriculture at Kasetsart University at Bangkhen just outside of Bangkok, I was introduced to a young man who had prepared a display relating to some work he was doing with entomogenous fungi. His professor, A. Manjikul, was obviously proud of his student, and since he was the only one at the university interested in the diseases of insects, he was anxious that I should see his student's exhibit and evaluate it. It was a fine display, and I was impressed with it and with the student. Little did he or I realize then that a few years later he would be coming to Berkeley as one of my graduate students, receiving both his M.S. and Ph.D. from the University of California. I am speaking, of course, of Sothorn Prasertphon, who returned to his country in 1967 to be one of the first (along with Boonsom Meksongsee) and leading insect pathologist of mainland Southeast Asia.

Having an opportunity to hitch a ride to Tokyo, Usinger and I left Bangkok on 30 November. This was my first visit to the Land of the Rising Sun, and I found it entrancing even though I confined my activities primarily to Tokyo and visits with silkworm pathologists at nearby sericulture stations. Bob's objectives were to visit the famous entomologist (hemipterist) Professor Asaki, then very ill in Fukuoka, as well as entomologists at Nagoya and Nara who specialized in aquatic insects.

Tokyo had already made an unbelievable recovery from World

War II. Its progress was evident on every hand. The people were gracious and friendly, apparently devoid of any bitterness from the war. I remember my first impression of how out of place Americans appeared, both in size and in general appearance. Most Americans, including servicemen and their dependents, seemed to be a demanding and discourteous group. I could not help wondering what the Japanese *really* thought of our occupying forces.

It was during this visit to Tokyo that I first met Keio Aizawa, the well-known Japanese insect pathologist. He and Jyuichi Kuwana came to the hotel and took me out to the Sericultural Experiment Station, where the station's director, T. Yokoyma had arranged a staff meeting. We talked the rest of the morning and through lunch hour. Afterwards we visited all the laboratories and experimental processing rooms in the station. I was particularly honored that N. Ishimori, then retired, had made a special trip from his home for this occasion. We discussed the work he had done relating to immunity in insects, and how he had come to work with Metalnikov in Europe. I was also pleased to have the opportunity to meet Professor Hisao Aruga of the University of Tokyo. Elsewhere I have discussed an extended relationship I was to have with him and his colleagues. [This was to have been in chapter 4 but was not written. MCS.] I was extremely impressed with the great interest and ability the Japanese showed in silkworm pathology, and the obvious desire of some of them to be able to extend this interest to other insects.

We left Tokyo the morning of 7 December, stopping for refueling on Midway Island, where we made some first-hand observations of the gooney birds (albatrosses) and their fascinating behavior. I was returning home from Thailand and Japan with a new realization and appreciation of the potential for insect pathology specifically, and biological sciences in general, in Asia. It was up to America to stretch both of its arms so that we might do what we could to stimulate our branch of science in Asia as well as Europe, and hope that eventually the tide of progress in the biological sciences could engulf Africa and South America as well.

Attending the Tenth Pacific Science Congress, 21 August to 6 September 1961, held in Honolulu was a less momentous occasion from a personal standpoint, but the program in insect pathology reflected the increase in interest and sophistication the field had attained during the intervening four years. The Symposium on Insect Pathology was chaired by Y. Tanada on 23 August. Joe had the difficult task of attempting to fit nine papers plus discussions into a period of only three and one half hours allotted him. Nonetheless, he did a magnificent job with a well-diversified program. It, together with the sessions in biological control, did much to establish insect pathology and microbial control as legitimate branches of entomological activity among Pacific entomologists.

For some years preceding the congress I had become aware of the fact that when one spoke of disease in insects and other invertebrates, or when one alluded to "insect pathology," most entomologists and invertebrate zoologists—indeed, most insect pathologists—automatically thought only of diseases caused by microorganisms and that the word pathogen was synonymous with microorganism or infectious agent. Accordingly, I thought this might be a good opportunity to point out the degree to which noninfectious diseases were being neglected by insect pathologists and others and that more attention and study be given to genetic diseases, teratologies, injuries caused by poisons, diseases caused by deranged physiology and metabolism, nutrition, noninfectious neoplasms, wounds and other injuries caused by physical agents, etc. I proceeded to do so, and later published the remarks as an editorial in the March 1962 issue of the *Journal of Insect Pathology*. From the comments made at the congress and from numerous letters received regarding the editorial, my intentions—to some degree at least—achieved their goal. Since then, it seems to me, expressions of concepts of insect pathology have frequently included noninfectious diseases, more papers on noninfectious diseases have appeared in the literature, and such scientists as geneticists and taxonomists, for example, have recognized that certain mutations, abnormalities, and teratologies they see are manifestations of what properly comes under the designation "disease." Not for a moment am I suggesting that my own single paper brought about this awareness, but I know that it has been pointed to by a number of workers as justification that their work in the areas I mentioned may be considered under the subject heading of insect pathology.

[EAS had intended to include a section on the Eleventh Pacific Science Congress. MCS.]

V

One evidence of the rivalry and politics involved in European science was the subsequent refusal by Western Europeans to recognize the Prague Conference in 1958 as the *first* international conference of insect pathology and biological control. The field of biological control had already been blighted with schism in a kind of four-way manner. After a 1948 meeting of a group of entomologists held at Stockholm under the auspices of the International Union of Biological Sciences, Western European countries had organized to form the Commission Internationale de Lutte Biologique contre les Ennemis des Cultures (CILB), now L'Organisation Internationale de Lutte Biologique (OILB), but this did not include the countries of Eastern Europe. The United States, in spite of the well-intended and well-founded Resolution No. 25 of the Food Conservation Conference held at Honolulu in 1924, maintained its unorganized independence from the European organization essentially because, as I was told by American leaders in biological control at the time, if they joined with the Europeans it would be a case of the Europeans running the show but the Americans paying for it. In the meantime, the Commonwealth Institute of Biological Control went its efficient way within the British Empire, later cooperating with governments and agencies about the world. At the Prague conference it became clear that the little ecumenical spirit prevailing among the several biological control groups was not of sufficient strength to bring them together into any world, or truly international, organization. Moreover, and to me this was most discouraging, the same splintering appeared to be beginning in insect pathology. During my efforts to investigate the matter, I met with resistance along the same lines as those which divided biological control for the unfortunate reason, in my opinion, that the basic science of insect pathology at that time appeared to be inextricably linked with biological control, as though microbial control were the only application of the field of insect pathology.

At any rate the next international meeting involving insect pathology was the Colloque International sur la Pathologie des Insectes et la Lutte Microbiologique held in Paris, 16–24 October 1962.[4] It was "Organisé par la Commission Internationale de Lutte Biologique contre les Ennemis des Cultures (CILB) Sous l'egide de l'Union Internationale des Sciences Biologique (UISB) et le haut patronage de M. Ministre de l'Agriculture avec le concours de l'Institut Pasteur de Paris, de l'Institut National de la Recherche Agronomique et de la Direction des Affaires Culturelles et Techniques du Ministère des Affaires Estrangères et le parrainage du Centre National de la Recherche Scientifique." A gesture toward true internationalism was evident in the composition of the Scientific Committee, the members of which were also chairmen of the scientific sessions, as well as the organizers. Included in this committee were representatives from Japan, France, Switzerland, the Federal Republic of Germany, the Union of Soviet Socialist Republics, England, and the United States. The secretaries were B. Hurpin of France and E. Müller-Kögler of Germany. The colloquium was held at the Centre de Conférences Internationales in Paris. During the conference, visits were made to the Pasteur Institute and the Laboratoire de Lutte Biologique, La Minière (near Versailles), of which P. Grison was director. Beginning on Sunday, 21 October, there were excursions to points in southern France including Avignon, Camargue via Arles, Montpellier (where a symposium on tissue culture was held), and St.-Christol-les-Alès. At the last-named location the group visited the Laboratoire de Cytopathologie directed by C. Vago.

The general program was officially opened with an address by Professor A. S. Balachowsky, then president of CILB. Shortly thereafter the first session began under the direction of Jost Franz of Germany. Few times have I experienced such embarrassment, frustration, and irritation at one and the same time. The meeting had commenced a bit late and opening ceremonies took longer than planned. I had the first paper. There was no rostrum so I was advised to remain seated at the bench. I had expected to be able to speak in a standing position, which I find preferable. My talk was scheduled for thirty minutes. Soon after introducing me, the moderator wrote a note asking me to finish in fifteen minutes in order to make up time. A continuous flow of notes reminding me how much time I had, trying to delete half of my address while I talked, and not being able to show my slides caused me to make a complete shambles of my speech, while the audience was completely unaware of the limitations imposed.

The colloquium, in general, was a marked success. As is frequently the case, the unscheduled conferences and the informal meetings and discussions among the participants were probably of greater long-lasting importance than were the formal presentations. As at the Prague conference, those who arranged the program accomplished the task of delicately balancing the honors of being chairmen of the several sections and being reviewers who, by invitation, gave the introductory addresses at the beginning of the sessions—all of this balanced according to nationality, Eastern and Western nations, and as to the several branches or segments of insect pathology and microbial control. The official languages of the colloquium were English, French, German, and Russian. The evenings were devoted to attending receptions by various French organizations, including the Louvre, and a final banquet was held in a restaurant of the Eiffel Tower during the early afternoon on the final day of the meetings in Paris. The excursions to southern France followed. Reports from those who were able to go were enthusiastic, and for those interested in insect tissue-culture work, it was especially worthwhile.

The proceedings of the colloquium, published in 1964, reveal the continued vigor of the field of insect pathology and attempts to use microorganisms to control insect pests. They were published as a supplement (Mémoire No. 2) of *Entomophaga,* the publishing organ of CILB. In order to have the 566-page volume produced less expensively, the manuscripts were submitted to a printer in Spain. Apparently he was in no hurry to accomplish the job, and much to the distress of its editors, and some of the authors, virtually two years passed before the proceedings appeared in printed form. From its introductory pages we find that 25 countries with a total of 111 official delegates were represented. Actually, more than this number of delegates were in attendance at some of the meetings; to me, these visitors gave additional testimony to the increasing interest in the subject of the colloquium and the dynamism of insect pathology.

As all good conferences should, this one in Paris ended with a list of resolutions—nine in this case. Again, they possessed the inevitable platitudinous quality except that they revealed a more genuine desire to have better international cooperation and to work together toward common goals. It is a pity that woven through these and other similar meetings were the elements of politics and nationalism which had to be dealt with diplomatically. I was amused by a bit of a stir that was caused when it was discovered that the Scientific Committee, formed to "fix the place and time of the next corresponding congress and [to] promote international cooperation," had on it two representatives from the United States, Art Heimpel and myself. When Jost Franz explained the difficulty, I told him the solution was simple—just drop my name from the committee. For reasons I never understood, this did not seem to be the solution Franz and his CILB associates were looking for. The situation was finally resolved by leaving both Heimple and myself on the committee, but my presence was to be justified by the fact that I was the editor of the *Journal of Insect Pathology.*

Actually, being a member of the Scientific Committee meant little in the way of responsibility, at least for the North American members (Angus, Heimpel, and myself), although we were consulted concerning the next colloquium to be held in the Netherlands and given an opportunity to offer suggestions. (Incidentally, from the time of the Paris meetings until the time the next colloquium was organized, four additional European members were added to the membership of the committee.) The next colloquium was held 5–10 September 1966, at Wageningen, the Netherlands, and was organized in cooperation with the International Agricultural Centre, Wageningen, "under the protection of the U.I.S.B. with the participation of the O.I.L.B." Of the organizing committee, J. G. ten Houten was president with P. A. van der Laan as secretary. Van der Laan was also editor of the *Proceedings* which were published in short order under the title "Insect Pathology and Microbial Control."

The Holland conference was different in some respects from former meetings in that the number of participants was restricted —invitations went out to about 200 potential delegates. As stated in the preface of the *Proceedings*, "we also broke with the convention of covering the whole field of our discipline. The Scientific Committee put forward eight topics and asked eight colleagues to give reviews on these. Other participants were invited to send in papers on any one of these topics." During the colloquium a symposium on the standardization of microbial-control preparations was held during two long afternoon sessions. With regard to the standardization of viruses two committees were appointed, both chaired by Art Heimpel of the U.S. Department of Agriculture, but including other American members (H. Dulmage, C. Ignoffo, and I. M. Hall). Group discussions on the standardization of entomogenous fungi chaired by P. A. van der Laan, and of *Bacillus thuringiensis*, chaired by H. D. Burges, were also held. The symposium produced a set of resolutions which can be found in the appendix of the *Proceedings* (van der Laan, 1966). Also, between scientific sessions, the customary excursions and receptions were held.

The colloquium was held at the Institute of Phytopathological Research in Wageningen. Ninety-two delegates from fifteen countries attended (although about one hundred from seventeen countries had accepted invitations to attend). From Canada and the United States nineteen participated.

VI

In the meantime, in 1964, there took place a significant event in the programming of the Twelfth International Congress of Entomology held in London from 8 July through 16 July of that year. By this I mean that for the first time in any international general entomological congress, insect pathology was recognized in its own right and not as subservient to one of its applications—usually microbial control. Instead, the program committee designated for insect pathology its own section, separate and apart from biological control. The title of the section, Section 11, was "Insect Pathology and Relationships between Insects and Micro-organisms," with L. ("Bill") Bailey and N. W. Hussey serving as organizers. This kind of recognition of insect pathology and insect microbiology as encompassing a branch of entomology was, up to that time, unique and a meaningful corner to turn. Although this decision was taken in committee by discussion with interested individuals, primarily it was one made by the congress secretary, Paul Freeman, with the advice of L. Broadbent and N. W. Hussey.

The subject matter of the section's program was divided into five parts: experimental techniques in insect pathology, influence of environmental factors on insect disease, diseases of insects of medical importance, the importance of *Bacillus thuringiensis* in insect pathology, and "contributed papers." Although most of the reports were in the nature of summaries or reviews of pertinent aspects of these areas, the focus on some of the leading problems (e.g., use of insect tissue cultures, latent infections and the role of stress in insect disease, and the toxins of *Bacillus thuringiensis*) was well made, and the cross-fertilization of ideas did well in the entomological setting of the congress.

[EAS intended to discuss the Thirteenth International Congress of Entomology held in Moscow in August of 1968 at this point. MCS.]

VII

An opportunity at an international level to champion the cause of invertebrate pathology as an important segment of comparative pathology came with the Eighth International Congress of Comparative Pathology held in Beirut, Lebanon, 11–23 September 1966. In addition to regular delegates to the congress, many countries had national committees to represent them at the congress. These national committees are, in a sense, subcommittees of the International Committee for Comparative Pathology. The origins lie with the French Société de Pathologie Comparée that sponsored the First International Congress of Comparative Pathology in Paris in October 1912. (Worthy of note is the fact that the program of this first congress, as with certain subsequent ones, included a substantial segment concerned with the diseases of plants, as well as with vertebrate animals.) The American National Committee of Comparative Pathology in 1966 was comprised of thirteen members. As a member of this committee chaired by J. K. Frenkel, I prepared to attend because I felt strongly that this particular international assemblage would be an appropriate forum at which to emphasize the importance of invertebrate pathology in comparative pathology. Unfortunately, at almost the last minute the officials in charge of the congress changed the dates of the congress, extending them for a longer period of time in order to accommodate some of the numerous excursions between the sessions. Apparently, inadequate numbers of delegates had been signing up for the excursions, originally placed after the completion of the scientific sessions, thus causing financial difficulties. By sandwiching in the excursions between scientific sessions it was thought that more delegates could be persuaded to take part in the outings. In my case, however, it meant that the day of my scheduled paper was extended forward several days, which then caused a conflict with inescapable responsibilities here at home.

It had been my hope that at the congress I could make a point as well as a plea to which the delegates would be attentive. The point I wished to make was that, in general, comparative pathologists (let alone so-called medical or clinical pathologists) are ignoring the vast possibilities for obtaining new knowledge concerning disease as offered through the study of the maladies and abnormalities of invertebrates—the ninety-five to ninety-seven percent of known animal species on earth! My plea was to have been to those who control the activities and destinies of the organizations, the laboratories, the journals, and the congresses of comparative pathology to encourage, support, and otherwise "open the doors" to invertebrate pathology. I tried to emphasize, as I had on previous occasions elsewhere, that pathology—considered in its broad definition—was in essence a basic biological science and that I agreed with such authorities as R. W. Leader and P. Rous in believing that only through biological unity could comparative pathology have real value. Whether once again "taking the stump" on these matters would do any good was questionable. Even though I could not attend, the secretary general, Louis Grollet, kindly saw to it that my planned talk was published in the *Revue de Pathologie Comparée* (see Steinhaus, 1967). The requests for reprints were of sufficient number to indicate that perhaps mine was not totally a voice crying in the wilderness. (See p. 304.)

The Society of Comparative Pathology, in which the International Committee for Comparative Pathology is based, is a rather old organization. In the words of one of the American delegates attending the Eighth International Congress, the society "is run by a group of French scientists, physicians, and professors at or near retirement age. The organization appears to be influenced greatly by consideration of personal friendship and allegiance . . . the president of the Congress . . . did not attend the Congress. The program on Pathology of overpopulation [was] based largely on interests of long-time associates to the Congress, rather than on what we mean by pathology, whether understood comparatively or geographically. . . . Attendance at the various sessions varied between 40 and 80, of which about one third were Lebanese, a third French, and a few individuals from other central European countries, Turkey, five from the United States and two from Canada." From this it would seem that organizationally, comparative pathology at an international level was not very dynamic or scintillatingly viable.

VIII

One intriguing aspect of the development of insect pathology at this particular period in its history was the fact that one in opening his mail frequently found that he was being invited to a conference or meeting of some type in which the diseases of insects were to be considered. This fact, if no other, gave evidence to the increasing interest in the subject and its possible applications. Thus it was on a day in February 1962 that I received a letter from Marshall Laird, then chief of environmental biology of the Division of Environmental Health of the World Health Organization (WHO) in Geneva, Switzerland. In his letter he informed me that preparations were being made for a "Scientific Group on the Utilization of Biotic Factors in Vector Control" to meet in Geneva in the fall of 1962. I was very interested in the unique subject matter of the proposed conference and was pleased to accept his invitation to attend.

Although the meeting was called for 29 October to 2 November, I arrived in Geneva early on 22 October, in order to accomplish some writing and reference work at the WHO library. Thus, I arrived on the same day that President Kennedy addressed the nation, and the world, announcing that he had issued orders for the establishment of a naval quarantine on Cuba because of the proof the U.S. government possessed that ballistic missiles had been installed on the island. A clash of some kind with the Soviet Union seemed inevitable. Although the Swiss press and radio undoubtedly were reporting the developments objectively and as fully as they could, I found it extremely difficult to assess precisely how serious the situation was. Inasmuch as another member of our panel was to be E. F. Knipling, chief of Entomology Research Branch of the U.S. Department of Agriculture, I decided to wait to see whether or not he would arrive. If Knipling did not show up for our meeting, I was to interpret this to mean that the situation was critical enough for me to return home at once.

By 28 October when I had reached the point that I felt I could wait no longer I heard over the radio that Premier Khrushchev had informed the United States that the weapons our country viewed as offensive would be removed from Cuba as soon as possible. Shortly thereafter I learned that Knipling had arrived. To say the least, the relief of the tension that had built up within me was great.

In addition to personages associated with WHO, Knipling, and myself, the active participants at the meeting were B. P.

Beirne of Canada, J. M. Franz of Germany, P. Grison of France, I. A. Rubtzov of the USSR, F. J. Simmonds of the United Kingdom (from Trinidad), and J. Weiser of Czechoslovakia. Knipling was chosen to serve as chairman, with Marshall Laird and J. W. Wright of WHO serving as the secretariat. The size of the group, the setting of the meeting, and the arrangements for the occasion were ideally appropriate.

It soon became even more clear as to why the WHO had called us together. As explained in the mimeographed report of the meeting made to the director general, and dated 20 December 1962, one of the major obstacles to malaria eradication that had arisen in the preceding few years was that of resistance shown by the mosquito victors to most of the established chemical insecticides. Moreover, in some parts of the world, the toxicity to vertebrates of such insecticides was creating problems. Accordingly, it became incumbent upon WHO to explore every possible means of control, and the methods of biological control, with their outstanding success with certain agricultural pests, could not be overlooked when it came to the suppression of the vectors of agents causing diseases in man and other vertebrates. Accordingly, the several days of our meeting were devoted to discussing the current status of research on the biological, cultural, and integrated control of insects of medical importance, the existing facilities for research and coordination, and what priorities should be established, and recommendations made, with regard to furthering methods of ecological control. So far as insect pathology and microbial control were concerned, the group made several worthwhile recommendations which, unfortunately, have only partially been implemented. Because of their basic and continuing value, however, the recommendations made by the Scientific Group, although directed to vector control, are applicable to all pest control, and are worth repeating in abbreviated form:

1. Compilation and dissemination of current information relating to the use or potential use of biotic agents for vector control.
2. Encouragement and facilitation of a broad survey of biotic agents potentially useful for vector control.
3. Designation of a reference center or centers for the diag-

nosis of microbial agents (i.e., the diagnosis of diseases and the identification of pathogens) potentially useful for vector control.

4. Encouragement and facilitation of follow-up investigations on biotic agents potentially useful for vector control.
5. Assistance in the field appraisal of biotic agents potentially useful for vector control.
6. Examples of research on biotic agents that could be facilitated by WHO. (Several examples were listed.)
7. Encouragement, facilitation and support of research fundamental to a better understanding and eventual practical use of biotic agents for vector control.

Among the ways of implementing the aims of the second resolution, the feasibility of providing collecting kits and instructions for collecting and preserving pathological material in preparation for diagnosis was discussed. With support from WHO, a collecting kit, with accompanying instructions, was designed by the Insect Pathology Laboratory of the U.S. Department of Agriculture. In the course of preparation, prototypes were sent to collaborating scientists in several parts of the world for field testing, criticism, and comment (Cantwell and Laird, 1966). In connection with the third resolution, the Department of Insect Pathology on the Berkeley campus of the University of California and the Laboratory of Insect Pathology of the Czechoslovak Academy of Sciences in Prague were cited as examples of organizations that might serve in this capacity. At Berkeley, we already had a functioning diagnostic service, which, I assured the group, could take on the added work if we could obtain the funds required for the necessary expansion. (At the time I did not realize that before another year had passed, I would be moving to the Irvine campus, leaving behind the diagnostic service, and thus would be unable to carry out any plans with WHO. Subsequently John D. Briggs at Ohio State University assumed some of this responsibility.)

While telling of the WHO Scientific Group meeting, it is logical to go on to tell of another meeting sponsored by WHO of significance to insect pathology and microbial control. I refer to the Symposium on Culture Procedures for Arthropod Vectors and Their Biological Control Agents held in Gainesville, Florida, from 30 September through 4 October 1963. The name of the symposium fairly well describes its subject matter. Again, the meeting was generated by Marshall Laird and his colleagues in WHO at Geneva. Of the twenty-nine participants, half a dozen of them—including J. D. Briggs, J. P. Kramer, C. Vago, and J. Weiser—concerned themselves with matters relating to insect pathology and microbial control. Among the several interesting summaries was one by O. P. Forattini of Brazil, who emphasized that it is "essential that the work of invertebrate pathologists and ecologists be co-ordinated," and he then listed the "stages" of collaboration he felt should take place between the two. To one who has always felt the need for greater emphasis on the ecological aspects of invertebrate pathology, it was gratifying to me to see this call for a "much fuller ecological baseline."

Of course, strictly speaking, any agent that is toxic for, kills, or causes disease in, an insect may properly be considered to be a pathogenic agent and may properly be studied from the standpoint of insect pathology. Accordingly, other parts of WHO's and FAO's program of insect control and its conferences and symposia called to consider this subject deserve at least mention here. A 1963 meeting of the organization's Expert Committee on Insecticides, a 1964 Scientific Group on Genetics of Vectors and Insecticide Resistance, and a 1966 FAO Symposium on Integrated Pest Control are cases in point. Matters relating to the resistance of insect vectors to insecticides permeate virtually every conference relating to insect control. And, bringing the focus in sharply, we must acknowledge that much in the way of pathology is involved in such methods of control as those using chemosterilant-attractant combinations, cytoplasmic incompatibility, hybrid sterility, genetic lethal factors, and the like. Their role in programs of integrated control was well emphasized by Laird and Wright (1966) in their report (which cites other references) of the FAO Symposium on Integrated Pest Control.

Marshall Laird's interest in microbial control, and invertebrate pathology, very probably stems from a titillating experiment in the Tokelau Islands of the South Pacific, where he introduced the fungus *Coelomomyces stegomyiae* in an apparently successful attempt to control the *Wuchereria* vector *Aëdes polynesiensis* (see Laird, 1960). Laird, well known for his work as a parasitologist and medical entomologist, became interested in the ecology of *Anopheles* and other mosquitoes of the South Pacific while he and Mrs. Laird were making an anopheline survey for the Royal New Zealand Air Force and the New Zealand Department of Scientific and Industrial Research from 1952 to 1954. He carried this interest with him when he accepted his post with WHO, and continued it and his work on *Coelomomyces* when he accepted the position as head of the Department of Biology at Memorial University of Newfoundland. In any event, apropos the subject of this chapter, the meetings he and his associates arranged while he was with WHO did much to further interest in the use of microbial pathogens in the possible control of insects of medical veterinary importance.

IX

As a member of the Invertebrate Consultants Committee for the Pacific (ICCP) of the Pacific Science Board, I attended one meeting of the committee at which the potentials of microbial control were discussed at some length and with considerable seriousness of purpose. It was held in Honolulu on 1–2 March 1963, with C. E. Pemberton chairing the two-day meeting. Among matters discussed at this meeting was the intriguing, but frustrating, work of A. R. Mead of the University of Arizona. I say "frustrating" because for some time I had been hoping that Mead would score a breakthrough with his work on the diseases of snails, especially the destructive giant African snail, *Achatina fulica*, which had spread from its East African home to the Indo-Pacific regions including Hawaii, so as to give greater impetus to this area of invertebrate pathology. I had been one of the referees on his applications to the National Institute of Health for research grants to support his work and in this capacity strongly supported him, providing he would collaborate with a microbiologist or virologist in the hope that the phenomenon he called disease could be pinned down as to its etiology. By the time of the ICCP meeting in Honolulu he was able to report that he had a virologist collaborating with him. Unfortunately, no specific etiological agent could ever be associated with the malady (characterized by leukodermic lesions through melanophore destruction, dermal atrophy, tentacular distortion, and "tissue alterations in the kidney, hepatopancreas and other elements of the viscera," and a tendency to become epizootic in older populations). Mead recognized the possible roles of genetic factors, nutrition, and stressors, as well as possible infectious agents, in the disease which he believed to be cyclic and to function as a natural regulator of snail populations.

3. MEETINGS IN THE UNITED STATES

I

So much for large international congresses or meetings outside the United States in which North American insect pathologists participated. In the present section I should like to relate some of the more interesting aspects of regional or special meetings in which North Americans were involved.

I shall not attempt to cover every seminar-type session relating to the diseases of insects and other invertebrates held in North America because no doubt some such meetings or discussions have been held under other names or under unsuspecting aegis. A case in point: In 1945, 11–14 September, the Committee on Grasshopper Research of the American Association of Economic Entomologists sponsored a conference held in Lincoln, Nebraska. Attending the conference were thirty-five research men "and others closely concerned with grasshopper problem" representing eleven states and the Canadian and United States departments of agriculture. From the recommendations of this conference (*Journal of Economic Entomology*, *39*, 111–3, 1946) we read:

> It is felt that special study of diseases is needed, but that progress in it will not be satisfactory until a special laboratory is established which will facilitate a concentrated, cooperative attack on the problem by pathologists as well as entomologists. . . . The Conference recommends that when experiments in control are started, priority should be given to the artificial use of diseases.

The committee was chaired by K. M. King and had as members F. A. Fenton, C. Wakeland, H. D. Tate, R. H. Handford, and J. H. Pepper. Unfortunately, few tangible results came directly from these recommendations so far as a sustained program in the study of the diseases of grasshoppers was concerned. Indeed the Entomology Research Division of the U.S. Department of Agriculture has no record of any implementation of any of the recommendations of the conference.

On 16–17 November 1959, the Conservation Foundation, headquartered in New York, sponsored informal discussions on the implications of the increasing use of chemical pesticides. Fairfield Osborn, president of the foundation, invited a panel of twenty-two authorities working in different areas of insect control to meet together with six representatives of the foundation. I was surprised and taken aback to find that although there was a goodly number of chemical control men at the meeting, I was virtually the sole representative for biological control. While I knew that I could speak to the subject of microbial control, I felt that those concerned with entomophagous insects deserved to be heard. I did the best I could, coming off fairly well by championing the idea of "integrated control" as then advocated by Ray F. Smith, V. M. Stern, and others. I took this liberty because I had advocated this approach with chemicals and insect-pathogenic microorganisms since about 1955 and had speculated about its possibilities in some of my classes. I was pleased that Stephen W. Bergen, who prepared most of the resulting mimeographed report, "Ecology and Chemical Pesticides," picked up a passing comment I made to the effect that in the United States as a whole the research program in economic entomology is out of balance so far as support is concerned, with funds for the research and development of chemical insecticides being Gargantuan compared to the financial support furnished for the development of methods of biological control—a fact that had been impressed upon me by Harry S. Smith. I had hopes then that perhaps the Conservation Foundation might help remedy this imbalance, but so far as I have been able to determine, few tangible results were forthcoming from that particular conference. Fortunately, now, ten years later, as I write these lines both state and federal governments have considerably improved their posture with regard to the support of nonchemical methods of controlling pests, without threatening the appropriate and essential use of chemical poisons.

Early in 1960 I was invited to serve as moderator for an interesting and one of the most unusual conferences it has been my pleasure to attend—I am sure that I would have enjoyed it more, however, had I not been moderator. The Conference on Biological Control of Insects of Medical Importance was called to meet on 3–4 February 1960 in the Dart Auditorium of the Armed Forces Institute of Pathology and was sponsored by the American Institute of Biological Sciences, The Armed Forces Pest Control Board, The Office of Naval Research, and The Army Chemical Corps. The proceedings were published by AIBS under the technical editorship of D. W. Jenkins in November 1960.

Following an address of welcome by Hiden T. Cox, then executive secretary of AIBS, Ralph W. Bunn, the executive secretary of the Armed Forces Pest Control Board, explained the objectives of the conference. His opening remark is pertinent: "Oliver Wendell Holmes is credited with saying, ' . . . no man can be truly called an entomologist, sir; the subject is too vast for any single human intelligence to grasp.' I could never be more fully in agreement with Dr. Holmes than at a time like this when confronted with a program involving not only entomology but the interrelationships of entomology with bacteriology, mycology, protozoology, virology, parasitology, and nematology." And he might have added a few more -ologies, including psychology!

Approximately sixty delegates participated in the two-day meeting. Most of the papers presented dealt with microbial diseases of insects of medical importance (especially Diptera), but other types of biological control were represented both by formal papers and in the very active discussions between papers. The conference ended with a summary statement by John R. Olive of the AIBS, and a series of twelve general recommendations and five specific ones formulated by an ad hoc committee headed by Dale Jenkins. These recommendations acknowledged that "pathogens, parasites, and some predators offer a real potential for natural control of medically important insects," and that there was need for more field and laboratory research, as well as for type-culture collections, and, of course, the appropriation of more funds to carry out these, and other, objectives including "scholarships, teaching, and establishing programs in insect pathology and biological control at universities. More laboratory space is urgently needed."

But, following the reading of the recommendations, it remained for Robert Traub to ask the all-important question, "How are these recommendations to be implemented?" Those responding to this cogent question suggested that the proceedings of the conference be sent to "granting groups, universities, and federal groups where research may be initiated and supported." Others suggested the designation of "an ecological year" or "a biological decade," during which biological control would be appropriately emphasized. (Interestingly enough, in 1965 biologists in America began planning, through a committee formed under the aegis of the National Research Council, U.S. participation in an International Biological Program (IBP); included were projects on insect pathology and biological control. The research phase of the IBP was planned to begin in July 1967 and to extend over a period of five years. However, no special emphasis was given to the biological control of insects of medical importance.)

II

From 31 March through 3 April 1964, the Entomology Research Division of the Agricultural Research Service of the U.S. Department of Agriculture held a special planning and training conference on the subject of insect pathology and biological control. The conference was held at Excelsior Springs, Missouri; it was the fourth of its kind on some aspect of the division's research program. In the words of the director, E. F. Knipling, such meetings were organized as "in-house planning and training programs to help develop maximum effectiveness and coordination in the over-all Division program." A. M. Heimpel was chairman of the committee (H. Baker, P. B. Dowden, and A. S. Michael) that planned this particular meeting on insect pathology and biological control. Although the conference was of the "in-house" variety and not open to those outside of the U.S. Department of Agriculture, its importance in the progress of insect pathology is of significance because of the consequential role played by the federal government's division of entomology in this field. Although no general conclusion or recommendations by the conference have been recorded, from the mimeographed record of the papers presented, it is clear that the Entomology Research Division covered well most of the principal thrusts insect pathology and microbial control were making at the time. To its credit, also, is the fact that the importance of the study of the diseases of beneficial insects, especially the honey bee, was brought out--an aspect of insect pathology all too often forgotten when a conference bears such a title as "insect pathology and microbial control." Incidentally, in passing, it is worthy of note that in a conference on insect nutrition and rearing held the preceding year (4–8 March 1963), several federal insect pathologists took part, and papers relating to the rearing of insects for use in experimental insect pathology and on the control of diseases in insect rearing were presented.

On 25 and 26 February 1965, a "work conference" was held on the Berkeley campus of the University of California, sponsored by the Division of Invertebrate Pathology. The subject of the conference was the utilization of nucleopolyhedrosis virus for the control of the bollworm or corn earworm, *Heliothis zea*. The meeting was attended by thirty-seven invited scientists from the university, from industry, and from state and federal agencies. The looming possibility that a nucleopolyhedrosis virus could be used to control *Heliothis* generated the two-day conference, where not only the virus and its action were discussed, but matters pertaining to its production, standardization, and registration were also on the agenda. Like all good conferences, this one, too, ended with a set (eight) of resolutions and recommendations. These have been published, as called for in one of the resolutions, in "an appropriate journal" (Falcon, 1965).

Another Berkeley meeting that involved insect pathologists was held on 24 and 25 April 1968, under the auspices of the Tropical Medicine and Parasitology Study Section of the Division of Research Grants of the National Institutes of Health. Its title was "Workshop on Integrated Control and Pest Management." Thirty or so invited delegates attended, in addition to members of the Study Section. The latter undoubtedly gained, as indeed did everyone, a greater appreciation of the goals and potentialities of integrated control (i.e., control using any and all methods and techniques in at least a relatively compatible manner); moreover, it served as an occasion to call for further attention to specific problems and to important research needed in integrated insect control programs—including those involving methods of microbial control. The speakers, led off by a smooth and clarifying introduction by Ray F. Smith, were especially articulate and well chosen; the discussions were penetrating and informative.

But a pall of sadness hung over the meeting. Bob Usinger, who was to have been moderator, had been only a few days before taken to the hospital with a terminal illness. It seemed so incredibly unfair and incongruous that Bob, who had given so much and still had so much to give to science, should be dying. Although Bob officially was not allowed visitors, his wife, Martha, arranged that my wife and I could go and visit him. That visit was one of the saddest hours of my life. I could not keep tears from coming as I said goodbye to a wonderful friend—great biologist, but more important, this great man—as he courageously prepared to meet and accept the spiritual mysteries of the universe.

But emphasis was being given to microbial control not only in entomology circles. I was surprised to be invited to participate in a symposium titled "Microbial Insecticides" held at the annual meeting of the American Society of Microbiology on 7 May 1964, in Washington, D. C. This particular symposium was held under the sponsorship of the Division of Agricultural and Industrial Microbiology, with Harlow H. Hall as convener. My contribution at the symposium was to highlight that what may be possible regarding the use of microorganisms in the control of insects may also be possible with regard to invertebrates other than insects. In doing this I made an effort to emphasize the economic importance of invertebrate animals of twelve phyla and to further emphasize that the study of infectious disease can be important in controlling invertebrate pests as well as (by the suppression of disease) in maintaining the health of the beneficial species (Steinhaus, 1965).

Other speakers at this symposium and their topics were T. A. Angus (bacterial pathogens of insects), R. A. Rhodes (milky disease of the Japanese beetle), D. Pramer (fungus parasites of insects and nematodes), A. M. Heimpel and J. C. Harshbarger (immunity of insects), and K. M. Smith (insect viruses). The papers presented by most of these participants were published in *Bacteriological Reviews* for September 1965. While in no wise comparing with the recognition given insect pathology by the Entomological Society of America, considering the greater size of the American Society for Microbiology and its preoccupation with so many other dynamic and well-established branches of its discipline, this symposium under the aegis of the latter organization was a significant happening.

III

The increasing number of special symposia or "workshops" being conducted in the United States makes the line between what one would ordinarily call a scientific "meeting" or "conference" and a specialized instructional course somewhat blurred. However, even though they are called "workshops" or even "courses," and are frequently of several weeks duration, basically they have many characteristics of a conference and hence might properly be considered in the present chapter.

One of these courses deserves special mention—not because it deals exclusively with the diseases of invertebrates (it does not), but because its general nature is such that it is germane to invertebrate pathology as well as to other areas of pathology. Designated as a "Course in Comparative Pathobiology," and held annually since 1964 at Aspen, Colorado, the federal-grant-supported workshop "was initiated as a result of interest displayed by a subcommittee on comparative pathology of the National Research Council and National Academy of Science. It was previously given impetus by a conference on comparative pathology held at Portsmouth, New Hampshire, sponsored by the Pathology Training Committee, National Institutes of Health." However, as a member of this subcommittee, I can testify to the fact that the moving spirit and working person of this course and its organization was Donald W. King of the University of Colorado, later of Columbia University. Notable about the course—the program of which changed from year to year—was the basic approach taken towards pathological phenomena and the involvement of the basic sciences. Having, at times, felt quite alone (although I was not) in advocating the basic biological approach including all levels from atomic and molecular to population and environmental to malfunctioning life systems, I was delighted that a course, so oriented, could gain the support of the National Research Council—National Academy of Sciences, as well as funding agencies. Ringing out loud and clear in the introduction to the first conference was the statement:

> We must now accept the fact that a "study of disease" is in reality a "study of the extremes of biology." Disease merely represents a variation of normal biological processes, either an excess or deficiency of metabolites produced by reactions on ultrastructural organelles, all fairly well described over the past three decades. . . . The cell's ability to maintain its homeostatic regulating mechanisms by degrading, sequestering, extruding or otherwise compensating for, and controlling detrimental factors in association with continued ability to resynthesize destroyed structural, enzymatic or metabolic moieties is a truer and a more modern concept of pathology.

Not quite the way I had been saying much the same thing,

nor the exact phraseology I would use, but certainly in the same ball park—and what a delight to have it coming from one in the community of practitioners and instructors of human medicine!

At this moment of writing, the fifth annual course is about to begin. It is to last seven days. It will be conducted by a "faculty" of twenty-one speakers, and is open to one hundred participants. The subject matter this year is succinctly divided under the headings: nucleic acid, genetics, proteins, lipids, carbohydrates, differentiation, and neurobiology. Its success, as in previous years, is assured by the quality of speakers and the seriousness of purpose of the entire endeavor.

In addition to the course in comparative pathology presented at Aspen, the subcommittee on Comparative Pathology, of the National Research Council's Division of Medical Sciences' Committee on Pathology has taken other action and has given moral support to the promotion of a wider and more dynamic base for comparative pathology. Most of this support, however, has been aimed at problems, programs, or projects relating to the pathology of vertebrates rather than invertebrates. Occasionally, as did King's course at Aspen, something like J. K. Frenkel's informative list of forty-five centers that have, in some measure, training or teaching programs in some phase of comparative pathology would generate assurance that the wide-spectrum concept of pathology was not entirely forgotten.

Another specialized symposium, and a highly significant one, was that titled "Neoplasia of Invertebrate and Primitive Vertebrate Animals," held in Washington, D.C., on 19, 20, and 21 June 1968. It was attended by approximately one hundred scientists from twelve or more countries. Although held under the aegis of the Registry of Tumors in Lower Animals of the Museum of Natural History of the Smithsonian Institution, it was conceived and initially organized by Clyde J. Dawe of the National Cancer Institute. Dawe, quiet and unassuming, a man of exceptional talents and wide breadth of vision in his view of oncology and of the study of neoplasia in all forms of life, greatly impressed me since I first came to know him while serving on the NRC-NAS subcommittee on comparative pathology in the early 1960s. He was joined in the planning and implementation of the symposium by John C. Harshbarger, who, after being independently associated with me on an NIH grant at Irvine, accepted the directorship of the registry in September 1967. The energy and ability with which he participated in arranging the symposium as well as simultaneously taking over the duties of director of the registry, initially and temporarily headed by G. E. Cantwell, was exemplary, to say the least.

The symposium covered neoplasia, and neoplasialike, pathologies in invertebrates from nearly every known standpoint. By bringing in foreign delegates, an international flavor was provided the conference. And while other commitments prevented me from accepting Dawe's kind invitation to chair a session, and to attend, I had a feeling of vicarious accomplishment in the occurrence of this meeting.

For years I had been distressed that more support was not being given the study of tumors in insects and other invertebrates. And whatever powers of persuasion I might have possessed were tested when I convinced a group of scientists making a site visit in response to a research grant application I had submitted to the Public Health Service that they should approve funds for the study of neoplasms in invertebrates. Fortunately, the application was approved, and part of the funds forthcoming enabled me to enlist John Harshbarger and Ronald L. Taylor in the quest for more knowledge relating to the occurrence of tumors and tumorlike formations in lower animals. At a meeting of the subcommittee on comparative pathology in the National Academy of Sciences building in Washington, D.C., on 19 September 1963, I submitted a formal statement which included the following recommendation: "Intensive research on tumors and other neoplastic diseases in invertebrates should be initiated. The same holds true for genetic diseases, nutritional disease, and teratologies found in invertebrates. These areas have been especially neglected." I am sure that Dawe would have proceeded with his plans for the Neoplasia Symposium without any such comment from me, but this subcommittee meeting—which he also attended—enhanced our common interest and I was greatly encouraged by his sympathetic appreciation of neoplasms in lower animals. The papers presented during the symposium were published in the *Journal of the National Cancer Institute.* Invertebrate oncology had arrived!

The growing significance of, and appreciation for, invertebrate pathology including insect pathology, in experimental biology became evident at the Fifty-first Annual Meeting (17–21 April 1967, in Chicago) of the Federation of American Societies for Experimental Biology. At that meeting the American Society of Experimental Pathology held a symposium, chaired by Frederick B. Bang, titled "Defense Reactions in Invertebrates." To be sure, comparative pathology had been the subject of symposia at previous federation meetings, but in this case an entire session was devoted to subject matter relating to disease in invertebrate animals. It undoubtedly will not be the last such session.

IV

Beginning in the late 1950s serious mortalities in oysters along the east coast, particularly in the Delaware Bay, of the United States caused the Bureau of Commercial Fisheries, and others, to take an interest in the diseases of this mollusk and thus to initiate a new dimension to invertebrate pathology in North America. To better meet the problems involved and to accelerate information as rapidly as possible, annual shellfish mortality conferences were started in 1958 at the instigation of the bureau. The bureau had begun to support research on the disease of oysters at Rutgers late in 1957, and itself entered active research on the mortalities in 1958, while continuing to support, with contracts, research at Rutgers and at the University of Delaware. Thus began not only a fascinating history of research on a practical marine problem, but a series of conferences or meetings that were to be one of the principal foci bringing together certain of the noninsect invertebrate pathologists.

Active in these conferences from the very first was Carl J. Sindermann, director of the bureau's biological laboratory at Oxford, Maryland. Sindermann attended most of the conferences, including the first, and it was from him and Al Sparks of the University of Washington that I learned some of the unrecorded facts pertaining to them. The conferences were informal; no minutes were taken until the 1963 meetings. At most of the other meetings, brief summaries were prepared, or individuals brought mimeographed papers or abstracts. A file of virtually all of this material, and much associated correspondence is held by Sindermann at the Oxford laboratory, as well as probably by others who attended the conferences. The early conferences were primarily concerned with reports on the mortality and pathology of oysters and with the cause of most of this mortality, a protozoan originally designated as "MSX" (multinucleate spherical plasmodium). Later the pathogen was identified as a haplosporidian and named *Minchinia nelsoni* by H. H. Haskin, L. A. Stauber, and J. G. Mackin (1966).

Toward the end of a paper by Farley (1967) on the life cycle of *M. nelsoni*, there is a brief addendum relating to the history of the shellfish mortality conference. Part of it bears quoting here:

> In 1959, the following institutions embarked on a cooperative program of research on the causes of oyster mortality, especially *Minchinia nelsoni* (MSX): Galveston Marine Laboratory, Agricultural and Mechanical College of Texas; Hiram College, Hiram, Ohio; Natural Resources Institute of the University of Maryland, Solomons, Maryland; New Jersey Oyster Research Laboratory, Bivalve and Cape May Court House, New Jersey; University of Delaware, Marine Laboratory, Lewes, Delaware; the U.S.F.W.S. Bureau of Commercial Fisheries Biological Laboratory, Oxford, Md.; and Virginia Institute of Marine Science, Gloucester Point, Virginia. Annual workshop conferences, with participants from the institutions named, were held. Because of the urgent need to rehabilitate the oyster industry there was, thru the year 1965, free exchange of information. Several formal presentations of ideas and data on possible life cycles were made as early as 1961. The best documented instance was during the 1965 annual conference when a panel discussion of the life cycle was held and each participant distributed mimeographed copies of his paper.

As the annual conferences progressed, the scope of the subject matter broadened. With regard to the disease caused by *M. nelsoni*, papers pertaining to the epizootiology of the disease were presented, as were suggestions on the prevention and cure of the disease. By the time of the eighth annual conference held at the University of Delaware in January 1966, not only were MSX and additional haplosporidia subjects discussed, but so were diseases of oysters caused by other pathogens; moreover, increased interest in immunological phenomena was evidenced by the program. Indeed, we find a paper on a bacterial disease of the American lobster listed on that program.

In 1967, an interesting and useful dichotomy took place: two meetings were held. The Ninth Annual Shellfish Pathology Conference took place on 27 and 28 January at Adelphi Suffolk College in Oakdale, New York, and the First Pacific Coast Oyster Mortality Workshop was held at the University of Washington in Seattle, on 9–11 May. (The latter meeting began on the day following the organizational meeting of the Society for Invertebrate Pathology, which we shall discuss in the next section of this chapter.) The Pacific Coast meeting was appropriate because for some years it had been observed that Pacific oysters suffered heavy losses for which, usually, no causative agent was known. Indeed, at the 1963 Shellfish Mortality Conference, C. Woelke and C. Lindsay reported that nearly every year at least one spectacular mass mortality, and in some years several, were encountered in the Puget Sound area.

Incidentally, worthy of note is a statement contained in a mimeographed announcement by J. D. Andrews relating to the then up-coming seventh conference, in 1965: "Dr. [Victor] Sprague has suggested that we change our name to 'Shellfish Pathology Conference' [from "Shellfish Mortality Conference"] in line with the recent action of the J. Insect Pathology changing to J. Invertebrate Pathology." Thus we do, indeed, find that the heading on the program of the subsequent conferences reads "Annual Shellfish Pathology Conference."

V

I should like to devote this last section to the Society for

Invertebrate Pathology, which was born in North America with international aspirations. It seems appropriate to end this chapter with reflections on how this organization came into being because—at least as I write this in the summer of 1968—it hopefully may be the future prime mover of the field of endeavor about which we are concerned in this book.

The thought of invertebrate pathologists having their own society or professional home had flickered through my mind—and perhaps those of others—since the late 1940s. Just as through the *Journal of Invertebrate Pathology* I had hoped to bring together the common interests of those concerned with disease in insects with those concerned with disease in oysters, snails, annelids, and all other invertebrates, so I thought it would be well to bring insect pathologists, already well integrated with entomologists, into closer association with other types of invertebrate pathologists. But to do so one had to proceed with great diplomacy and caution lest he appear to be simply an iconoclast or a self-interested promoter. To accomplish the intended goals it would be inadvisable and undesirably sad to alienate in any way the entomological societies and organizations which had hospitably received and sheltered insect pathology. Nor, with care, was there any reason why this should happen. Nevertheless, there was this other dimension of invertebrate pathology from which the insect pathologist was inadvertently isolating himself, and from which there was much he could learn.

Quietly, I approached a number of Europeans who were leading the destiny of the Commission Internationale de Lutte Biologique (later to become the Organisation Internationale de Lutte Biologique)—commonly designated as the CILB or the OILB—as to whether they would be interested in expanding their insect pathology activities into a separate, truly international society concerned with basic invertebrate pathology. My overtures met with little encouragement. Clearly, at that time, the Europeans felt quite self-sufficient and were satisfied with the tie-in of microbial (biological) control with insect pathology without much concern about invertebrate pathology generally. Then three developments or happenings took place which induced me to take action.

In 1963, while serving as president of the Entomological Society of America, I appointed a special committee to review the arrangment, content, and names of the sections of that society. The committee did a fine job. I was personally disappointed in the names they gave to the six sections, but as president I did not feel I could say so. Feeling strongly that insect pathology should stand equally with physiology, morphology, systematics, toxicology, and the rest, I was certainly surprised to find that the committee presented to the governing board a set of sectional names which buried pathology under a descriptive subheading "physiological aspects of life phenomena"—a quite inadequate umbrella for a subject that also included morphological and ecological aspects in its purview! So as not to hold up or endanger the other excellent recommendations of the committee, I decided to call the attention of the governing board to the matter, but otherwise to "hold my spit." The board, tired from long hours of deliberations, accepted and approved the report as it was submitted. Clearly, to me, this action meant that while the Entomological Society of America recognized pathology as subject matter, this discipline would have to look elsewhere to find its place in the societal or organizational sun.

Some time after this action by the Entomological Society of America, I began receiving vague inquires about the feasibility of some sort of an organization for those broadly interested in invertebrate pathology. Some of these inquiries came in association with my decision to change the name of the *Journal of Insect Pathology* to the *Journal of Invertebrate Pathology*, beginning with the 1965 volume (volume 7) of the periodical. Because I knew that the Europeans were planning an international colloquium to be held in Wageningen, the Netherlands, in September 1966, I put off such inquiries with the request that I be allowed to try once more to ask the OILB to sponsor a branch or society devoted to basic invertebrate pathology. Some months before the colloquium, while he was visiting the United States, I asked the secretary of the colloquium, P. A. van der Laan, in person, as I had in correspondence, if I could present such a proposal from the floor of the convention. He politely asked me not to because he felt it would engender misunderstanding and was "politically not feasible." Efforts by John Briggs and others who explored the situation while at Wageningen came to naught.

It was now clear that unilateral action of some kind was needed. In my talk prepared for the International Congress for Comparative Pathology, as well as in an editorial in the March 1967 issue of the *Journal of Invertebrate Pathology*, I openly proposed the formation of an international society for invertebrate pathology. In the meantime, I had been approached by several oyster pathologists—in particular A. K. Sparks of the College of Fisheries of the University of Washington, Seattle, and Carl J. Sindermann of the U.S. Department of Interior's Bureau of Commercial Fisheries at Oxford, Maryland—with the suggestion that "we get together" because if we did not, the oyster pathologists would in all probability go ahead to form their own society. Accordingly, I promised to prepare a questionnaire that was circulated over the names of Sparks and Steinhaus to 560 invertebrate pathologists, of which approximately 400 were insect pathologists while 160 were concerned with the pathology of invertebrates other than insects.

The results of the questionnaire, of which a surprising forty-four percent had been received at the time of tallying, were published in an editorial report in the June 1967 issue of the *Journal.* In response to the question, "Are you in favor of a truly international society of invertebrate pathology?" a total of 229 answered yes while nineteen (twelve from one laboratory) replied no. The mandate was clear: most invertebrate pathologists interested enough to answer the questionnaire wanted a scientific society of their own. The greatest number of enthusiastic returns came from pathologists in North America, Japan, and Eastern Europe (including the USSR). The Western Europeans were less enthusiastic about the idea obviously because they felt that an international society for invertebrate pathology posed a threat of some kind to OILB. It did not, of course; and, in any case, OILB was not truly international but rather a consortium of European national representatives. It was "affiliated with IUBS and has consultative status in the FAO," and, as its name implies, was oriented primarily to matters of biological control and not toward basic invertebrate pathology broadly conceived.

At this point it might be well to mention the number and principal types of animal pathology societies or organizations in existence at the time (ignoring for this purpose, those general societies having special sections or programs devoted to pathology subject matter). As mentioned earlier, the International Committee for Comparative Pathology is a fairly old international society, largely dominated by the French. The International Society of Geographic Pathology (1929), the International Society of Clinical Pathology (1948), and the International Academy of Pathology (1916) are essentially oriented toward human pathology in their activity and interests. The last-named organization is a member of the International Council of Societies of Pathology (1962), the original purpose of which was to distribute the information developed from World Health Organization Reference Centers on the definitions of tumor types. Its official "Purposes" are listed in its rules as: (1) to provide an international medium for the cooperation of societies of pathology and for the exchange of information among pathologists, (2) to aid and to cooperate in the development of uniformity in the criteria for the definition and diagnosis of disease, (3) to encourage research and education in the field of pathology, (4) to promote relations with international organizations concerned with problems of health. The council had its inaugural meeting on 8 July 1962 at the University of Zurich, Switzerland. However, neither then nor since has there been much overt expression of interest in comparative pathology let alone invertebrate pathology. Hopefully, however, it could be a world umbrella under which all societies of pathology (be they human, other vertebrates, invertebrates, or plants) could form some sort of fruitful liaison or affiliation.

At the first meeting of the ICSP, thirty-one countries and thirty-six pathology societies or institutions were represented. This included six societies or associations from North America, none particularly concerned with the pathology or disease of invertebrate animals: Canadian Association of Pathologists,

Asociacion Mexicana de Patologos, American Society for Experimental Pathology (perhaps the nearest to accommodating the needs of invertebrate pathology), American Society of Clinical Pathologists, College of American Pathologists, and American Association of Pathologists and Bacteriologists. I have listed these organizations concerned with pathology to indicate another reason why, especially in the United States and Canada, there was developing a ground swell for a society or association dedicated to the promotion of invertebrate pathology.

But back to the formation of the Society for Invertebrate Pathology. On the basis of the nominations made by those responding to the questionnaire, an organization committee (later the first governing council) was formed consisting of the following members: Thomas A. Angus, Arthur M. Heimpel, Mauro E. Martignoni, Carl J. Sindermann, A. K. Sparks, and Victor Sprague. This committee held its first meeting on 9 May 1967 at the University of Washington, Seattle. It chose as the first officers of the society Heimpel as secretary-treasurer, Sparks as vice-president, and me as president. It was decided to establish the society as an incorporated, nonprofit organization through a charter legally derived in the state of Maryland. (There were two reasons for the latter action: we found the legal fees to be less in Maryland than in, say, California, and it was the resident state of the secretary-treasurer, who had to handle most of the details involved.) To the extent agreed to by its publishers, Academic Press, the already existing *Journal of Invertebrate Pathology* was to be published under the auspices of the society beginning with the first (February) issue of 1968. Other decisions made at this organizational meeting related to matters of framing a constitution, deciding—initially at least—to affiliate as an "adherent society" with the American Institute of Biological Sciences, deciding out of deference to worried Western Europeans not to include the word "international" in the name of the society but to emphasize that membership was open to pathologists anywhere in the world, and deciding to establish the categories of founding member ($5, one time only), charter member ($4), regular member ($4) student member ($2), emeritus member, honorary member, and sustaining member. Art Heimpel, very graciously arranged for me, as stated in a letter dated 17 May 1967, to "have the honor of being the first founding member." Art certainly could have been the first one, so I very much appreciated this distinction. Upon a recommendation initiated by Mauro Martignoni, the well-known insect pathologist, Enrico Masera, was elected in April, 1968, the first honorary member of the society.

With Heimpel hard at work establishing the legality of the society (the articles of incorporation were certified in Baltimore on 26 October 1967), working with the council on redrafts of the constitution, and receiving an unexpectedly encouraging number of membership applications, I proceeded with the initial necessary appointments. These were: nominating committee (chaired by Leslie A. Stauber), membership committee (chaired by Carlo M. Ignoffo), program committee (chaired by John Briggs), and the society's representative to the governing board of AIBS (Victor Sprague). With the enthusiastic and energetic activity of these committees and the members of the council, the society appeared to be off to an encouraging start. This was the spirit we attempted to echo in the first *Newsletter* of the society, published 20 December 1967. The first general annual meeting of the society was planned to be held with the AIBS at Columbus, Ohio, 3–7 September 1968, with the council meeting during the afternoon and evening of 2 September.

It is to be expected that initiating a new scientific society "from scratch" is a difficult and frustrating undertaking. Because of world tensions existing at the time, attempting to place the society on an international footing appeared somewhat ambitious even though, except for Western Europe, the questionnaire we had circulated showed world-wide interest in such an international organization. As the months of 1967 went by, it became apparent that at the insect pathology centers in France and Germay, real resistance to joining the society had taken place. Inquiries of some of the leaders in those countries revealed that in spite of our protestations to the contrary, these leaders and most of their colleagues feared that for some reason (which I could never really understand) the new society threatened the insect pathology and microbial control activities of the European OILB, and that, in any case, they would not enjoy adequate representation in the government of the society. Somehow these fears had to be laid to rest. The manner in which our European colleagues mixed their scientific activities and international "politics" was quite foreign to most American insect pathologists and presented us with an enigma.

A break came early in January 1968 when it was learned that C. Vago, one of the concerned European leaders, would be having a short stop-over in New York during a trip he was making to Central America. Heimpel called me, and over the phone I authorized him to go to New York on behalf of the society to attempt to learn at first hand the reasons for the apprehensions that Vago and others were feeling with regard to the society. The discussion that these two had did much to clear the air, although I was not sure whether Vago had modified his view, about which he wrote to me in March of 1967, stating that each country should have its own society with the several societies then meeting internationally under the aegis of a permanent committee for international congresses. In view of the French dominance of the International Committee for Comparative Pathology, I could understand his fears concerning the international practicality of an international society originating in the United States. Heimpel advised me that psychologically it would be well officially to ask Vago to advise the council concerning the further internationalization of the society. Accordingly, I invited him to assume this role, as I subsequently asked Weiser for eastern Europe and Aizawa for Asia; as soon as possible we should, I thought, also invite similar advice from representatives of Africa, South America, and Australia. In the meantime, I carried on considerable correspondence with several other non-American members, notably van der Laan of the Netherlands. I was beginning to learn a bit as to how the continental European scientist went about the matters of "politics" and prestige. It was not too long after Vago's return to France, for example, that the applications for membership began coming in—to be charter members they had to reach the secretary-treasurer's office by 1 March (later extended to 15 March) 1968.

Then, in late February 1968, another development took place which further intensified the matter of how the society was to implement its desire to function on an international basis. The nominating committee, knowing the council's desire to give the society international parameters, submitted to the membership a slate containing three Europeans as candidates for vice-president (who two years hence would appear on the ballot as the sole nominee for president). The elections were held by mail ballot during the month of April. The "race" did credit to all three (Franz, Vago, and Weiser), but Vago received the most votes and was therefore elected to follow A. K. Sparks as the second vice-president, and hence he would be the third president of the society. We had achieved our goal of forming an international society for invertebrate pathology.

[Discussions of the first meeting of the society held in Columbus, Ohio, August 1968, and the second meeting in Burlington, Vermont, August 1969, were to have followed this section had EAS completed his writing. MCS.]

1. In a 1968 letter William E. Ogilvie, secretary-general manager of the International Live Stock Exposition, stated,

> The "Corn Borer Committee" [the Administrative Committee], of which Mr. [A. G.] Leonard was chairman, raised some $100,000 to finance these investigations. Business interests of the day who were concerned with the welfare of the Cornbelt farmer contributed to the fund.
>
> The reason that the Union Stock Yard & Transit Company spear-headed the project in the name of the International Live Stock Exposition was that the European corn borer was at the time regarded as a serious threat to the welfare of the Cornbelt farmer's major feed grain crop; and the Cornbelt farmer was the chief patron of the Chicago Union Stock Yards' livestock market as well as an annual contributor of major importance to the annual International Live Stock Shows.
>
> European research was apparently thought by this committee to have uncovered methods for control [i.e., biological control] of the infestation not yet discovered by American institutions; so the work that was sponsored by the so-called "International Corn Borer Investigations" was carried on in Europe under the direction of Tage U. Ellinger, a Dane, whose father at the time was president of an agricultural research institution in Denmark.

While Mr. Ogilvie's comments no doubt accurately represent the industrial concept of the investigations and the associated conferences and publications, as we shall shortly see, subtler influences were at work which made the endeavor essentially a European one.

2. Concerning this meeting, Dr. Jaroslav Weiser wrote to me on 4 April 1956

as follows: "I am pleased that you have received the microfilm of the index to the papers read in Leningrad, and I understand your surprise. I believe that we both had a finger in the origin and history of this meeting. After Pospelov died, his laboratory in the Union Institute of Plant Protection in Leningrad was without a leader, and only two women workers remained: A. A. Jevlachova and O. I. Schwetzova. Accordingly, there was no one who could effectively fight for the continuation of studies on the diseases of insects in an organized way. The first help to correct this situation came with the translation of your first book, *Insect Microbiology*, although this inspired work more along lines of medical entomology than insect pathology. Following this, the paper by Jevlachova and Schwetzova appeared, and after this the Russian translation of your second book, *Principles of Insect Pathology*, with its appended list of Soviet papers, was published. This was the second 'impulse.' In October of 1954, Professor Pawlowskij, the head of all zoology in the Academy of Sciences, was in Prague. He spent about three hours in my laboratory, during which I informed him in detail about our work. That was October 11th. Six weeks later the conference in Leningrad was held. I think it was in direct connection with his visit here. Other than Jevlachova and Schwetzova, most of those attending were from laboratories of apiculture and sericulture. I believe that the meeting was held primarily to give support to the resolution that special training in insect pathology be introduced in apiculture, sericulture, and plant protection research. . . . Among the more promising laboratories, so far as developing aspects of insect pathology are concerned, are those of Gerschenson in Kiev, Pojarkov (and in other laboratories, Chachanov, Michailov, and Dikasova) in Taschkent, Talalajev in Irkutsk, and Smirnov in Voronezh."

3. [A new International Committee on Nomenclature of Viruses (ICNV) was formed at the International Congress for Microbiology, in Moscow, in 1966. P. Wildy is president and C. Vago is a member of the executive committee. Vago is also the chairman of the Invertebrate Virus Subcommittee, with K. Aizawa, C. Ignoffo, M. E. Martignoni, L. Tarasevich, and T. W. Tinsley as members. The objective of ICNV was to find a universal toxonomic system for all viruses. The first report of ICNV and of its subcommittees was published in 1971 (volume 5 of *Monographs in Virology*). MEM.]

4. Strictly speaking, one must antedate this colloquium with a session, Symposium III, titled "Insect Microbiology," conducted during the Eighth International Congress of Microbiology held at Montreal, Quebec, 19–25 August 1962. In August 1960, I received a request for suggestions as to the makeup of the symposium from E. H. Garrard, chairman of the agricultural Microbiology Subcommittee of the congress. He also invited me to chair the symposium, but I had to refuse since I would not be attending the congress; I suggested that he invite L. Bailey, the well-known English expert on the diseases of the honey bee. Under Bailey's chairmanship, five interesting reviews were presented: A. Krieg spoke on crystalliferous bacteria pathogenic for insects; D. M. MacLeod and T. C. Loughheed reviewed the entomogenous fungi; Anton Koch dealt with the symbiotes in wood-destroying insects; C. Vago discussed the pathogenesis of insect viroses; and J. P. Kramer's paper was on entomophilic Microsporidia. Chairman Bailey had asked the participants to address themselves largely to the nature and ecology of the entomogenous microorganisms concerned; for the most part the presentations were in accordance with this request. (For the proceedings of this and other symposia, see Gibbons, 1963.)

It was good to have Koch speak on the mutualistic relationships between insects and their microbial symbiotes because, until recently, so little attention had been given them by North American biologists. Indeed, my own reviews of the subject (Steinhaus, 1946, 1949) were perhaps the first general consideration of the matter in the English languge; yet, spearheaded by Paul Buchner and his school, Europeans were well aware of this group of fascinating biological relationships which had been studied under the name *endosymbiosis*.

Unexpressed ideas are of no
more value than kernels in
a nut before it is cracked.

Author Unknown

6

Of Books and Journals

1. THOSE WHO PAVED THE WAY

I

This chapter is being written at a time of great transitions in the art, and science, of communication. The technologies involved in recording, retrieving, and transmitting information are daily becoming more complex and varied. Commercial combines of book publishers, film makers, television producers, computer companies, and the like, are forming on every hand. Even so, most scientists are still looking to the printed word and the publications and meetings of their professional societies for their basic informational sources.

At any rate, in attempting to trace, in a semihistorical manner, the development of insect pathology, we cannot ignore the role that journals and books have played in this development. In insect pathology, as in other disciplines and in daily life, John F. Kennedy's comment is appropriate: "Print is the true international currency of the modern world. The printer's art has built a vast array of secular and religious thought, of technical and scientific achievement. . . . To affect the thinking of man [through printing] is to influence the course of history." Certainly the printed word has influenced the course of the history of insect pathology. And, undoubtedly, as is already happening, the new technologies of communication will be adding their influence to the developments still to come.

II

In previous chapters I have commented upon some of the earliest writings on the diseases of insects (primarily the silkworm and the honey bee) and how this took the form of poetry as well as prose. Although P. H. Nysten's 1808 *Recherches sur les maladies des vers à soie*, was one of the, if not *the*, first books published solely on the disease of an insect, and Kirby and Spence (1826) published the first generalized chapter on the "Disease of Insects," still my fancy is taken by what I believe is the first Japanese publication I have been able to find devoted solely to the diseases of the silkworm. I have described this cloth-covered folder containing a single, loose sheet of mitsumata paper printed with woodcuts in chapter 1. I still recall the feeling of exaltation I felt when, with the help of my friends K. Aizawa and T. Hukuhara, I found this rare and unique document in the library of the Tokyo National Sericultural Experimental Station.

Passing down through Pasteur's 1870 classic *Études sur la maladie des vers à soie* and Omori's 1899 monumental *Modern Treatise on Japanese Silkworm Disease*, we must pay tribute to André Paillot's *L'infection chez les insectes* published in 1933. Although this was not a broad treatise on insect pathology, it was a magnificent portrayal of microbial disease in insects as known at the time and centrally pivoted on Paillot's own investigations. The subtitle of the 535-page book was "Immunité et symbiose", and went on to include a section on the role of arthropods in the transmission of pathogens of man and other vertebrates.

It is extremely difficult even to attempt to select the first significant North American publications relating to the diseases of insects. To be sure early federal and state bulletins on apiculture and sericulture in the United States contained sections or passages relating to the diseases of the honey bee and the silkworm. However, most of this was based on information obtained from abroad. Certainly the bulletins and circulars written by the Floridians on the diseases of citrus insects and those by Snow, Forbes, and others on the "chinch bug fungus" have to be hailed as pioneering highlights so far as early publications on the diseases of destructive insects are concerned. Thaxter's 1888 monograph on the Entomophthoraceae is a monument in itself.

One can take his choice, but as one who has read probably every piece of early American literature relating to the microbiology and diseases of insects, I find it difficult not to feel that the first American really to visualize the broad scope of the phenomenon of disease in insects was probably S. A. Forbes of Illinois. His work and interests, so far as insect disease is concerned, went far beyond the efforts of him and others to control the chinch bug by means of *Beauveria.* I catch his feeling of vision especially as I read his 1888 paper, published in *Psyche,* titled "On the Present State of Our Knowledge Concerning Contagious Insect Diseases." My conviction is enforced by the extensive and comprehensive "bibliographic record" appended to the article. Prior to and immediately following this article by Forbes, American publications relating to the diseases of insects began to appear about specific diseases or about diseases of specific insects. Admittedly not a great deal was known about insect diseases at that time, but limited as was the knowledge, Forbes's review indicates that he was seeing the forest as well as the trees.

As the twentieth century began, papers and bulletins on particular projects continued to appear. Thaxter proceeded with his "contributions toward a monograph of the Laboulbeniaceae" (chapter 2), and more was written about the fungi on citrus insects and on the chinch bug, grasshopper, and other insects. But then the subject matter began to vary. D'Herelle, beginning in 1911, published a series of papers on *Coccobacillus acridiorum* and its use in the control of certain grasshoppers. About this same time, R. W. Glaser and J. W. Chapman (1912) started publishing on "the wilt disease" of the gypsy moth. From then until his death in 1947, Glaser was to publish alone and with others, a steady stream of papers—most of them on some phase of insect microbiology or insect pathology.

The U.S. Department of Agriculture's Bureau of Entomology bulletins and circulars by G. F. White on the diseases of the honey bee were a monumental set of publications. Although he published a report on this subject in 1904 for the state of New York and did a thesis on the subject at Cornell University, it was his classical bureau bulletins (1906, 1912, 1917, 1920*a*, *b*) that can be considered great not only for their scientific content but for their literary clarity and excellence. Moreover, the high quality of White's work and writings ensured a continuation of concern with the diseases of the honey bee by the U.S. Department of Agriculture so that we find excellent papers by C. E. Burnside, J. I. Hambleton, and associates following those by White. In Canada similar contributions were being made by Lochhead, Katznelson, and others.

Although he is probably best known to most insect pathologists for his work on *Coelomomyces* infections in mosquitoes and other insects, J. N. Couch published not only a number of papers on *Septobasidium* (those semiparasitic fungi associated with certain scale insects) but an outstandingly superb treatise of 480 pages titled, simply, *The Genus Septobasidium.* Largely taxonomic in nature, it remains today, as it was in 1938 when published, one of North America's notable contributions to the literature in insect microbiology-pathology.

Chronologically, I have skipped over a publication which influenced me greatly as a student and, therefore, one to which I must pay personal homage. I refer to the seventy-page paper by H. Glasgow, published in 1914 and titled "The Gastric Caeca and the Caecal Bacteria of the Heteroptera." One day, as a graduate student in 1937, I came upon this paper quite by accident while searching for a reference on rickettsiae. That certain insects possessed special ceca, which contained specific bacteria, and that these bacteria were transmitted from generation to generation struck me as profoundly miraculous. (About this same time I was lead to the magnificent 1930 monograph on symbiosis by Paul Buchner of Germany. That certain living cells of insects regularly harbored living bacteria and yeasts, intensified my wonderment concerning the insect-microbe relationships that existed.) As I read Glasgow's paper and admired his drawings and illustrations, the desire grew within me to tell everyone about this fascinating subject. So, to me, Glasgow's paper is one of America's great publications in insect microbiology. Unfortunately, because I found that none of my teachers or fellow students were aware of Glasgow's and Buchner's publications, I naïvely set about to tell America about their work—as well as other insect-microbe relationships—in a review that, in all honesty, I think *Bacteriological Reviews* and science could have done without. For better or worse, however, preparing the review did get me started on a tangent that eventually ended up in the preparation of a book, through the production of which I was initiated into some of the mysteries and pitfalls of book publishing.

2. VENTURES IN WRITING

I

While pursuing the goals of my Muellhaupt Scholarship at Ohio State University, I began a collection of pertinent annotated references regarding the biological relationships between insects and bacteria. These I took with me a year later when I accepted an offer of a position from the U.S. Public Health Service at Hamilton, Montana. After spending numerous winter evenings at my reference collection and loaned references, I was bold enough to send a manuscript off to Burgess Publishing Company, which, in 1942, published it under the exaggerated title *Catalogue of Bacteria Associated Extracellularly with Insects and Ticks.* This small 206-page offering did not amount to much, but it did seed in me the desire to collect and study in a similar fashion, the references to all types of biological relationships between all types of microorganisms and insects.

Soon after the appearance of my so-called catalogue I tried to enlist, but was turned down a number of times by the army and navy for physical disabilities, I turned my attention to whatever research in medical entomology and vaccine production the government requested of our agency. More or less as relaxation, I spent many evenings—with the able assistance of my wife—continuing to dig out of the literature whatever I could find relating to insect (and arachnid) microbiology generally. My cabinet of annotated reference cards grew gradually to a surprisingly great number. Then a peculiar set of circumstances took place that almost inadvertently led to the eventual publication of *Insect Microbiology*.

Late in 1943 I was contacted by a Seattle-based admiral and asked if I would lead a team of specialists to the islands of the South Pacific to study scrub typhus (tsutsugamushi disease), which had become a serious problem as our armed forces occupied areas inhabited by the mite that transmitted this rickettsial malady. I responded eagerly, especially when the admiral assured me that he could have my physical disqualifications waived. Director Parker suggested that since "going to war" meant an uncertain future, I should do something about all the reference material I had been collecting on insect microbiology. By "do something" he meant the information should be published so that it would be available to others—even though I felt that my compilation of information was both incomplete and disorganized.

The pressure of expectation of doing military service left no time for careful consideration of the matter. Accordingly, I had the material typed up, and contacted the Comstock Publishing Company as to their possible interest in publishing the material in book form. My approach to Comstock Publishing Company (now a subsidiary of Cornell University Press) was purely on the basis of their then historic interest in publishing works of entomological interest, although in truth I had no idea at the time whether entomologists or microbiologists, or anyone, would be interested in a book of the kind I had put together. I say "put together" advisedly, because I did not consider the manuscript to be in any sense a scholarly, critical review of the literature. The latter had been my original aspiration, but now suddenly confronted with leaving for overseas duty I took the advice of Dr. Parker and submitted the material to the publisher in a "such as it is" state.

Matters concerning this manuscript then took an ironic twist. After signing the contract with Comstock and completing my preparations for leaving, I received a call from the admiral in Seattle. It seemed that although he had sufficient influence to have my physical disabilities cleared at the local and regional levels, he was not successful at the top level (i.e., the surgeon general's office) in Washington. Not too long after this, I left the Rocky Mountain Laboratory to accept a position on the faculty of the University of California at Berkeley.

To my great disappointment the publisher at Comstock, for one reason or another, delayed processing and printing the book for a period of more than two years from the time he received the manuscript. This delay made the book somewhat out of date when it finally appeared in August of 1946. Being a young, naïve, first-time author of a major manuscript made me a rather anxious and impatient author. I must admit that I became pretty indignant at times, especially when I was paying for all the revisions which had become necessary due to the long delay in publishing. In addition, I was to receive no royalties from the book.

So it was that *Insect Microbiology* was born after a most agonizing period of gestation and labor. In spite of the deficiencies caused by too-rapid compilation, publishing restrictions imposed by the war (the book had to be kept under 800 printed pages), and no real scholarship, the book did surprisingly well—primarily, I think, because it was the first and only compilation of its kind and served as a reference work for medical entomologists and for those few just becoming interested in the field of insect microbiology as such. In general, the reviews were kind and complimentary, but it was now clear that it was the entomologists who appreciated the effort rather than the microbiologists. In fact, the only unkind review was by a rather eminent bacterial physiologist who clearly completely missed the point of the book and thus revealed the narrowness of his view of what the published compilation could mean to those for whom it had been intended. Although his criticisms were, in general, valid from the viewpoint of a bacterial physiologist, the apparent viciousness of his comments at first dispirited me even though complimentary reviews appeared elsewhere at the same time. But this was a rewarding experience because it was this individual who first opened my eyes concerning the idiosyncrasies of book reviewers. I shall have more to say about this later, and will only comment now that, generally speaking, my advice is to ignore them. Do not take too seriously the bouquets and compliments tossed at your work, nor the derogatory remarks. Profit from the latter insofar as they are constructive criticisms, but biting, nasty, cruel, and condescending reviews are, as frequently as not, written either by people having these same qualities of personality or by those who themselves have never written a book.

II

My next venture into book writing was a much more enjoyable one, even if more hectic. After organizing the Laboratory of Insect Pathology on the Berkeley campus, it became clear that the next phase of our plans—teaching a course in insect pathology—should be implemented. Inasmuch as such a full-blown course had never been a part of the curriculum of any university or college, no textbook or other adequate teaching materials were available. At first I thought perhaps I could teach the course by giving lectures, providing mimeographed notes and hand-outs, and distributing loose-leaf instruction for the laboratory sessions. And, indeed, this is pretty much the way we began when I first gave the one-semester formal course in insect pathology beginning on 4 March 1946. The four-unit course (three lectures and one afternoon of laboratory work a week) was given in the curriculum of the Department of Entomology and Parasitology.

After the first presentation of the course, it was clear to me that not only was there a need for an up-to-date reference work devoted entirely to the diseases of insects, but a textbook on the subject was a real necessity. (The best I could hope for my just-published *Insect Microbiology* was that it serve as a sort of general reference work.) Accordingly, I decided to attempt such a reference-textbook combination by assuming a crash program of writing. By making good use of weekends and holidays, I was able to complete the manuscript for *Principles of Insect Pathology* by 1 May 1948. In this case, the publisher (McGraw-Hill Book Company) was most pleasant and considerate regarding the author's idiosyncracies. They lost no time in processing the manuscript, and although a book on such a specialized subject as insect pathology could, at best, mean but a marginal profit to the publisher, there was no indication of their giving the book a low priority on their publishing schedule. Yet, for a 757-page technical work of this kind, quality production could not be made hastily. It was not until 15 August 1949 that the volume was published.

A small part of the time taken by the publisher to process the book was the result of one of those once-in-a-million happenings which I am sure McGraw-Hill will not mind my relating briefly. For a time after his acceptance of the manuscript there was a strange quiet on the part of the publisher. On a visit to New York I took the occasion to stop by the offices of the McGraw-Hill Book Company to see how matters were going. Mr. Hugh Handsfield, editor of the college department, cordially greeted me, but it was obvious that he was greatly relieved about something that must have happened shortly before my arrival. Mr. Handsfield explained to me the reason for the period of quiet and of cryptic responses to my recent letters. For the first time in the company's history, they had lost a manuscript—mine! And only that morning had it been found. The manuscript had been sent to one of their out-of-town editors; somewhere along the way the mails had mislaid the package, delivering it only after several weeks, during which I was being protected against the anguish of their frantic search. Had the manuscript remained lost it would indeed have been distressful to me because in order to attempt to have the book ready for classes I had cut corners by submitting the first draft of the manuscript, with written-in corrections, additions, and deletions, and hence no second copy of the manuscript as submitted existed. (This was in the days before easy copying by Xerox!)

Another incident might be cited which reflected the courteous concern and good humor shown by this large publishing house. In one piece of advertising for the book, one of their copywriters made a rather impossible statement concerning the attributes of the book. After calling this to their attention, Mr. Stuart Dorman, manager of their college department, apologized and sent me a signed, specially drafted certificate reading: "This is to certify that all false claims made in advertising Professor Edward A. Steinhaus' PRINCIPLES OF INSECT PATHOLOGY are the fictitious products of the oftimes fevered brains of McGraw-Hill copywriters." Obviously part of this book company's great success in the publishing world is knowing how to handle its authors.

I remember how, during the book's production—the reading of proofs and all—I grew apprehensive about its reception inasmuch as I had literally thrown it together in much too short a time. Fortunately, those who reviewed the book were generous in their evaluations, and, probably because it was the first book of its kind in English, it was well received as an attempt to provide the broad discipline we could now call "insect pathology" with a reference base or rallying point. As a former student once said to me, "I figured that if the subject were important enough to write a book about, it was worth my looking into."

In 1950 the USSR published a Russian translation of *Insect Microbiology*, and in 1952 did the same with *Principles of Insect Pathology*. I understand that both titles sold well in Slavic countries and in those parts of the world where Russian was a common language. Shortly thereafter the Chinese Republic published a Chinese translation of the latter title using the Russian copy. These translations sold for but a fraction of the price of the original. As late as 1967, long after both *Insect Microbiology* and *Principles of Insect Pathology* were out of date, a reprint publisher (Hafner Publishing Company) reprinted both titles. In a sense this was flattering, but books which had ostensibly served their purpose deserved to die gently and to rest in peace.

Inasmuch as the books just referred to were used as reference works or textbooks for classes in insect pathology, mention should be made of the laboratory manual we prepared and used for our elementary course in insect pathology. As might be expected, in the first year or two the class was given loose-leaf directions for laboratory exercises. To the extent possible, the order of these exercises was in approximately the same order as the subject matter presented in the lectures. After a few years, Kenneth M. Hughes, who assisted in teaching the laboratory section of the course, prepared a regular mimeographed manual of laboratory exercises. This manual was augmented and modified as it was used through the years until 1959 when Mauro E. Martignoni, who had then begun teaching the laboratory sessions, and I completely revised the manual.

During the spring meeting of the American Society of Microbiology, while passing through the commercial exhibits, I passed by that of the Burgess Publishing Company of Minneapolis, and there I found Mr. C. S. Hutchinson, executive vice-president of the company, who had treated me so graciously when his company had published by catalogue on entomogenous bacteria back in 1942. I explained to him our problem of not having a laboratory manual of first-class appearance and asked him if he would be interested in publishing ours, even though the field concerned and the few courses given in the subject would mean not a great many sales. He jovially responded that for old times' sake he would seriously consider it. On 20 May 1960, I received a letter from him expressing definite interest.

Mauro and I got busy and polished up the manuscript, using Mauro's unique arrangement of the contents, and submitted it to Burgess promptly. The contract was signed by the middle of June, and, after an unexpected production delay, *Laboratory Exercises in Insect Microbiology and Insect Pathology* by Martignoni and Steinhaus appeared early in 1961. At last, courses in insect pathology, slowly emerging across the country, had a commercially published laboratory manual available--and so did we.

III

The publication, in 1963, of the two-volume work titled *Insect Pathology: An Advanced Treatise*, which I edited for Academic Press Inc., was an endeavor that gave me new insights into insect pathology and insect pathologists. It made clear how rapidly the field had grown since the publication of *Principles of Insect Pathology* fourteen years previously. Although the thirty-five authors who prepared the thirty-four chapters comprising the two volumes tried, in general, to limit their consideration of their assigned subjects to the previous ten or fifteen years (allowing of course for root attachments to historical aspects involved), the treatise totaled 1,350 pages—in spite of the fact that not all areas were covered (e.g., teratologies, Laboubeniales). Also it was a rare privilege to observe how, at least in their original manuscripts, the thirty-five pathologists viewed and approached their respective fields. The speed with which these two volumes were written and produced was unusual and was accomplished largely because of the exceptional cooperation of the authors involved.

Sometime in 1959 I chatted briefly with a representative of the Academic Press as to the possibility of publishing an editorial treatise to be used primarily as a reference work. I was not inclined to revise my own *Principles* because I had arrived at the conviction that insect pathology had reached the point at which a single author could not do justice to an in-depth treatment of the subject even if he were to cover only the developments of the past ten or fifteen years. However, I told the press representative that I could not undertake such a project until I had returned from a sabbatical leave that I had planned in the Orient and Europe during 1960–61. Moreover, I wanted time to circulate a questionnaire to representative insect pathologists as to the nature of, and contents most desired by them in, a treatise of the kind proposed. (The questionnaire, circulated and returned during May 1960, heartily endorsed in general the idea of producing a reference-type work arranged, for the most part, according to the types of diseases suffered by insects.)

I returned to the States, with my family, from Asia and Europe via Washington, D.C., in January 1961, where I remained for several months taking advantage of the library facilities in that area. As soon as I arrived back in California, I set about outlining the chapters of the treatise and inviting prospective authors. By 1 May, this part of the planning was virtually complete. However, knowing that I would not have time to spend for writing and editing beyond 1962, I requested of the authors that they have their manuscripts in my hands by 1 December 1961. In most cases, this allowed the authors only nine months in which to prepare their chapters. Some expressed consternation at being allowed so little time, but from my standpoint it was either that or nothing. But, using all the tricks of cajoling, persuasion, and periodic encouragement I could muster, we met our deadline for manuscripts with remarkable punctuality.

My experience with editing the *Annual Review of Entomology* aided me considerably in knowing how to apply this cajoling process, especially with reluctant authors who were "too busy right now," as well as with the inevitable number of procrastinating authors who have little regard for deadlines no matter how their delay may hold back the project and penalize other authors who take their assignment seriously and have their manuscripts in on time or ahead of time. (I, myself, had worked diligently to meet a deadline for a chapter on insect diseases to be included in a book on biological control, only to have it sit and age three years while the editor waited for slow and procrastinating authors to submit their contributions. I was determined that such would not happen with the treatise I was editing.)

The preface of the *Treatise* explained some of the problems involved, the difficulty in selecting one author for a particular subject when admittedly there were more than a dozen others competent to write on the subject, why some subjects had to be omitted, the difficulty of editing into smoother English the manuscripts received from non-English-speaking authors, etc. (With regard to this last point, colleagues have often asked me why I willingly rewrite papers or, in this case, chapters, submitted in poor English that can be "saved" by some rewriting. Admittedly it is frequently a most onerous task to go through a manuscript sentence by sentence, interpret the meaning, and then attempt to cast it into reasonably smooth-flowing English. There are several reasons, not all altruistic: In the *Advanced Treatise*, for example, it was important to obtain an international prospective and cross-section of recent development. If we depended on only those who wrote fluent English, such an objective would not be obtained. Therefore, it is worth the hypercriticism of some to have in English—for the convenience of the all-too-common monolingual English-language scientist—the thoughts and ideas of a foreign colleague, even though his heavily edited article may not read as smoothly or in as precise a form of English as the critic might be able to write. Also, sometimes, I have been pleased to rewrite a colleague's article purely for friendship's sake. Frequently because of the wide international usage of English as a scientific language, a colleague not skilled in the English language would like to have his contribution published in a more widely read language than his own. Rewriting his paper for him may be doing more for science and mankind than writing an original paper of my own. To say the least, it is ungracious and inconsiderate for an English-reading scientist to be critical of awkward phraseology and syntax in an article by a non-English scientist, when otherwise the contents of that article might not be available to him except through translating services.)

A while back I alluded to the preface of the *Treatise*. It reminds me of another point I should like to make. I have never understood why those who review books for journals so frequently fail either to read the author's preface or, if they do, ignore it and the qualifications and disclaimers stated in it. In the case of the *Treatise*, there were generally favorable reviews and the two volumes, although expensive, were well received. One exception, however, humorously stood out. The journal *Science* had asked an entomologist—not an insect pathologist—to review the *Treatise*. (A habit, incidentally not infrequently indulged in by that periodical at that time—that of asking nonauthorities to review certain works.) In this particular instance the reviewer, apparently feeling he should criticize something, condemned the book for omitting a subject to which, in actuality the *Treatise* had devoted an entire chapter. Then he criticized the work—perhaps justly—for omitting a favorite subject of his, ignoring the fact that the preface had clearly stated its regret in not being able to include all subjects that, had space been available, deserved inclusion in one of the two volumes. Presumably, many of our scientific brethren who review works of their colleagues, but ignore qualifying words, phrases, or explanations in the preface or introduction of a work, rely on that medieval devil, Tutivillus, whose duty it was to collect and carry to hell all the words and phrases skipped over or mutilated by priests in performing the services or in reciting the paternoster. If ever one seeks extremists he has but to read through the book review sections of our scientific journals (and the same applies to literary periodicals); here he will find the spectrum of criticisms all the way from the perfunctory pablum of undeserved praise to the condescending and frequently unfair cavils of the pontifical critic carrying on because he has found a misplaced comma or a typographical error. One can be almost certain that the latter has himself never gone through the agonies of writing a book.

But to return to the actual publication of the *Treatise*. As I have said, what with innumerable letters, memos, telegrams, cables, and long-distance phone calls, most of the manuscripts met the deadline of 1 December 1961. For the next two months I devoted virtually every free moment to editing the manuscripts. Shortly after 1 February 1962, I was able to deliver the edited manuscripts to the publisher. Again, since the *Treatise*, because of its technical nature in a very limited field, was obviously not going to be a best seller, the publisher could not make a "rush" job of its production. Nevertheless, early in 1963, volume 1 appeared, to be followed a few months later by volume 2.

I regret that there is insufficient space on these pages to present some of the interesting aspects of the promotion and sales programs engaged in by publishers of books and journals—from the viewpoint of the author or the editor. The game seems to be one in which the author is rarely satisfied with the publicity given his opus, while the publisher has charts, tables, and figures to prove that certain types of advertising are better than others and that beyond a given point the promotion of any given book reaches diminishing returns long before the author would believe.

3. ADVENTURES OF AN EDITOR

I

There was a time when serving as an editor of a journal, series, or multiauthored treatise was considered somewhat prestigious in the scientific community. Even serving on an editorial board was worthy of citation in one's curriculum vitae. Indeed, major universities considered the duties of an editor to be equivalent to other scholarly activities, and even provided reduced teaching and administrative loads to compensate for the time involved. Today, such is not often the case. Indeed, there is a tendency in some of the more snobbish areas of science to assume that the editorship of journals and periodic reviews should be left to effete old professors or to scientists who are "over the hill" and have nothing better to do. The cause of this attitude in some quarters is difficult to discern—perhaps the illusion that there are as many scientific journals these days as there are scientists tends to demean the art and tasks of editorship to a status of commonness.

My own experience has been quite otherwise. To the extent that I found time to indulge in serving as an editor or as a member of an editorial board, I have found it to be an exciting, creative, and rewarding activity. Of course, I carry a bias in this regard—and have done so since a boy of twelve. As a youth I both worked in the print shop of our local weekly newspaper and published a small news magazine with equipment my father purchased and gave to me in return for printing handbills and advertising matter for his store. And ever since this exposure, which lasted through high school, I have carried printer's ink in my veins. To this day I am one of those who would like to run his own small newspaper, but who also knows that to do so would mean utter bankruptcy.

I do not recall having really contributed much to the activities of the several editorial boards (*Journal of Economic Entomology*, *Virology*, *Life Sciences*, and others) on which I served, but I always found what tasks there were to be stimulating. Indeed, I must at this point admit that of the innumerable university committees on which I, as do most academicians, had to serve in the course of my university career, the only one I felt accomplished something and was rewarding was the editorial committee. This committee, at the time I served on it during the 1950s, had to approve every monograph, book, or publication that was entitled to bear the imprint of the University of California Press. Moreover, to my knowledge, it was the only committee that was funded directly from the president's office.

Although there was, of course, satisfaction in reviewing manuscripts pertaining to the work in one's own or related disciplines, the real intellectual thrill came in joining in or listening to the discussion and evaluations made by experts in areas quite removed from one's own. I could give many examples, but one fellow committee member who comes to mind today is Robert A. Scalapino, the political scientist whose expertise and knowledge of China and the Far East is widely recognized. Time after time I listened in fascination as Bob articulately presented his critiques of manuscripts. Our discussions were always interesting and frequently quite animated and lively. Each member had his own style of presenting his reviews or critiques—one or two were solemn and pedantic, another almost highly excited and flamboyant, some terse and scholarly, others detailed or perhaps even verbose. Foster Sherwood, another political scientist and later a vice-chancellor at UCLA, frequently found himself most comfortable pacing the floor and presenting arguments for or against the manuscript under discussion somewhat in the manner of a lawyer presenting his case. One of my perennial tasks was to explain to new committee members that the word "revision" in the titles of treatises in systematic biology was a term pertaining to the modification of taxonomic groups and did not mean that we had before us a manuscript in which the author was announcing that the work itself had been revised. Indeed, the indoctrination of a new member of the committee was always an interesting occurrence—ordinarily guided by the patient skills of the director of the university press, August Frugé.

It was while serving on the university editorial committee

that the first volume of the *Annual Review of Entomology* appeared—in January 1956. And thereby hangs a tale!

Having had little formal training in entomology I found myself in a constant state of self-education as far as insects were concerned. Indeed, except for certain taxonomic catalogues, I eventually came to feel that I had read and studied everything ever written on insects. This experience caused me to have a great appreciation for the need of a review publication of some sort in the field of entomology generally.

As part of my self-education in entomology I attended as many local, regional, and national meetings as I could. During these meetings I would find occasion to talk with the officers of the group and the editors and members of the editorial boards of their publications, asking them to consider initiating a review publication. For the most part my pleas went unheeded. It is important to mention that at a meeting of the old Association of Economic Entomologists, I approached Ernest N. Cory, who at that particular time gave me no encouragement whatever. Later, he was to save the day.

About the time I had given up the idea as a lost cause, the American Association of Economic Entomologists was amalgamated with the Entomological Society of America, the new organization taking the name of the latter society. This took place in 1953, with Charles E. Palm of Cornell as the new society's first president. Although Palm was submerged in an avalanche of new and onerous duties that accompanied the amalgamation, I approached him on the subject of a review publication, more or less in the spirit of "what harm could it do?" To my surprise—and to entomology's debt—Palm responded by saying he thought it was a good idea (Robert L. Metcalf had recently written an editorial for the *Journal of Economic Entomology* recommending a similar venture), and would I be willing to be chairman of a committee on the matter. Needless to say, I accepted—this could be the break I had been awaiting.

As chairman of the committee, I more or less assumed the responsibility of personally investigating all the possibilities (whether we should have a review journal, a quarterly review, an intermittent periodical review, or an annual review), and contacting the several publishing houses known to publish review publications. It soon boiled down to the decision that the style of publication best suited for our needs was that of the well-known *Annual Review of Biochemistry*, published by Annual Reviews, Inc., a nonprofit organization then located in Stanford, California, and later moved to Palo Alto.

On 2 July 1953, I wrote to J. Murray Luck, managing editor and founder of Annual Reviews, Inc., asking if his organization would be interested in publishing an *Annual Review of Entomology*. (Luck was also Professor of Biochemistry at Stanford University.) Dr. Luck invited me to appear before the Board of Directors of Annual Reviews to discuss the matter with them. I gathered data of all sorts to support our case—unknown to me, Annual Reviews was making a similar survey and exploration, typifying the thorough manner in which, I was to learn, that Murray Luck went about all matters in his role as manager of the organization. To give me moral as well as substantive support I asked E. Gorton Linsley to accompany me to the meeting with their board. Moreover, although my credentials as an honest-to-goodness entomologist could be questioned, those of Gort Linsley certainly could not.

The meeting went off quite well in spite of the board's apprehension concerning the relatively few book-buying entomologists that they calculated were in the world as compared with biochemists. Our request did not receive a carte blanche response. Annual Reviews was willing to undertake the publication of the *Annual Review of Entomology* providing the society would pledge subsidizing support not to exceed $2,500 for each of four years. In return, Annual Reviews agreed to give the society a certain amount (depending on the amount of sales) for each copy of the *Review* sold through the society's office and to continue this arrangement indefinitely beyond the four-year subsidy period. This, it seemed to me, was a very reasonable arrangement and one which, in the long run, would benefit the society financially as well as prestigiously. Accordingly, I now had a selling job to do to the governing board of the Entomological Society of America.

At the national meetings of the society held in Los Angeles in December 1953, I presented my report to the society's governing board. Although I presented Annual Reviews' proposition as convincingly as I could, there were two members of the board who spoke out against undertaking the venture at this time when the newly amalgamated society was just getting under way. I left the room prior to the vote being taken, but my heart sank because I knew that my request could not withstand the attack of the two prominent members of the board who opposed it. I was right.

But I had not considered one factor: Dr. Cory, the outgoing secretary, had not attended the meeting at which I had made my presentation. (Remember, it was Dr. Cory whom I had found earlier not too sympathetic to my ideas regarding a review publication for entomology.) Upon seeing Cory the next morning I expressed my disappointment over the previous evening's action of the board. Thanks to his keen business sense he reacted immediately by commenting, in effect, that the board was throwing away a golden opportunity to guarantee itself a regular income on a publication following the four-year period of paying for any losses (up to $2,500) incurred in publishing the *Review*. Accordingly, he returned to the board meeting on the morning of 8 December and convinced them of the financial merits of the proposal I had presented the previous evening. The board then rescinded its previous action and voted in favor of the arrangement with Annual Reviews, Inc. How ironical, I thought, was the fact that the man who I thought might least approve of the idea turned out to be the one to save it, albeit largely on an economic basis rather than one of scholarship. (I prepared other, more diplomatic, explanations of the agreement with Annual Reviews for publication in the *Bulletin of the Entomological Society of America* [vol. 1, No. 2, 1955], and in the preface to volume 4 [1959] of the *Annual Review of Entomology*.)

I happily reported to Murray Luck the action taken by the society. From the standpoint of the society, the committee appointed by Charlie Palm to look into the matter had performed its service and could be discharged. (Later, Herbert H. Ross, who followed Palm as president of the society, asked the committee to reassemble and to function as a liaison group between the society and Annual Reviews.)

Two or three weeks later Dr. Luck drove up from Stanford and dropped in for a visit. After exchanging a few pleasantries, I was utterly amazed when he asked if I would be editor of the new *Review*. I immediately protested explaining that there were any number of entomologists who could fill the post of editor better than I. After assuring me that I could select my own associate editor and suggest the members (with the society's approval) of the first editorial committee, we parted with my promise to think it over. On 14 January 1954, he wrote me, formally asking me to accept the editorship. I accepted, but not with a great deal of confidence in my ability to meet the challenge such a new undertaking involved.

I gave a great deal of thought to my selection of an associate editor, but the choice was not difficult to make. Inasmuch as he assumed the editorship when, because of the pressure of other duties, I relinquished the position after the publication of volume 7, and has carried on ever since with obvious skill and diligence, it is obvious I selected the right man. Of course, I am speaking of Ray F. Smith, who, since 1962, has not only supervised the editing of the *Review* but has successfully carried out matters pertaining to the form of references and indices we had planned during the first seven years we worked together. Ray's broad knowledge of entomology and incisive skills as an editor were such that beginning with volume 5 (1960) of the *Review*, I asked that he be made a full co-editor. When he took over the helm in 1962, with Thomas E. Mittler as associate editor (and later as co-editor), he did so with such consummate skill as to ensure the complete success of the *Review*.

I shall not go into the mechanics of how the *Review* functions (for this, see the preface to volume 3, 1958) except to say this: each year the editors meet with the *Review*'s editorial committee. The committee (selected with the confirmation of the Entomological Society of America), using a rather involved formula, appraises the needs for reviews of subjects falling within all branches of entomology. Topics and authors are selected, and the final result is usually a high quality review of pertinent subjects at an appropriate time.

The first editorial committee was comprised of A. W. A. Brown, H. M. Harris, R. L. Metcalf, C. D. Michener, C. B. Philip, and the editor. Ordinarily the committee meets at the time (around the first of December) of the regular annual meeting of the Entomological Society of America. However, this first committee had to meet early in the spring of 1954 in order to get matters under way. Accordingly, through the help of Harlow Mills—a member of the society's original committee to consider the feasibility of having a review publication and director of the Illinois Natural History Survey—arrangements were made to meet at the Natural History Museum in Urbana. Such a site for our first meeting had the advantage of being on the home grounds of the society's president, Herb Ross, and president-elect, George Decker. Dr. A. B. Gurney, the society's secretary, would come from Washington, D.C., to attend, and the location was convenient for Dr. Luck, who, being a biochemist, obviously wanted to meet and observe some entomologists, to talk with the society's officers, and to see that the editorial committee stepped off with the right foot. There is no doubt that the meeting was a definite success. The enthusiasm over having an *Annual Review of Entomology* was shared by everyone; the officers of the society were pleased with Dr. Luck and the straightforward, businesslike manner with which he approached the contractual agreements between the society and Annual Reviews, Inc.; and Dr. Luck appeared pleased with what he saw of entomologists and the scholarly, professional way they proceeded to launch the new undertaking.

From a purely personal standpoint, I thoroughly enjoyed my association with Annual Reviews and its personnel. Dr. Luck was always pleasant and was tolerant and understanding of our requests and innovations. He was the talented and skillful manager of a publishing corporation which grew from his first love, the *Annual Review of Biochemistry* (which he initiated in 1932) to the publication of a prestigious annual review in each of thirteen different major fields of science. (I am writing these words in the spring 1968.) I appreciated the opportunity that the *Review* provided for a creative association with so many other entomologists—readers, authors, members of the editorial committee, and especially with my co-editor, Ray Smith. I sincerely regretted that the pressure of other obligations forced me to abandon the editorship in 1962 after almost ten years (including more than two years preparatory work) of association with it. [The editors of the *Annual Review of Entomology* in a notice placed between the preface and the contents in volume 15, 1970, stated: "Edward A. Steinhaus . . . conceived the idea of an *Annual Review of Entomology*, encouraged its publication and then served as its editor for the first seven volumes (1956–1962). MCS.]

I realize that I have spent a somewhat longer time than I should telling of the beginnings of the *Annual Review of Entomology*. However, this publication relates in two ways to insect pathology. The *Review* offered an excellent outlet for high-quality reviews of different aspects of insect disease, of which there have been some truly outstanding examples. Secondly, the experience in encouraging and otherwise handling authors gained as an editor of the *Review* stood me in good stead when I came to find a publisher for the *Journal of Insect Pathology* and to edit the two volume work *Insect Pathology: An Advanced Treatise.*

II

As insect pathology became established in North America after World War II there was revealed a growing need for a journal of its own. Superficially, at least, two experiences were most often encountered, especially by those engaged in basic or fundamental research in insect pathology. At that time (in the 1940s and 1950s) few, if any, entomology journals used paper of a high-quality gloss or used fine-screened halftones which would portray photographs of sectioned tissue or of electron micrographs to the satisfaction of most pathologists. Secondly, a common experience of those working in basic insect pathology was to be told by entomological or zoological journals that their papers more properly belonged in journals of bacteriology or microbiology; contrariwise, microbiology journals would frequently suggest that the author send his manuscript to an entomology journal because basically his work concerned insects. A frustrating situation indeed!

Moreover, as I was later to write in the introduction of volume 1, number 1, of the *Journal of Insect Pathology*,

> Whenever a branch of science, consisting of overlapping disciplines, finds difficulty in having its papers published in journals devoted to the classical fields of knowledge; whenever a frustrating diffusion is caused by a scattering of papers in numerous, sometimes obscure, journals, because they have 'no place of their own' to go; whenever the science itself can be stimulated, dignified, made more intimate, and given greater coherence, form, and substance by having its own journal; whenever these needs appear to be felt by workers in an increasingly active field all over the world; whenever these situations prevail—as they do with insect pathology—surely this is justification enough for the type of scientific periodical we hope the *Journal of Insect Pathology* will prove to be.
>
> . . . the competition for publication space in biological journals has increased to a point that separate and more specialized publishing outlets have been required.
>
> Insect pathology is a field of endeavor utilizing the techniques and knowledge of such disciplines as entomology (in most of its various branches), pathology, bacteriology, mycology, protozoology, virology, nematology, physiology, histology, and immunology. It may also encompass certain aspects of the general field of insect microbiology and certain of the biological relationships existing between insects and microorganisms not pathogenic to them. Because of the numerous disciplines involved, papers and reports pertaining to insect pathology, insect microbiology, the prevention and therapy of diseases of useful and beneficial insects, and the use of microorganisms in the control of insects (i.e., microbial control), are being published not only in a variety of entomological and microbiological journals, but also in journals, bulletins, transactions, and periodicals of almost every conceivable type. This state of affairs is not necessarily bad insofar as it draws the attention of biologists in other fields to the problems and activities of insect pathology, but it does leave a virile and active field without a publishing organ of its own. The *Journal of Insect Pathology* is designed to help fill this void, and to provide a common medium for all phases of insect pathology, and eventually, we hope, the broader field of invertebrate pathology.

As these frustrations became more manifest, I decided to attempt to relieve them by seeking a high-quality publication outlet. I proceeded with some hesitation, at first, however, because I saw dangers in isolating papers in insect pathology and microbial control from papers in other branches of entomology or from general entomology. If insect pathology were to be accepted as a legitimate branch of entomology, in its own right, papers concerning the subject should mingle with those of other areas of entomology. If, when an entomologist perused his regular entomological journal, he were forced at least to notice papers dealing with the diseases of insects, he would come to accept insect pathology on the same basis as he accepted insect physiology, insect taxonomy, insect morphology, insect toxicology, and the rest. However, the more I thought about it the more I felt that, while there was justification in this apprehension, to accept the status quo would not solve the problem. Moreover, eventually basic insect pathology should also relate to other areas of invertebrate pathology, and to comparative pathology generally. Thus, it would be expected that a journal for invertebrate pathology would some day be appropriate.

There was another matter that bothered me. Entomology and entomologists had treated my intrusion into their midst with kindness and generosity. I would not want to start a journal which, if successful, in any sense harmed or stole papers from other entomology journals—especially the two principal publishing organs of the Entomological Society of America. Accordingly, I conferred with the society's executive secretary, Robert H. Nelson, and with the society's managing editor, F. W. Poos. Both of these gentlemen not only encouraged me, but Poos, who was at the time overburdened with a large backlog of manuscripts and attempting to cope with an increasing flood of new manuscripts, said he would welcome anything that took away some of the load of manuscripts.

Somewhat in the same spirit with which I had proceeded to find a publisher for the *Annual Review of Entomology*, I set forth with my hopes concerning the journal idea. Except this time I did not believe that the situation could wait so long for fruition. Moreover, I was a bit more audacious now than when I was a relative newcomer among entomologists.

At any rate, I carefully studied the nature of the journals being published by the leading journal publishers. Because the journal would not be backed by a society (the *Annual Review*, in spite of its remarkable success, had taught me the difficulties in taking this route), I had to find a publisher satisfied in assuming a risk (the initial sales were certain to be low) and yet willing to produce a superior product on high-quality paper, with high-quality printing and illustration. An inquiry at our University of California Press was discouraging because they sought to limit the number of journals they published. Of the two or three remaining possibilities, I decided to approach Academic Press, Inc., of New York.

To my pleasant surprise, Academic Press showed a genuine interest in my proposal, largely, I believe, because they saw it as a long-range investment in a newly developing field. Nevertheless, I was surprised at their decision to go ahead with the project because the results of their sampling of opinions of selected scientists were anything but encouraging. The general concensus was that there was not enough being written in insect pathology to fill a journal. (How ironic this seems now, in 1968—ten years later—when, because we find a backlog of almost 100 manuscripts on hand, we are negotiating an increase in the number of pages to be published each year!) However, I took this information as a good omen because I had been told that an annual review of entomology could not succeed; and before that I had been told by doubters that the development of the field of insect pathology was ridiculous, and other such similar experiences began to cause me to feel that the surest sign of success was to have the idea condemned or ridiculed.

In the present instance, however, I am afraid that even my closest associates thought I had taken on an impossible venture, or that at least it showed undue temerity on my part. In any case, I assured the vice-president of Academic Press, Mr. Kurt Jacoby—a grand gentleman—that somehow I would solicit an adequate supply of manuscripts if they would be willing to give the proposed journal a three-year try. Undoubtedly there are insect pathologists who may remember the pleading, yet Fifth-Avenue-style, letters they received from me requesting manuscripts. To those who responded (with encouragement and promises if not with manuscripts) the *Journal* will always be indebted. Not every author is willing to entrust one of his brain children to an unknown periodical the fate of which is so uncertain!

I received the "go-ahead" signal from Mr. Jacoby late in 1958. No formal contract was involved because my service as editor was to be a service of love. It was reward enough to feel that by an extra bit of work I could help provide insect pathologists with somewhat of the kind of journal they needed. I was given a free hand in determining the policy of the *Journal*, the standards required, and the make-up of the editorial board. Aside from a free copy of the *Journal*, and the satisfaction of witnessing its existence, my only--but very real—compensation was the right to spout off at the beginning of each volume (originally each volume consisted of four numbers) in an editorial.

I was anxious to have the *Journal* begin publication in 1959. However, as a quarterly, it was already too late to solicit and edit the manuscripts in time for a March, 1959 number. Accordingly, the press was willing for the first number to appear in May, 1959, with succeeding numbers appearing in August, October, and December. The following year, and every year thereafter, until 1968, they appeared in March, June, September, and December of each year.

After the first year of the *Journal's* publication, there was never any doubt in my mind about there not being enough manuscripts to fill the *Journal*. The size, or number of pages, comprising the periodical was determined by Academic Press, which, for economic reasons, could not afford to allow the *Journal* to run much over 450 to 500 pages per year during the early years of operation. Again, it was a case of a private publishing company attempting to sell a first-class product in a field having but a limited number of scientists, few of whom could afford a costly publication. As planned by Academic Press, subscriptions from the libraries of universities, research laboratories, and other institutions carried the product, which, happily, was generally well received by insect pathologists.

During the second and third year of the *Journal's* publication

I began the eventually successful negotiations with the publishers to provide separate subscriptions, at a reduced price, to individuals promising that theirs was a personal subscription and not one for their institution. At first the press was hesitant to agree to such an arrangement. However, they were sympathetic to my pleas on behalf of the individual scientists who wanted to have their own copies (but who could not afford the ever-increasing high price being charged institutions), and finally agreed to give the idea a try. I believe this was the first of their *Journals* with which Academic Press tried this dual subscription rates. It worked to their satisfaction, i.e., it did not reduce the number of institutional subscriptions, and was eventually used with a number of their periodical publications.

Another initial decision that should be mentioned is the fact that my original proposal was to name the publication the *Journal of Invertebrate Pathology*. However, those at the press felt that inasmuch as it was the pathology (diseases) of insects that was then receiving most of the attention, in order for the endeavor to succeed it would be necessary to capitalize upon this "wave of interest" and to reflect this in the name of the journal. Hence, the name *Journal of Insect Pathology* was selected as the initial title for the periodical. However, from the very first editorial in the very first number our aspirations concerning the broader field of invertebrate pathology were made clear.

Choosing the first editorial board was a difficult task. So many aspects of the selection seemed to work at cross purposes. For reasons I do not recall, an arbitrary number of twelve was decided on as an appropriate number to be on the board. Yet this was a small number when one tried to represent (but never could) the several different areas of insect pathology activity, several different countries to indicate the international tone we hoped for, and men with qualifications to assist in reviewing manuscripts and promoting the *Journal*. All this plus a logical system of rotation to keep these several factors in balance. Nevertheless, I made a stab at it and all twelve of those selected willingly agreed to join in setting the sails on an unknown ship setting out to unknown seas. They were Aizawa (Japan), Angus (Canada), Bailey (England), Bergold (then of Canada), Briggs, Hall, Mains, and Thompson (United States of America), Franz (Germany), Masera (Italy), Vago (France), and Weiser (Czechoslovakia).

Although the editorial board was of international complexion, the *Journal* could publish its articles only in English. The reason for this is interesting but still little understood. Scientific periodicals of limited circulation, published by private publishing companies, depend upon subscriptions from individual scientists even though their mainstay is library and institutional subscriptions. At least during the first ten years of the *Journal*'s existence the world economic situation was such that only a few individual European scientists appeared to have the money with which to purchase expensively produced journals, even if they were multilingual, and American scientists refused, for the most part, to invest in a publication unless the articles were printed in English. Therefore, although it seemed only fair and sensible for an international journal to publish its articles in the several leading languages of the science of the times, it was economic suicide to do so. It was difficult to convince our European colleagues, used to multilingual periodicals, of this fact, and they undoubtedly considered our "publish in English only" policy to be a parochial act. But it was in truth a matter of economics; had we published the *Journal of Insect Pathology* with the articles in different languages, it would not have been able to remain solvent. Fortunately for us, English has become pretty much the universal language in science. Even so, had the publisher been willing and had we the expertise, I would have liked to have had foreign language summaries.

A word or two need be said concerning the change, in 1965 (vol. 7), of the name of the journal from *Journal of Insect Pathology* to *Journal of Invertebrate Pathology*. I have already explained why we did not use the latter name to begin with when the *Journal* was initiated in 1959. By 1964, it was clear that the study of disease in other invertebrates was gaining strength; papers on the diseases of shellfish were increasing in number, some of them being submitted to and published by our *Journal of Insect Pathology*. Accordingly, I decided to go to New York to request of Academic Press that (1) the name of the *Journal* now be changed to indicate the coverage of the pathology and microbiology of all invertebrates and (2) the page size of the *Journal* be increased and the format be changed to the more-easy-to-read double-column make-up. Mr. Jacoby, still the "boss" of this fantastically rapid growing publishing company, even at his advancing age, readily agreed to the requests. He authorized the changes to be made beginning with the first number of volume 7 (1965). Inasmuch as I had declared our aspirations regarding the name from the introduction of the very first number of the *Journal*, I am sure that the name change did not come as any great surprise to our readers, although it may have confused some librarians at first.

As our editorial appearing in the first number of the "renovated" *Journal* declared, "The new name is not merely a device to provide an appropriately named medium of communication for a broader spectrum of pathological endeavor; it is an overt attempt to intensify and to accelerate interest and activity in invertebrate pathology generally and in comparative pathology." Future events were to justify this aspiration.

4. RAMBLINGS OF AN AUTHOR

I

The compensations of writing technical books for a very limited group of specialists or for editing treatises of journals for this same group are anything but monetary in nature. But compensations there are. Some are psychological in nature, some involve the satisfaction of "doing one's duty as he sees it" when the need is there, and some involve the pure sense of fulfillment from helping one's colleagues and one's fellow man, as corny as that may sound. Writing, and having published, a book that some of your associates say is "premature" (which may be translated to mean "you're not the guy to do it"), founding and editing a review publication or journal that your elders or peers tell you cannot succeed, and grinding through and rewriting manuscripts that are of excellent scientific merit but submitted in poor English or with complete disregard to format—the only reason I can personally account for this type of activity is to say that it gives one a feeling of creativity and service, as well as fulfillment. But funny and interesting things happen along the way.

Although, to be sure, a plethora of new scientific nomenclature, terminology, and jargon is proposed and generated by individual scientific papers, nevertheless an author of a text or major reference work, or the editor of a review publication or a journal has a unique opportunity to strike a blow for what he considers to be correct or proper terminology and semantics. This may even extend to coining new terms and then encouraging their use.

One of my first experiences in this connection had to do with the use of the words *symbiosis* and *symbiont*. The results should encourage young scientists not to hesitate to enter where angels fear to tread. Perhaps the story is worth telling.

During 1942 and 1943 when I was compiling the information contained in *Insect Microbiology*, I was distressed by the fact that contrary to the meaning given the word *symbiosis* by its originator, de Bary, it was used throughout most of the literature to mean the relationship between organisms characterized by a mutually advantageous association. Even *Webster's Dictionary* indicated that the term is ordinarily used where the association is advantageous. But to my youthful mind this usage was wrong. I was fully aware of Humpty-Dumpty's admonition that a word "means just what I choose it to mean—neither more nor less," but de Bary's original concept made much more sense. Why was it violated? Such eminent biological writers as Marston Bates wrote that it was useless to attempt to change the tide of common usage. It may have been common usage but it was also uninformed usage. De Bary (1879) originally used the word *symbiosis* as a general term referring simply to the living together of dissimilar organisms. It was an umbrella term including mutualism, commensalism, parasitism, and other associations. But who was I to set the record straight?

Fortunately, while writing *Insect Microbiology*, I found a

decision (apparently generally ignored) by the Committee on Terminology of the American Society of Parasitologists (*Journal of Parasitology*, 1937, *23*, 326–29) that recommended that the term *symbiosis* be used as de Bary had originally defined it, and to mean simply the living together of dissimilar organisms without regard to the nature of this association. The term *mutualism* should be used to refer to the relationship between organisms that live in a mutually advantageous association. I was determined to follow this recommendation.

The use of the word *symbiote* rather than the then commonly used *symbiont* was a bit stickier. Regarding this the committee on terminology stated:

> *Symbiont* is the form coined by De Bary, *Webster's Dictionary* gives *symbiont* as the preferred form . . . whereas *symbiote* is listed as a synonym or variant. Meyer (1925) and others have maintained that symbiote is the correct form. . . . the philologist would prefer symbiote which has a definite Greek origin and is correctly formed in English. . . . The matter is apparently one of taste and usage rather than correctness.

Well, I decided to side with the philologist in this instance. Etymologically, symbiont was incorrectly formed; why not use the correctly formed (from the Greek) symbiote? And so I wrote the pertinent chapters in *Insect Microbiology* accordingly. Without wishing to take credit for what may have happened anyway, I did notice that after the book was published, other authors began to follow suit. Professor Harold Kirby, famous for his studies of the mutualistic protozoa in the gut of termites, chatted with me on one occasion, telling me that if I took the stand I had indicated in the manuscript of my book, he would support it in his writings. He did so, but unfortunately, his untimely death lost me a staunch colleague in this cause.

Although even today there are authors who misuse both *symbiosis* and *symbiote*, the point to be made is that as an author in a field relating to mutualistic associations between insects and microorganisms, for example, and as an editor of a journal which receives papers relating to such associations, it is possible, without being dictatorial about it, to guide at least a certain segment of the literature along lines of proper usage. To the few--very few—who prefer the "incorrect" usage, my answer has been one of permissiveness, but with an editorial request that they provide a footnote to their paper explaining why they prefer to use the word *symbiosis*, for example, in a manner contrary to that defined by de Bary and recommended by the Committee on Terminology of the American Society of Parasitologists. This request frequently settles the matter in favor of de Bary.

As already mentioned (chapter 3) the words *epidemic* and *epizootic* in relation to insects, to my thinking, also had been misused in the past, and it took some campaigning on my part to convince my associates that the proper usuage was insect *epizootic* and not insect *epidemic*.

By the time I had finished writing *Principles of Insect Pathology*, I realized that the almost universally used expressions *polyhedral disease* and *polyhedral viruses* used to designate those insect-pathogenic viruses occluded in polyhedral-shaped (Gr. *polyhedros*, with many sides) inclusion bodies were incorrect (chapter 3). Authors using these terms did not really mean to say that they were concerned with a many-sided disease or a many-sided virus (although later, true polyhedral-shaped viruses were discovered in insects). Accordingly, as an editor, I insisted that instead of an author's saying polyhedral disease or polyhedral virus he say polyhedrosis (as he would say tuberculosis or coccidiomycosis) and polyhedrosis virus as he would say tuberculosis bacillus or encephalitis virus. Gradually one could see the change take effect throughout the English literature in insect pathology, although one still occasionally finds misusages—but hopefully not in the *Journal of Invertebrate Pathology*.

Some of the terminology used in insect pathology, and for which I have been credited or blamed, is derived from certain of my scientific publications. Thus, for example, there is *granulosis* (*-es*, plural), *stressor*, *incitant*, and others. Believing that they added to the precision of discussions in which their use was pertinent, I was able, as an editor, to guide their usage for authors whose writing required such terminology or who were misusing it.

Today it is difficult to realize that during the renaissance years of insect pathology, just following World War II, much of the terminology and vocabulary of insect pathology lacked precision and was carelessly used. Of course, new terminology and jargon continued to emerge. But there is no denying that attention to the preciseness of meaning and the nuances of this new terminology in the writings of leading insect pathologists, as well as in carefully edited journals, has a steadying influence on the semantics of this branch of science, as well as any other. Also a great factor in establishing proper usage, without restricting innovation, is the attention given this matter in the teaching of courses in insect and general invertebrate pathology.[1]

5. RAMBLINGS OF AN EDITOR

I

Most editors of scientific publications inevitably develop a philosophy and a set of biases concerning editing. This is to be expected. Unfortunately, one of the faults of most editors is that they feel called upon, at one time or another, to express this philosophy and these biases. Being no better than the others, and probably a great deal worse, I shall take this opportunity and this time to express myself on these matters. Not that what I have to say on the subject is particularly new or original, but inasmuch as I have evaluated, edited, and otherwise handled approximately 2,000 of my colleagues' opuses, they may be curious—should they be reading these pages—to know why I handled their brain children as I did.

Most scientists of any experience at publishing are aware that although there may be standardizations of style, format, and copy editing, the attitudes and philosophies of subject editors and editors-in-chief vary widely. Accordingly, every viewpoint I am about to express will be objected to by certain others who, at one time or another, have had similar editorial experience. Moreover, a good case could frequently be made for each of these varying viewpoints or attitudes.

My fundamental, and usually prevailing, philosophy regarding scientific publishing is that every scientist has the right to have his data and opinions published—providing he presents his material in a form that can be understood clearly by his peers and colleagues. This right cannot be abridged no matter how radical, how "far out," how bizarre they may seem. So long as he arranges and presents his data and results in a logical form, and so long as he writes clearly and without ambiguity, he has a right to the printed page of the most respectable of journals.

Strangely, I have found that many a referee will recommend that a well-written paper not be published because "I do not believe the author could have obtained the results he claims" or "the author's results are not in accord with those obtained previously by others" or, believe it or not, "the author of this paper was fired from his last position and has a personal reputation that does not entitle him to publish in a reputable journal." Any editor susceptible to the pressure of such evaluations is, to my mind, unworthy of the trust placed in him. Even worse, he may live to see the day when that well-written but radical, nonconforming paper he rejected will be one that helped its author receive a Nobel Prize. When one's data are sound, when one has done his best and done it sincerely, when one has come as close as one can to saying exactly what he wants to say, it is nonsense to assume that many, or perhaps a critical few, will not find the paper worthwhile.

Actually, however, most quarrels between authors and editors have less to do with the basic philosophy of either than with details of grammar, format, adequacy of data to support the conclusions, number and size of illustrations, and the like. The variety of ways to handle such differences are so numerous and have been spelled out in so many articles and books that I shall not attempt to cover the subject here. In general, as editor, I try never to change, alter, or distort the "style" of an author's writing, especially when that author does in fact have a style which is unique or, at least, characteristic. Beyond this, whenever an author has wished to depart from standard form, a standard spelling, etc., I have usually solved the matter

—as I mentioned earlier in the case of terminology—by requiring the author to explain in a footnote (if not in the text of his article) why he chooses to depart from a standardized form. Usually, such a request causes the author to see that the point to which he is taking exception is of little or no relevance to the meat or thrust of his paper and that such a footnote will merely emphasize his idiosyncrasy—which subconsciously, at least, he recognizes is only that, an idiosyncrasy.

But to get away from the mechanical, physical, and technical parts of editing. (Although even here it should be emphasized that many an author has been saved from making a fool of himself, or has been made to read fluently and lucidly, because of the patience and alertness of a competent editor, copy editor, or referee.) One of the psychological adjustments a scientist-editor must make is not to overjudge a fellow scientist by his writing—especially if his writing is confused. As an editor meets his scientific colleagues periodically at scientific meetings, or otherwise, he has to fight to avoid dichotomizing his friends into "good scientists" and "bad scientists" because they are "good writers" or "bad writers." In my experience there is little if any correlation between the brilliance of a man as a scientist and his excellence as a smooth writer. I believe there is such correlation between his scientific talents and his ability to arrange his thoughts in logical fashion on paper. But many a brilliant scientist is a murderer of syntax, of grammar, and of clarity of expression. He frequently "hates to write up" his results and often has an inadequate understanding of the rudiments of smooth writing. If he could, he would like to avoid split infinitives, dangling participles, and long, ponderous, never-ending sentences.

On the other hand, more often than not, I believe, superior scientists are concerned that their formal scientific publications be accurate expressions of what they wish to report to the scientific community. Either this or I have been extraordinarily fortunate in dealing with invertebrate pathologists who are undoubtedly extraordinarily fine people. Throughout editorial experiences of several kinds, I have rarely found myself at an impasse with an author, providing I took the trouble—

which I tried always to do—to explain why, for example, a particular sentence structure was generally considered to be incorrect by most authorities on formal writing or why rewriting a paragraph, or his entire paper, would be to his advantage. My warning that "the printed word is more permanent than marriage" usually caused the author to take another look at his manuscript without enmity.

Perhaps the most difficult job of an editor is that of rejecting a manuscript, and maintaining the high quality of his journal. To be sure, not a single number appears but what some readers will wonder why particular papers were ever accepted. The reasons for this difference of opinion are, of course, legion. Only another editor would understand, and so it is useless to explain even a few of the reasons. But once convinced by the referees or editorial board that a manuscript is unredeemably unacceptable—usually because the author has not provided the basic data to support his conclusions—the editor has the difficult and unpleasant task of so informing the author. By convincing authors that publication would only "hurt" them and that the greatest of scientists have had papers rejected and have lived to be thankful, most with whom I have dealt have taken the rejection in stride.

Nevertheless, any way one slices it, the result ranges from disappointment to insult in the minds of the authors. In some cases where the author is unwilling to accept the rejection as a sound judgment of his contribution, he should by all means submit it to another journal. Since eventually he is almost certain to find a publisher, he really is not prevented from having his work presented to the world, but by the same token neither is he prevented from making a fool of himself or of truly ruining his scientific reputation. However, unlike some novels and poems, usually two or three rejections of a scientific manuscript should cause the author to stop and reconsider. Perhaps a few more experiments or observations are all that is needed to make his opus a first-rate publication.

The job of editing a journal or any other publishing venture is truly made more rewarding and certainly easier if there is a cordial relationship with someone in the publishing house.

We were indeed fortunate to have the talented and cooperative help of several during the years we were associated with Annual Reviews and that of Mrs. Roselle Coviello in my association with Academic Press. There are also those associates who never receive credit for the able assistance rendered in time of emergency or otherwise. Such was Gordon A. Marsh, whose loyal and faithful help has been invaluable all these years in most of my publishing ventures. This was especially true during the time I was Dean of Biological Sciences, for I could never have continued as editor of the *Journal of Invertebrate Pathology* without his able assistance. He not only read galley proof but he also prepared the annual index, and most of all his expertise in insect nomenclature was of inestimable value.

At this point I find myself on the verge of pontificating regarding the entire business of editing and publishing scientific works. To do so would be tragic for the book as well as for this particular chapter. Instead, may I suggest to any reader who may be interested in scientific editing in biology that he read issues of the *Newsletter* of the Council of Biology Editors, Inc. Not that he will find any definitive answers in this valuable mimeographed publication or in any other publications on the subject. But he will find plenty of diverging opinion and frustration together with some wisdom and worthwhile experience.

There is every indication that with the overwhelming mass of scientific literature being produced today, new and more sophisticated means of communicating information are upon us. There are those who, appreciating the electronic advances of today, predict the disappearance of the scientific book and journal. (In this connection, however, and without being reactionary about it, I remember that phonograph records were declared effete when radio came into virtually every home.) Be this as it may, so long as man is allowed to discover new facts and ideas he will also desire to inform others and to express his opinion. Whatever means he uses to accomplish these objectives, the scientists (and my thoughts are on the invertebrate pathologist in particular) should present his findings in such a manner that they can be best evaluated by his fellow scientists. And, in so doing, he should make his presentations with the modesty and dignity which, at least in the past, most enlightened men have associated with the word "science."

1. [It is of interest that at the time of his death Edward Steinhaus had just finished covering the terminology of invertebrate pathology for the *Dictionary of Comparative Pathology and Experimental Biology* by Robert and Isabel Leader (Philadelphia: W. B. Saunders Company, 1971). In their preface to the dictionary the Leaders comment, "The late Dr. Steinhaus, who . . . assisted us with this material, spent his career proselytizing the scientific community to the cause of invertebrate pathology. Perhaps we can help his cause creep a small distance further forward." MEM.]

Bibliography

Abdel-Malek, A., and E. A. Steinhaus, 1948. Invasion route of *Nosema* sp. in the potato tuberworm, as determined by ligaturing. J. Parasitol. 34:452–453.

Acqua, C. 1919. Recherche intorno alla malattia della flaccidezza nel baco da seta. Annali della Regia Scuola Superiore di Agricoltura in Portici 15: 1–6.

Ainsworth, G. C. 1956. Agostino Bassi, 1773–1856. Nature 177: 255–257.

Alexander, F. W. 1905. How to rid your apiary of black brood. Gleanings in Bee Culture 33: 1125–1127.

Alfieri, E. 1925. Opere di Agostino Bassi. Tipografia Cooperativa di Pavia. 673 pp.

Allaire, J. P. 1824. Remarks on certain entozoical fungi. Ann. Lyceum Natur. Hist. N. Y. 1: 125–126.

Allen, H. W., and M. H. Brunson. 1945. A microsporidian in *Macrocentrus ancylivorus*. J. Econ. Entomol. 38: 393.

Angus, T. A. 1965. Symposium on microbial insecticides. I. Bacterial pathogens of insects as microbial insecticides. Bacteriol. Rev. 29: 364–372.

Aristotle. 330–323 B.C. Historia animalium. Trans. 1878 by R. Cresswell. George Bell and Son, London, 326 pp. Trans. 1910 by D. W. Thompson, *in* The works of Aristotle, J. A. Smith and W. D. Ross, eds. Vol. 4. Clarendon Press, Oxford. n. pag.

Armour, R. 1966. It all started with Hippocrates. McGraw-Hill Book Co., New York. 136 pp.

Audouin, V. 1837*a*. Recherches anatomiques et physiologiques sur la maladie contagieuse qui attaque les vers à soie, et qu'on désigne sous le nom de muscardine. Ann. Sci. Natur. 8: 229–245.

———. 1837*b*. Nouvelles expériences sur la nature de la maladie contagieuse qui attaque les vers à soie, et qu'on désigne sous le nom de muscardine. Ann. Sci. Natur. 8: 257-270 *and* C. R. Seances Hebd. Acad. Sci. 5: 712–717.

———. 1839. Quelques remarques sur de la muscardine, à l'occasion d'une lettre de M. de Bonafous, faisant connaitre les heureux resultats obtenus par M. Poidebard, dans la magnanerie de M. le comte de Demidoff. L'Institut 7: 154, 199–200.

Aymard, A. 1793. Notes sur l'education des vers à soie. Quoted by Robin, 1853.

Bail, [C.A.E.T.] 1867*a*. Ueber Epidemieen der Insecten durch Pilze. Entomol. Z., ent. Verein Stettin 28: 445–462.

———. 1867*b*. Mittheilungen über das Vorkommen und die Entwicklung einiger Pilzformen. Osterprogramm Real-Schule I. Ord. zu St. Johann, Danzig. 46 (8): 45.

———. 1868. Vorläufige Mittheilung über eine durch pilze verursachte Epidemie der Forleule, *Noctua piniperda* L. Krit. Blätt. Forst.- Jagdwiss. 50: 244–250.

———. 1869*a*. Ueber Pilzepizootien der forstverheerenden Raupen. Schr. Naturfosch. Ges. Danzig. N. F. 2 (2): 1–22.

———. 1869*b*. Pilz-Epidemie an der Forleule *Noctua piniperda* (L.). Z. Forst-. Jagdwesen 1: 243–247.

———. 1870. Weitere Mitteilungen über den Frass und das Absterben der Forleule, *Noctua piniperda*. Z. Forst-. Jagdwesen 2: 135–144.

———. 1904. Eine Käfer vernichtende Epizootien und Betrachtungen über die Epizootien der Insekten im allgemeinen. Festschrift. z. Feier 70. Geburtstages P. Acherson, Verl. Gebr. Borntraeger, Leipzig. 209–215.

Bailey, L. 1957. The cause of European foulbrood. Bee World 38: 85–89.

———. 1963. The pathogenicity for honey-bee larvae of microorganisms associated with European foulbrood. J. Insect Pathol. 5: 198–205.

———. 1968. Honey bee pathology. Annu. Rev. Entomol. 13: 191–212.

Balbiani, E. G. 1882. Sur les microsporidies ou psorospermies des Articulés. C. R. Seances Hebd. Acad. Sci. 92: 1168–1171.

———. 1884. Lecons sur les sporozaires. O. Doin, Paris. 184 pp. [see p. 166.]

Balsamo, C. G. 1835. Osservazione sopra una nuova specie di mucedinea del genere *Botrytis*, et cetera. Bibl. Ital. 79: 79, 125.

Bartlett, K. A., and C. L. Lefebvre. 1934. Field experiments with *Beauveria bassiana* (Bals.) Vuill., a fungus attacking the European corn borer. J. Econ. Entomol. 27: 1147–1157.

Bary, A. de. 1878. [On a disease among small weevils.] In a letter to E. Metchnikoff dated 1 November 1878 and quoted in Metchnikoff, 1879.

———. 1879. Die Erscheinung der Symbiose. Karl J. Trübner, Strassburg. 30 pp.

Bassi, A. 1835. Del mal del segno, calcinaccio o moscardino, malattia che affigge i bachi da seta e sul modo di liberarne le bigattaie anche le più infestate. Parte prima. Teoria, Orcesi, Lodi. 67 pp. Rpt. *in* Opere di Agostino Bassi. Cooperativa di Pavia, 1925. pp. 11–67. Trans. by A. C. Ainsworth and P. J. Yarrow *in* Phytopathological classics, no. 10. Monumental Printing Co., Baltimore. 49 pp.

———. 1836*a*. Del mal del segno, calcinaccio o moscardino, malattia che affigge i bachi da seta e sul modo di liberarne le bigattaie anche le più infestate. Parte seconda, Pratica, Orcesi, Lodi. 60 pp. Rpt. *in* Opere di Agostino Bassi. Cooperativa di Pavia, 1925. pp. 87–152.

———. 1836*b*. Maladies des vers à soie: recherches sur la muscardine. C. R. Seances Hebd. Acad. Sci. 2: 434–436.

Beall, G., G. M. Stirrett, and I. L. Conners. 1939. A field experiment on the control of the European corn borer, *Pyrausta nubilalis* Hübn., by *Beauvaria bassiana* Vuill. II. Sci. Agr. 19: 531–534.

Beard, R. L. 1945. Studies on the milky disease of Japanese beetle larvae. Conn. Agr. Exp. Sta. Bull. pp. 491, 505–581.

Beauverie, M. J. 1909. Notes sur les muscardines: sur une muscardine du ver à soie. pp. 55–81. *In* Laboratoire d'etudes de la soie de la condition publique des soies de Lyon: rapport présenté à la Chambre de Commerce de Lyon par la Commission Administrative 1906–1907. Vol. 13. A. Rey and Cie, Lyon.

———. 1914. Les muscardines: le genre *Beauveria* Vuillemin. Rev. Gen. Bot. 26: 157–168.

Bechamp, A. 1867. Faits pour servi à l'histoire de la maladie parasitaire des vers à soie appelée pébrine et spécialement du dévelopment du corpuscle vibrant. C. R. Seances Hebd. Acad. Sci. 64: 873–875.

Beeson, R. J., and P. T. Johnson. 1967. Natural bacterial flora of the bean clam, *Donax gouldi*. J. Invertebr. Pathol. 9: 104–110.

Benton, F. 1895. Bulletin No. 1, U.S. Dept. Agr., Division of Entomology. Cited by Howard, 1930.

Berg, A. 1963. Miasma und Kontagium. Naturwissenschaften 50:389–396.

Berger, E. W. 1906. Report of the assistant entomologist. Florida Agr. Exp. Sta. Annu. Rept. 1906. Pp. xvii–xx.

———. 1910. Report of the entomologist. Florida Agr. Exp. Sta. Annu. Rept. 1910. pp. xxxv–xliv.

———. 1919. Work of the Entomological Department, State Plant Board. Florida State Hort. Soc. Quart., Proc. 32nd Annu. Meeting. Pp. 160–170.

———. 1921. Natural enemies of scale insects and whiteflies in Florida. Quart. Bull. State Plant Bd. Florida 5: 141–154.

———. 1923. How to introduce the red and yellow whitefly fungi (red *Aschersonia* and yellow *Aschersonia*). State Plant Bd. Florida Dep. Entomol. Bull. 4 pp.

———. 1932. The latest concerning natural enemies of citrus insects. Proc. Florida State Hort. Soc. 45: 131–136.

———. 1938. The mango shield scale, its fungus parasite and control. Florida Entomologist 21: 1–4.

———. 1942. Status of the friendly fungus parasites of armored scale-insects. Florida Entomologist 25: 26–29.

Bergold, G. 1943. Über polyederkrankheiten bei insekten. Biol. Zentralbl. 63: 1–55.

———. 1947. Die Isolierung des Polyder-virus und die Natur der Polyeder. Z. Naturforsch. 2b: 122–143.

———. 1948*a*. Bündelformige Ordnung von Polyederviren. Z. Naturforsch. 3b: 25–26.

———. 1948*b*. Über die Kapselvirus-Krankheit. Z. Naturforsch. 3b: 338–342.

Berliner, E. 1915. Über die Schlaffsucht der Mehlmottenraupe (*Ephestia kuhniella*, Zell.) und ihren Erreger *Bacillus thuringiensis*, n. sp. Z. Angew. Entomol. 2: 29–56.

Bevan, E. 1838. See Munn, W. A., 1870.

Billings, F. H., and P. A. Glenn. 1911. Results of the artificial use of the white-fungus disease in Kansas. Kansas Agr. Exp. Sta. Tech. Bull., Bur. Entomol. 107. 58 pp.

Billings, F. P. 1888. The germ of the southern cattle plague. Amer. Natur. 22: 113–128.

Blunck, H. 1952. Ueber die bei *Pieris brassicae* L., ihren Parasiten und Hyperparasiten schmarotzenden Mikrosporidien, *in* Proc. Int. Congr. Entomol., 9th, Amsterdam, 1952, 1. Pp. 432–438.

———. 1954. Mikrosporidien bei *Pieris brassicae* L., ihren Parasiten und Hyperparasiten. Z. Angew. Entomol. 36: 316–333.

———. 1957. *Pieris rapae* (L.), its parasites and predators in Canada and the United States. J. Econ. Entomol. 50: 835–836.

———. 1958. Is there a possibility of using microsporidia for biological control of Pieridae? *in* Proc. Int. Congr. Entomol., 10th, Montreal, 1956, 4. Pp. 703–710.

Bogoyavlensky, N. 1922. *Zografia notonectae* n. g., n. sp. Arch. Russian Protistol. Soc. 1: 113–119.

Boissier de [Sauvages], P. A. 1763. Mémoires sur l'éducation des vers à soie, divisés en trois parties. Troisième Mémoire, Nîmes. Second memoire, diseases, pp. 99–100 and 109–123; third memoire, muscardine, 74–84, and grasserie, 91–97.

Bolle, G. 1894. Il giallume od il mal del grasso del baco da seta. Communicazione preliminare. Atti. e Mem. dell' I. R. Soc. Agr. Gorizia 34: 133–136.

———. 1898. Die Gelb-oder Fettsucht Seidenspinners eine Schmarotzerkrankheit. Supplement to Der Seidenbau in Japan. A. Hartleben, Budapest, Wein, Leipzig. 141 pp.

Borah, W. 1943. Silk raising in colonial Mexico. Ibero-Americana 20: 47, 51, 63–66, 88. Univ. Calif. Press, Berkeley. 169 pp.

Bosanquet, W. C. 1894. Notes on a gregarine of the earthworm (*Lumbricus herculeus*). Quart. J. Microsc. Sci. 36: 421–433.

Bounhiol, J.-J., J. Bruneteau, and A. Couturier. 1947. La Pathologie des insects d'après l'oeuvre d'André Paillot. Société d'Etude et de Vulgarisation de Zoologie Agricole. Rev. Zool. Ar. Appliquée. [Second half-year, 1947.] Nos. 7–12. 15 pp.

Boyd, W. 1961. A textbook of pathology. Structure and function in diseases. 7th ed. Lea and Febiger, Philadelphia. 1378 pp.

Breasted, J. H. 1962. Ancient records of Egypt. 3: 138, 198, 252; 4: 54, 456. Russell and Russell, Inc., New York. 5 vols.

Brefeld, O. 1870. Entwicklungsgeschichte der *Empusa Muscae* und *Empusa radicans*. Bot. Ztg. 28: 161–166, 177–186.

———. 1871. Untersuchugen über die Entwicklung der *Empusa Muscae* und *Empusa radicans* und die durch sie verurasachten Epidemien der Stubenfliegen und Raupen. Abh. Naturforsch. Ges. Halle 12: 1–50.

———. 1877. Ueber die Entomophthoreen und ihre Verwandten. Bot. Ztg. 35: 345–355, 368–372.

———. 1881. Botanische Untersuchungen über Schimmelpilze. Untersuchungen aus dem Gesammtgebiete der Mykologie. IV. Heft. A. Felix, Leipzig. 191 pp.

Briggs, J. D. 1963. Commercial production of insect pathogens. Pp. 519-548 *in* E. A. Steinhaus, ed., Insect pathology: an advanced treatise, vol. 2. Academic Press, New York. 689 pp.

Brockett, L. P. 1876. The silk industry in America. The Silk Assoc. Am., Washington, D. C. 237 pp.

Brongniart, C. 1888. Les Entomophthorées et leur application à la destruction des insectes nuisible. C. R. Seances Hebd. Acad. Sci. 107: 872–874.

Brooks, F. T. 1949. Obituary, Mr. T. Petch. Nature 163: 202.

Brooks, W. M. 1968. The role of the holotrichous ciliates, *Tetrahymena limacis* (Warren) and *Tetrahymena rostrata* (Kahl), as parasites of the gray field slug, *Deroceras reticulatum* (Müller). Hilgardia 39: 205–276.

Brouzet, G. 1863. Recherches sur les maladies des vers à soie. Roget et Laporte, Nîmes. 81 pp.

Brown, F. M. 1930. Bacterial wilt disease. J. Econ. Entomol. 23: 145–146.

Bruner, L., and H. G. Barber. 1894. Experiments with infectious diseases for combating the chinch-bug. Nebraska Agr. Exp. Sta. Bull. 34: 143–161.

Bruyne, C. de. 1893. De la phagocytose observée, sur le vivant, dans les branchies des Mollusques lamellibranches. C. R. Seances Hebd. Acad. Sci. 116: 65–68.

———. 1895. Contribution à l'etude de la phagocytose. Arch. Biol. 14: 161–241.

Buchner, P. 1911. Über intrazellulare Symbionten bei zuckersaugenden Insekten und ihre Vererbung. Sitzber Ges. Morphol. Physiol. 27: 89–96.

———. 1930. Tier und Pflanze in Symbiose. Gebrüder Borntraeger, Berlin. 900 pp.

———. 1953. Endosymbiose der Tiere mit Pflanzlichen Mikroorganismen. Verlag Birkhäuser, Basel and Stuttgart. 771 pp.

———. 1965. Endosymbiosis of animals with plant microorganisms. Interscience Publishers, Div. of John Wiley and Sons, Inc., New York. 909 pp.

Bulloch, W. 1938. The history of bacteriology. Oxford University Press, London. 422 pp.

Bütschli, O. 1882. Gregarinida. Bronn's Klassen und Ordnungen des Thierreichs. 1: 503–589.

Calvert. G. 1655. The reformed Virginia silkworm. John Streater, London. 40 pp.

Cantwell, G. E., J. C. Harshbarger, R. L. Taylor, C. Kenton, M. S. Slatick, and C. J. Dawe. 1968. A bibliography of the literature on neoplasms of invertebrate animals, *in* Gann Monograph 5: Experimental Animals in Cancer Research, pp. 57–84. Maruzen Co., Ltd., Tokyo. 128 pp.

Carossa, H. 1951. Foreword *to* Paul Buchner: Leben und Werk. Dem Freunde und Lehrmeister Gewidmet von Seinen Schülern. Pp. 5–6.

Casas de las Gonçalo. 1581. Arte para criar seda desde que se rebiue vna semilla hasta sacar otra. Rene Rabut Ympressor de Libros, Granada. [Cited by Borah, 1943.]

Chapman, J. W., and R. W. Glaser. 1915. A preliminary list of insects which have wilt, with a comparative study of their polyhedra. J. Econ. Entomol. 8: 140–150.

Charles, V. K. 1941. A preliminary check list of the entomogenous fungi of North America. U. S. Dep. Agr., Bur. Entomol. and Plant Quarantine 21 (supplement to 9): 707–783.

Chauthani, A. R. 1968. Bioassay technique for insect viruses. J. Invertebr. Pathol. 11: 242–245.

Chavannes, A. 1862. Les principales maladies des vers à soie et leur guérison. Joël Cherbuliez, Geneva. 111 pp.

Cheshire, F. R. 1884. Foul brood (not Micrococcus but Bacillus), the means of its propagation and the method of its cure. Brit. Bee J. 12 (151): 256–263. Also, The health exhibition literature 5: 173–218; published for the International Health Exhibition by Wm. Clowes and Sons, London.

Cheshire, F. R., and W. W. Cheyne. 1885. The pathogenic history and history under cultivation of a new bacillus (*B. alvei*), the cause of a disease of the hive bee hitherto known as foul brood. J. Roy. Microsc. Soc. Ser. II, 5:581–601.

Chinese Repository. 1834, 2: 485–486; 1847, 16: 224–231; 1849, 18: 306–309. Printed for the Proprietors, Canton, 1832–1851. 20 vols.

Chorine, V. 1930. On the use of bacteria in the fight against the corn borer. Pp. 94–98 *in* T. Ellinger, ed., International Corn Borer Investigations, vol. 3. International Live Stock Exposition, Union Stock Yards, Chicago. 174 pp.

Cist, J. 1824. Notice of the *Melolontha* or may bug. Am. J. Sci. and Arts 8: 269–271.

Clark, E. C. 1955. Observations on the ecology of a polyhedrosis of the Great Basin tent caterpillar *Malacosoma fragilis*. Ecology 36: 373–376.

———. 1956. Survival and transmission of a virus causing polyhedrosis in *Malacosoma fragile*. Ecology 37: 728–732.

Clark, E. C., and C. G. Thompson. 1954. The possible use of microorganisms in the control of the Great Basin tent caterpillar. J. Econ. Entomol. 47: 268–272.

Clausen, C. P. 1940. Entomophagous insects. McGraw-Hill Book Co., Inc., New York and London. 688 pp.

———. 1954. Biological antagonists in the future of biological control. J. Agr. Food Chem. 2: 12–18.

Clay, J. 1929. Fighting the corn-borer. Live Stock Markets 39 (12): 1 and 7. June 6, 1929.

Clinton, G. P. 1937. Roland Thaxter. Nat. Acad. Sci. Biographical Memoirs 17: 55–68.

Cobb, J. H. 1832. A manual containing information respecting the growth of the mulberry tree, with suitable directions for the culture of silk. Carter and Hendon, Boston. 68 pp.

Cohn, F. 1855. *Empusa muscae* und die Krankheit den durch parasitische Pilze charakterisiten Epidemieen. Verh. kaiserl. Leopold-Carolin. Naturforscher 25: 300–360.

———. 1870. Ueber eine neue Pilzkrankheit der Erdraupen. Beitr. Biol. Pflanz. 1 (1): 58–86.

Coker, W. C. 1920. Notes on lower Basidiomycetes of North Carolina. J. Elisha Mitchell Sci. Soc. 35: 113–182.

Collinge, W. E. 1891. Note on a tumour in *Anodonta cygnaea* Linn. J. Anat. Physiol. Norm. Pathol. 25: 154k.

Comstock, J. H. 1879. Report upon cotton insects. Commissioner of Agri., Wash., D. C., pp. 217–218.

Cooke, A. H. 1895. Molluscs. Cambridge Natur. Hist. 3: 1–459.

Cooke, M. C. 1875. Fungi: their nature and uses. D. Appleton and Company, New York. 299 pp.

———. 1892. Vegetable wasps and plant worms. Soc. for Promoting Christian Knowledge, London. 364 pp.

———. 1893. Romance of low life amongst plants. Soc. for Promoting Christian Knowledge, London. 320 pp. [Entomogenous fungi covered pp. 225–246.]

Cornalia, E. 1856. Monografia del bombice del gelso (*Bombyx mori* Linneo). Memorie dell' I. R. Instituo Lombardo di Scienze, Lettere ed Arti 6: 1–387 [Parte quarta: Patologia del baco. pp. 332–366.]

Couch, J. N. 1938. The genus *Septobasidium*. University of North Carolina Press, Chapel Hill, N.C. 480 pp.

———. 1945. Revision of the genus *Coelomomyces*, parasitic in insect larvae. J. Elisha Mitchell Sci. Soc. 61: 124–136.

Couch, J. N., and H. R. Dodge. 1947. Further observations on *Coelomomyces*, parasitic on mosquito larvae. J. Elisha Mitchell Sci. Soc. 63: 69–79.

Crawley, H. 1903. The polycystid gregarines of the United States. Proc. Acad. Natur. Sci. Philad. 55: 41 and 632.

Cuny, H. 1967. Louis Pasteur. The man and his theories. Trans. by P. Evans, Fawcett Publications, Inc., Greenwich, Conn. 192 pp.

Cutbush, E. 1808. On the prevention and cure of certain diseases of the silk-worm. The Philad. Med. Phys. J. 3 (Part I): 20–24. [Collected and arranged by B. S. Barton.]

Dandolo, V. 1830. L'art d'élever les vers à soie. Chez Bohaire, Libraire, Lyon. 392 pp.

Danysz, J. 1893. Destruction des animaux nuisibles à l'agriculture (rongeurs et insectes) par les maladies contagieuses. Ann. Sci. Agron. 1: 410–493.

———. 1900. Un microbe pathogène pour les rats (*Mus decumanus* et *Mus ratus*) et son application à la destruction de ces animaux. Ann. Inst. Pasteur 14: 193–201.

Dean, G. A., and R. C. Smith. 1935. Insects injurious to alfalfa in Kansas. 29th Bienn. Rept. Kansas State Bd. Agr. 202–249.

Degeer, C. [K.] 1776. Memoires pour servir à l'histoire des insectes. Vol. 6. Pierre Hesselberg, Stockholm. [See page 75.]

———. 1782. Abhandlungen zur Geschichte der Insekten. Trans. by J. A. E. Goeze. 200 pp. [See p. 38.]

De Kruif, P. 1926. Pasteur: microbes are a menace!, pp. 57–104; Metchnikoff—the nice phagocytes, pp. 207–233, *in* Microbe hunters. Blue Ribbon Books. New York. 363 pp.

Derry, T. K., and T. I. Williams. 1961. A short history of technology from earliest times to A.D. 1900. Oxford Univ. Press, Oxford. 782 pp.

Dobell, C. 1932. Antony van Leeuwenhoek and his little animals. Harcourt Brace Co., New York. 2nd ed. 1958, 435 pp.

Doolittle, G. M. 1889. Scientific queen-rearing as practically applied. Thomas G. Newman and Son, Chicago. 169 pp.

Dorfeuille, J. 1822. Natural History [An insect plant]. The Western Quarterly Reporter of Med. Surg. and Natur. Sci. 1 (4): 398–400. [Of Cincinnati.]

Doutt, R. L. 1964. The historical development of biological control. Pp. 21–73 *in* Paul DeBach and E. T. Schlinger, eds., Biological control of insect pests and weeds. Chapman and Hall, Ltd., London. 844 pp.

Dresner, E. 1949. Culture and use of entomogenous fungi for the control of insect pests. Contrib. Boyce Thompson Inst. Plant Res. 15: 319–335.

———. 1950. The toxic effect of *Beauveria bassiana* (Bals.) Vuill. on insects. J. N.Y. Entomol. Soc. 58: 269–278.

Dubos, R. J. 1950. Louis Pasteur: free lance of science. Little, Brown and Co., Boston. 418 pp.

Dufour, J. 1892. Note sur le *Botrytis tenella* et son emploi pour la destruction des vers blancs. Bull. Soc. Vaudoise Sci. 28: 106.

Dufour, L. 1826. Recherches anatomiques sur les carabiques et sur plusiers autres insectes colèptères. Ann. Sci. Natur. 8: 5–54.

———. 1828. Note sur la gregarine, nouveau genre de ver qui vit en troupeau dans les intestins de diver insectes. Ann. Sci. Natur. 13: 366–368.

Duggar, B. M. 1896. On a bacterial disease of the squashbug (*Anasa tristis* de G.). Illinois State Lab. Natur. Hist. Bull. 4: 340–379.

Dunn, P. H. 1960. Control of houseflies in bovine feces by a feed additive containing *Bacillus thuringiensis* var. *thuringiensis* Berliner. J. Insect Pathol. 2: 13–16.

Dunn, P. H., and B. J. Mechalas. 1963. The potential of *Beauveria bassiana* (Balsamo) Vuillemin as a microbial insecticide. J. Insect Path. 5: 451–459.

Dutky, S. R. 1940. Two new spore-forming bacteria causing milky diseases of Japanese beetle larvae. J. Agr. Res. 61: 57–68.

———. 1942. Method for the preparation of spore-dust mixtures of type A milky disease of Japanese beetle larvae for field inoculation. U. S. Dep. Agr. Entomol. and Plant Quarantine Bull. ET-192, 10 pp.

———. 1963. The milky diseases. Pp. 75–115 *in* E. A. Steinhaus, ed., Insect pathology: an advanced treatise, vol. 2. Academic Press, New York and London. 688 pp.

Dzierzon, Johannes. 1880. Dzierzon's rational bee keeping or the theory and practice of Dr. Dzierzon. Trans. by H. Djeck and S. Stutterd. Ed. by Charles N. Abbott. London. 350 pp.

Earle, F. S. 1889. Orchard notes. Alabama Exp. Sta. Bull. 106: 163–176.

Eckert, J. E., and F. R. Shaw. 1960. Beekeeping, successor to "Beekeeping" by Everett F. Phillips. Macmillan Co, New York. 536 pp.

Edelstein, L. 1967. Ancient medicine. O. Temkin and C. L. Temkin, eds., Selected papers of Ludwig Edelstein. The John Hopkins Press, Baltimore. 496 pp.

Ellinger, T., ed. 1928. International Corn Borer Investigations, vol. 1. International Livestock Exposition, Union Stock Yards, Chicago. 237 pp.

———. 1929. International Corn Borer Investigations, vol. 2. International Livestock Exposition, Union Stock Yards, Chicago. 183 pp.

———. 1930. International Corn Borer Investigations, vol. 3. International Livestock Exposition, Union Stock Yards, Chicago. 174 pp.

———. 1931. International Corn Borer Investigations, vol. 4. International Livestock Exposition, Union Stock Yards, Chicago. 96 pp.

Escherich, K., and M. Miyajima. 1911. Studien über die Wipfelkrankheit der Nonne. Naturw. Z. Forst- u. Landw. 9: 381–402.

Essig, E. O. 1931. A history of entomology. The Macmillan Co., New York. 1029 pp.

———. 1945. Silk culture in California. Circular No. 363, College of Agr., University of California, Berkeley. 15 pp.

Evans, J. 1806–1813. The bees: a poem in 3 vols. with notes, moral, political, and philosophical. J. and W. Eddanes, Shrewsbury.

Falcon, L. A. 1965. Report of work conference on the utilization of nucleopolyhedrosis virus for the control of *Heliothis zea*. Bull. Entomol. Soc. Am. 11: 84.

Farley, C. A. 1967. A proposed life cycle of *Minchinia nelsoni* (Haplosporida, Haplosporidiidae) in the American oyster *Crassostrea virginica*. J. Protozool. 14: 616–625.

Farrington, B. 1947. Greek science: its meaning to us. Harmondworth, Middlesex, England. 143 pp.

Fawcett, G. L. 1915. Apuntes botanicos y micologicos. El hongo moscadino l'n enemigo de los insectos que atacan la cana. Revista Indust. Y Agracola de Tucuman 5: 497–498.

Fawcett, H. S. 1908. Fungi parasitic on *Aleyrodes citri*. Univ. Florida Spec. Studies No. 1. 41 pp. Master's thesis.

———. 1912. The effects of spraying. Florida Agr. Exp. Sta. Rep. pp. lxxiii–lxxvii.

———. 1944. Fungus and bacterial diseases of insects as factors in biological control. Bot. Rev. 10: 327–348.

———. 1948. Biological control of citrus insects by parasitic fungi and bacteria. Pp. 628–664 *in* L. D. Batchelor and H. J. Webber, eds., The citrus industry, vol. 2. University California Press, Berkeley and Los Angeles. 933 pp.

Fenner, F., and M. F. Day. 1953. Biological control of the rabbit in Australia. Sci News 28: 7–22.

Filippi, F. de. 1851. Alcune osservazioni anatomico-fisiologico sugl' insetti in gnerale, ed in particolare sul bombice del gelso. Accademia d'Agricolture, Turin, Annali 5: 1–26.

———. 1852. Anatomisch-physiologische Bemerkungen über die Insecten im Allgemeinen und über den *Bombyx mori* (*bombice del gelso*) im Besondern. Ent. Ztschr. (Stettin) 13: 258–267; 14: 124–132; 15: 7–11.

Fisher, F. E. 1947. Insect disease studies. Annu. Rpt. Florida Agr. Exp. Sta. for year ending June 30, 1947. p. 162.

———. 1948. Diseases of citrus insects. Annu. Rpt. Florida Agr. Exp. Sta. for year ending June 30, 1948.

———. 1950*a*. An advantage of using wettable sulfur. Citrus Magazine 12: 17, 20.

———. 1950*b*. Entomogenous fungi attacking scale insects and rust mites on citrus in Florida. J. Econ. Entomol. 43: 305–309.

———. 1950*c*. Two new species of *Hirsutella* Patouillard. Mycologia 42: 290–297.

———. 1951. An *Entomophthora* attacking citrus red mite. Florida Entomologist 34: 83–88.

———. 1952. Diseases of scale insects. Citrus Magazine 15: 25–26.

———. 1954. Diseases of citrus insects. Florida Agr. Exp. Sta. Annu. Rep. for 1953. p. 191.

Fisher, F. E., and J. T. Griffiths, Jr. 1950. The fungicidal effect of sulfur on entomogenous fungi attacking purple scale. J. Econ. Entomol. 43: 712–718.

Fisher, F. E., W. L. Thompson, and J. Griffiths. 1949. Progress report on the fungus diseases of scale insects attacking citrus in Florida. Florida Entomologist 32: 1–11; also, *in* Univ. Florida Agr. Exp. Sta. Progress Rpt. December, 1948. 10 pp.

Fisher, R., and L. Rosner. 1959. Toxicology of the microbial insecticide, Thuricide. Agr. Food Chem. 7: 686–688.

Fleming, W. E. 1968. Biological control of the Japanese beetle. U. S. Dep. Agr. Res. Ser., Tech. Bull. 1383. 78 pp.

Florenzano, G. 1949. Prospetto sistematico delle specie batteriche segnalate come entomosimbionti. Redia 34: 125–159.

Florey, H. W., ed. 1962. The history and scope of pathology. Pp. 1–39 *in* General pathology. 3rd ed. Lloyd-Luke Ltd., London. 1104 pp.

Forbes, S. A. 1882. Bacterium a parasite of the chinch bug. Am. Natur. 16: 824–825.

———. 1883. Studies on the chinch bug. (*Blissus leucopterus* Say.) I. 12th Rpt. State Entomologist of Illinois for the year 1882. pp. 32–63.

———. 1886*a*. The entomological record for 1885. Illinois State Entomologist Miscellaneous Essays. pp. 5–9.

———. 1886*b*. Studies on the contagious diseases of insects. Bull. Illinois State Lab. Natur. Hist. 2 (article 4): 256–321.

———. 1886*c*. A contagious disease of the European cabbage worm, *Pieris rapae*, and its economic application. Proc. Soc. Promotion Agr. Sci: 26–32.

———. 1888. On the present state of our knowledge concerning contagious insect diseases. pp. 3–12. Bibliographical record. Contribution of American bibliography of insect diseases. pp. 15–22. Psyche 5 (nos. 141–142).

———. 1890. Studies on the chinch bug. (*Blissus leucopterus* Say.) II. 16th Rpt. State Entomologist on Noxious and Beneficial Insects of the State of Illinois. pp. 1–57.

———. 1895*a*. Preliminary note on a contagious insect disease. Science 2: 375–376.

———. 1895*b*. Experiments with the muscardine disease of the chinch-bug, and with the trap and barrier method for the destruction of that insect. Univ. Illinois Exp. Sta. Bull. 38: 25–86.

———. 1895*c*. On contagious disease in the chinch-bug (*Blissus leucopterus* Say). 19th Rpt. State Entomologist on Noxious and Beneficial Insects of the State of Illinois. Pp. 16–176.

———. 1898*a*. The spontaneous occurrence of white muscardine among chinch-bugs in 1895. 20th Rpt. State Entomologist on Noxious and Beneficial Insects of the State of Illinois. Pp. 75–78.

———. 1898*b*. Note on a new disease of the army worm (*Leucania unipuncta* Haworth). 20th Rpt. State Entomologist on Noxious and Beneficial Insects of the State of Illinois. Pp. 106–109.

———. 1899. Recent work on the San Jose scale in Illinois. Illinois Exp. Sta. Bull. 56. pp. 241–247.

Foscarini, G. M. 1821. Sperienze ed osservazioni sulla malattia de' bachi, conescuita sotto il nome di calcinetto. Bibl. Ital. 22: 59–83.

Fracastoro, G. 1546. Cited by A. Berg, 1963.

Franz, J. 1953. *Laricobius erichsoni* Rosenhauer (Col. Derodonidae), ein Rauber an Chermesiden. Z. Pflanzenkrankr. Pflanzenpathol. 60: 2–14.

Franz, J. M. 1961 *a*. Biologische Schädlingsberkämpfung. Pp. 1—302 *in* P. Sorauer, ed., Handbuch der Pflanzenkrankheiten. Paul Parey, Berlin and Hamburg. 627 pp.

———. 1961 *b*. Biological control of pest insects in Europe. Annu. Rev. Entomol. 6: 183–200.

Frauenfeld, G. 1849. [Untitled report.] I. Versammlungsberichte. Berichte über die Mittheilungen von Freunden der Naturwissenschaften in Wien. (Gesammelt und herausgegeben von W. Haidinger.) 5: 169–174.

Freeman, J. 1852. Life of the Reverend William Kirby, M. A., F. R. S., F. L. S., etc. Rector of Barham. Longman, Brown, Green and Longmans. 506 pp.

Freeman, P., ed. 1965. Proceedings XII International Congress of Entomology, London, July 8–16, 1964. Royal Entomol. Soc. London, London.

Fresenius, G. 1856. Notiz, Insekten-Pilze betreffend. Bot. Ztg. 14: 882–883.

———. 1858. Ueber die Pilzgattung *Entomophthora.* Abhandl Senckenberg. Naturforsch. Ges. 2: 201–210.

Fukushima, T. 1828. Propagation of knowledge of nine diseases of sericulture [in Japanese]. Tokuhoin, Oshu. One folded page.

Gadd, C. H. 1949. Article on T. Petch. Tea. Quart. 20: 1–2.

Gardner, E. J. 1965. History of biology. 2nd ed. Burgess Publishing Co. 376 pp.

Garrison, F. H. 1929. An introduction to the history of medicine. W. B. Saunders Co., Philadelphia. 996 pp.

Giard, A. 1890. Emploi des champignons parasites contre les insectes nuisibles. Rev. Mycol. (Toulouse) 12: 71–73.

———. 1892*a*. *L'Isaria densa* (Link) Fries, champignon parasite due Hanneton commun. (*Melolontha vulgaris* L.) Bull. Sci. France et Belg. 24: 1–112.

———. 1892*b*. Nouvelles études sur le *Lachnidium acridiorum*, Gd. champignon parasite du criquet pèlerin. Rev. Gén. Bot. 4: 449–461.

———. 1893. Sur le *Cordiceps militaris*. Bull. Soc. Entomol. 67: 344.

———. 1894. Sur une affection parasitaire de l'huître (*Ostrea edulis* L.) connue sous le nom de maladie du pied. C. R. Seances Soc. Biol. 46 (10s. 1): 401–403.

———. 1896. Exposé des titres et travaux scientifiques (1869–1896) de Alfred Giard. Générale Lahure, Paris. 390 pp.

———. 1897. Sur une cercaire sétigère (*Cercaria lutea*) parasite des Pélécypodes. C. R. Seances Soc. Biol. 49: 954–956.

Gibbons, N. E., ed. 1963. Recent progress in microbiology. VIII International Congress for Microbiology, Montreal, August 20–24, 1962. University of Toronto Press, Toronto, Canada. 721 pp.

Glaser, R. W. 1914. The economic status of the fungous diseases of insects. J. Econ. Entomol. Dec. 1914, 1–4.

———. 1918. A systematic study of the organisms distributed under the name of *Coccobacillus acridiorum* d'Herelle. Ann. Entomol. Soc. Am. 11: 19–42.

———. 1926. The green muscardine disease in silkworms and its control. Ann. Entomol. Soc. Am. 19: 180–192.

Glaser, R. W., and J. W. Chapman. 1912. Studies on the wilt disease or "flacheria" of the gypsy moth. Science 36: 219–224.

———. 1913. The wilt disease of gypsy moth caterpillars. J. Econ. Entomol. 6: 479–488.

———. 1916*a*. The nature of the polyhedral bodies found in insects. Biol. Bull. Woods Hole 30: 367–391.

———. 1916*b*. Further studies on wilt of gypsy moth caterpillar. J. Econ. Entomol. 9: 149–167.

Glaser, R. W., and W. M. Stanley. 1943. Biochemical studies on the virus and the inclusion bodies of silkworm jaundice. J. Exp. Med. 77: 451–466.

Gray, G. R. 1858. Notices of insects that are known to form the basis of fungoid parasites. Privately printed, London. 22 pp.

Griffiths, J. T., Jr., and F. E. Fisher. 1949. Residues on citrus trees in Florida. J. Econ. Entomol. 42: 829–833.

———. 1950. Residues on citrus trees in Florida: changes in purple scale and rust mite populations following the use of various spray materials. J. Econ. Entomol. 43: 298–305.

Griffiths, J. T., Jr., and W. L. Thompson. 1950. The behavior of purple scale populations on citrus trees in Florida. Florida Entomologist 33: 61–70.

Grigarick, A. A., and Y. Tanada. 1959. A field test for the control of *Trichoplusia ni* (Hbr.) on celery with several insecticides and *Bacillus thuringiensis* Berliner. J. Econ. Entomol. 52: 1013–1014.

Guérin-Méneville, F. E. 1847. Muscardine. Mission confiée, Paris. 66 pp.

———. 1849. Études sur les maladies des vers à soie: observations sur la composition intime du sang sur les insectes et surtout sur les vers à soie en santé et en maladie, et sur la transformation des éléments vivants des globules de ce sang en rudiments du végétal qui constitute la muscardine. Mém. Soc. Agr. France 5. 187 pp.

Haberlandt, F. 1871. Der Seidenspinner des Maulbeerbaumes, seine Aufzucht und seine Krankheiten. Carl Gerolds Sohn, Wein. 247 pp.

———. 1872. Il giallume. La Sericoltura Austriaca 4: 49–52.

Haddow, A. J. 1924. The mosquito fauna and climate of native huts at Kisumu, Kenya. Bull. Entomol. Research 33: 91–142.

Haekel, E. 1862. De telis quibusdem Astaci fluviatilis über die Gewebe des Flusskrebses. Die Radiolarien. George Reimer, Berlin.

Hagen, H. A. 1879*a*. Obnoxious pests—suggestions relative to their destruction. Can. Entomologist 11: 110–114.

———. 1879*b*. Destruction of obnoxious insects, phylloxera, potato beetle, cotton-worm, Colorado grasshopper, and greenhouse pests, by application of the yeast fungus. University Press, John Wilson and Son, Cambridge. 11 pp. [Reprint, with revisions and additions, of the article in Can. Entomologist 1879.]

———. 1880. On the destruction of obnoxious insects by yeast. Can. Entomologist 12: 81–88.

———. 1882*a*. Experiments with yeast in destroying insects. Can. Entomologist 14: 38–39.

———. 1882*b*. [On the destruction of insect pests by the application of yeasts.] Am. Monthly Microsc. J. 3: 179.

Haggard, H. W. 1933. Mystery, magic, and medicine. Doubleday, Doran, and Co. Inc., Garden City, New York. 192 pp.

Hall, I. M. 1952*a*. Observations on *Perezia pyraustae* Paillot, a microsporidian parasite of the European corn borer. J. Parasitol. 38: 48–52.

———. 1952*b*. A new species of microsporidia from the fawn-colored lawn moth, *Crambus bonifatellus* (Hulst) (Lepidoptera, Cambidae), J. Parasitol. 38: 487–491.

———. 1953. The role of virus diseases in the control of the alfalfa looper. J. Econ. Entomol. 46: 1110–1111.

———. 1954. Studies of microorganisms pathogenic to sod webworm. Hilgardia 22: 535–565.

———. 1955. The use of *Bacillus thuringiensis* Berliner to control the western grapeleaf skeletonizer. J. Econ. Entomol. 48: 675–677.

———. 1957*a*. Methods of releasing fungi for control of the spotted alfalfa aphid. Pest Control Rev. (Feb.–March): 3.

———. 1957*b*. Use of a polyhedrosis virus to control the cabbage looper on lettuce in California. J. Econ. Entomol. 50: 551–553.

———. 1959. The fungus *Entomophthora erupta* (Dustan) attacking the black grass bug, *Irbisia solani* (Heidemann) (Hemiptera, Miridae), in California. J. Insect Pathol. 1: 48–51.

Hall, I. M., and L. A. Andres. 1959. Field evaluation of commercially produced *Bacillus thuringiensis* Berliner used for control of lepidopterous larvae on crucifers. J. Econ. Entomol. 52: 877–880.

Hall, I. M., and K. Y. Arakawa. 1959. The susceptibility of the house fly, *Musca domestica* Linnaeus, to *Bacillus thuringiensis* var. *thuringiensis* Berliner. J. Insect Pathol. 1: 351–355.

Hall, I. M., and J. V. Bell. 1960. The effect of temperature on some entomophthoraceous fungi. J. Insect Pathol. 2: 247–253.

———. 1961. Further studies on the effect of temperature on the growth of some entomophthoraceous fungi. J. Insect Pathol. 3: 289–296.

———. 1962. Nomenclature of *Empusa* Cohn 1885 vs. *Entomophthora* Fresenius 1856. J. Insect Pathol. 4: 224–228.

———. 1963*a*. The synonymy of *Empusa thaxteriana* Petch and *Entomophthora ignobilis* Hall and Dunn. J. Insect Pathol. 5: 182–186.

———. 1963*b*. Identification of an entomophthoraceous fungus isolated by Sawyer. J. Insect Pathol. 5: 272–275.

———. 1963*c*. Note on *Cordyceps sobolifera* (Berkeley) on the desert cicada, *Diceroprocta apache* (Davis). J. Insect Pathol. 5: 270–272.

Hall, I. M., and P. H. Dunn. 1957. Entomophthorous fungi pathogenic to the spotted alfalfa aphid. Hilgardia 27: 159–181.

———. 1958*a*. Susceptibility of some insect pests to infection by *Bacillus thuringiensis* Berliner in laboratory tests. J. Econ. Entomol. 51: 296–298.

———. 1958*b*. Artificial dissemination of entomophthorous fungi pathogenic to the spotted alfalfa aphid in California. J. Econ. Entomol. 51: 341–344.

———. 1959. The effect of certain insecticides and fungicides on fungi pathogenic to the spotted alfalfa aphid. J. Econ. Entomol. 52: 28–29.

Hall, I. M., R. I. Hale, H. H. Shorey, and K. Y. Arakawa. 1961. Evaluation of chemicals and microbial materials for control of cabbage looper. J. Econ. Entomol. 54: 141–146.

Hall, I. M., and J. C. Halfhill. 1959. The germination of resting spores of *Entomophthora virulenta* Hall and Dunn. J. Econ. Entomol. 52: 30–35.

Hall, I. M., and V. M. Stern. 1962. Comparison of *Bacillus thuringiensis* Berliner var. *thuringiensis* and chemical insecticides for control of the alfalfa caterpillar. J. Econ. Entomol. 55: 862–865.

Hamilton, E. 1942. Mythology. Little and Brown, Boston. 497 pp.

Harshbarger, J. C. 1966. Some effects of X-irradiation on larvae of *Galleria mellonella*. J. Invertebr. Pathol. 8: 277–279.

———. 1967. Responses of invertebrates to vertebrate carcinogens. Fed. Proc. 26: 1693–1697.

Harshbarger, J. C., and A. M. Heimpel. 1968. Effect of zymosan on phagocytosis in larvae of the greater wax moth, *Galleria mellonella*. J. Invertebr. Pathol. 10: 176–179.

Harshbarger, J. C., and R. L. Taylor. 1968. Neoplasms of insects. Annu. Rev. Entomol. 13: 159–190.

Hart, J. H. 1890. Report on sugar-cane blight. Agr. Res. Trin. 2: 127.

Haskin, H. H., L. A. Stauber, and J. G. Mackin. 1966. *Minchinia nelsoni* n. sp. (Halosporida, Halosporidiidae): causative agent of the Delaware Bay oyster epizootic. Science 153: 1414–1416.

Hawley, I. M., and F. G. White. 1935. Preliminary studies on the diseases of larvae of the Japanese beetle (*Popillia japonica* Newm.). J. N.Y. Entomol. Soc. 43: 405–412.

Hazarabedian, K. M. 1938. Practical silkworm culture and its possibilities in America. Privately printed. Fresno Bee, Fresno, Calif. 63 pp.

Headlee, T. H., and J. W. McColloch. 1913. The chinch bug (*Blissus leucopterus* Say). Kansas Agr. Exp. Sta. Bull. 191. 66 pp. (numbered 287–353).

Heimpel, A. M., and J. C. Harshbarger. 1965. Symposium on microbial insecticides. V. Immunity in insects. Bacteriol. Rev. 29: 397–405.

Herelle, F. de. 1911. Sur une épizootie de nature bactérienne sévissant sur les sauterelles au Mexique. C. R. Seances Hebd. Acad. Sci. 152: 1413–1415.

———. 1912. Sur la propagation, dans la République Argentine, de l'epizootie des sauterelles du Méxique. C. R. Seances Hebd. Acad. Sci. 154: 623–625.

Hergula, B. 1930. On the application of *Metarrhizium anisopliae* against *Pyrausta nubilalis*. Pp. 130–141 *in* T. Ellinger, ed., International Corn Borer Investigations, vol 3. International Livestock Exposition, Union Stock Yards, Chicago. 174 pp.

———. 1931. Recent experiments on the application of *Metarrhizium anisopliae* against the corn borer. Pp. 46–62 *in* T. Ellinger, ed., International Corn Borer Investigations, vol. 4. International Livestock Exposition, Union Stock Yards, Chicago. 96 pp.

Herrick, F. H. 1891. V. Alpheus: a study in the development of Crustacea. Appendix II, p. 462. Memoirs of the Nat. Acad. Sciences 5: 370–463.

Herrmann, G., and E. Canu. 1891. Sur un champignon parasite du talitre. C. R. Seances Soc. Biol. 43: 646–651.

Hill, S. B., Jr., W. W. Yothers, and R. L. Miller. 1934. Effect of arsenical and copper insecticides on the natural control of whiteflies and scale insects by fungi on orange trees in Florida. Florida Entomologist 18: 1–4.

Hittell, T. H. 1881. The California silk grower's instructor. California Silk Culture Assoc., San Francisco. 25 pp.

Hofmann, O. 1891. Die Schlaffsucht (Flacherie) der Nonne (*Liparis monacha*) nebst einem Anhang. Insektentötende Pilze mit besonderer Berücksichtigung der Nonne., P. Weber, Frankfurt. 31 pp.

Holloway, J. K., and T. R. Young, Jr. 1943. The influence of fungicidal sprays on entomogenous fungi and on the purple scale in Florida. J. Econ. Entomol. 36: 453–457.

Howard, L. O. 1901*a*. Insects as carriers and spreaders of disease. U. S. Dep. Agr. Yearbook, 1901. Pp. 177–192.

———. 1901*b*. Experimental work with fungus diseases of grasshoppers. U. S. Dep. Agr. Yearbook, 1901. Pp. 459–470.

———. 1921. A fifty-year sketch history of medical entomology. Pp. 412–438 *in* M. P. Ravenel, ed., A half century of public health—jubilee historical volume of the American Public Health Association. Am. Public Health Ass., New York. 461 pp.

———. 1930. A history of applied entomology. Smithsonian Misc. Collections 84: 1–564.

———. 1933. Fighting the insects: the story of an entomologist. The Macmillan Co., New York. 333 pp.

Howard, W. R. 1894. Foul brood, its natural history and rational treatment, with a review of the work of others. G. W. York, Chicago, Ill. 47 pp.

———. 1896. A new bee-disease—pickled brood or white fungus. Amer. Bee J. 36 (37): 577–578.

Hubbard, H. G. 1885. Insects affecting the orange. (Cited by Fisher, 1951).

Huber, J. 1958. Untersuchungen zur Physiologie insektentötender Pilze. Arch. Mikrobiol. 29: 257–276.

Huger, A. 1963. Granuloses of insects. Pp. 531–575 in E. A. Steinhaus, ed., Insect pathology: an advanced treatise, vol. 1. Academic Press, New York and London. 661 pp.

Hughes, K. M. 1950. A demonstration of the nature of polyhedra using alkaline solutions. J. Bacteriol. 59: 189–195.

———. 1952. Development of the inclusion bodies of a granulosis virus. J. Bacteriol. 64: 375–380.

———. 1953*a*. Conservation of the generic name *Borrelina* and designation of the type species. Int. Bull. Bacteriol. Nomencl. Taxon. 3: 134.

———. 1953*b*. The development of an insect virus within cells of its host. Hilgardia 22: 391–406.

———. 1957. An annotated list and bibliography of insects reported to have virus diseases. Hilgardia 26: 597–629.

———. 1958. The question of plurality of virus particles in insect-virus capsules and an attempt at clarification of insect terminology. Trans. Am. Microsc. Soc. 60: 22–30.

Hughes, K. M., and C. G. Thompson. 1951. A granulosis of the omnivorous looper, *Sabulodes caberata* Guenée. J. Infect. Diseases 89: 173–179.

Hume, E. D. 1923. Béchamp or Pasteur? Covici-McGee, Chicago. 296 pp.

Husz, B. 1928. Bacillus Thuringiensis Berliner, a bacterim pathogenic to corn borer larvae. Pp. 191–193 *in* T. Ellinger, ed., International Corn Borer Investigations, vol. 1. International Livestock Exposition, Union Stock Yards, Chicago. 237 pp.

———. 1929. On the use of Bacillus Thuringiensis in the fight against the corn borer. Pp. 99–105 *in* T. Ellinger, ed., International Corn Borer Investigations, vol. 2. International Livestock Exposition, Union Stock Yards, Chicago. 183 pp.

———. 1930. Field experiments on the application of *Bacillus thuringiensis* against the corn borer. Pp. 91–98 *in* T. Ellinger, ed., International Corn Borer Investigations, vol. 3. International Livestock Exposition, Union Stock Yards, Chicago. 174 pp.

———. 1931. Experiments during 1931 on the use of *Bacillus thuringiensis* Berliner in controlling the corn borer. Pp. 22–23 *in* T. Ellinger, ed., International Corn Borer Investigations, vol. 4. International Livestock Exposition, Union Stock Yards, Chicago. 96 pp.

Ignoffo, C. M. 1965*a*. The nuclear-polyhedrosis virus of *Heliothis zea* (Boddie) and *Heliothis virescens* (Fabricius). I. Virus propagation and its virulence. J. Invertebr. Pathol. 7: 209–216.

———. 1965*b*. The nuclear-polyhedrosis virus of *Heliothis zea* (Boddie) and *Heliothis virescens* (Fabricius). II. Biology and propagation of diet-reared *Heliothis*. J. Invertebr. Pathol. 7: 217–226.

———. 1965*c*. The nuclear-polyhedrosis virus of *Heliothis zea* (Boddie) and *Heliothis virescens* (Fabricius). IV. Bioassay of virus activity. J. Invertebr. Pathol. 7: 315–319.

———. 1966. Insect viruses. Pp. 501–530 *in* C. N. Smith, ed., Insect colonization and mass production. Academic Press, New York. 618 pp.

Ignoffo, C. M., A. J. Chapman, and D. F. Martin. 1965. The nuclear-polyhedrosis virus of *Heliothis zea* (Boddie) and *Heliothis virescens* (Fabricius). III. Effectiveness of the virus against field populations of *Heliothis* on cotton, corn, and grain sorghum. J. Invertebr. Pathol. 7: 227–235.

Ignoffo, C. M., and A. M. Heimpel. 1965. The nuclear-polyhedrosis virus of *Heliothis zea* (Boddie) and *Heliothis virescens* (Fabricius). V. Toxicity-pathogenicity of virus to white mice and guinea pigs, J. Invertebr. Pathol. 7: 329–340.

Ishikawa, K. 1940. Pathology of the silkworm [in Japanese]. 2nd edition. Meibundo, Tokyo. 512 pp.

Ishimore, N. 1934. Contribution a l'étude de la grasserie du ver à soie (*Bombyx mori*). C. R. Seances Soc. Biol. Soc. Franco-Japonaise Biol. 116: 1169–1170.

Iyengar, M. O. T. 1935. Two new fungi of the genus *Coelomomyces* parasitic in larvae of *Anopheles*. Parasitol. 27: 440–449.

Jaynes, H. A., and P. E. Marucci. 1947. Effect of artificial control practices on the parasite and predators of the codling moth. J. Econ. Entomol. 40: 9–25.

Jenkins, D. W., ed. 1960. Biological control of insects of medical importance. Amer. Inst. Biol. Sci. Tech. Bull. 144 pp.

Johnson, J. R. 1915. The entomogenous fungi of Porto Rico. Board of Comm. Agr. of Porto Rico (Insular Expt. Sta.), Bull. 10. 33 pp.

Johnson, P. T. 1966*a*. On *Donax* and other sandy-beach inhabitants. The Veliger 9: 29–30.

———. 1966*b*. Mass mortality in a bivalve mollusc. Limnol. Oceanogr. 11: 429–431.

———. 1968*a*. A new medium for maintenance of marine bacteria. J. Invertebr. Pathol. 11: 144.

———. 1968*b*. An annotated bibliography of pathology in invertebrates other than insects. Burgess Publishing Co., Minneapolis. 335 pp.

———. 1968*c*. Population crashes in the bean clam, *Donax gouldi*, and their significance to the study of mass mortality in other marine invertebrates. J. Invertebr. Pathol. 12: 349–358.

———. 1969*a*. The coelomic elements of sea urchins (*Strongylocentrotus*). I. The normal coelomocytes: their morphology and dynamics in hanging drops. J. Invertebr. Pathol. 13: 25–41.

———. 1969*b*. The coelomic elements of sea urchins (*Strongylocentrotus*). II. Cytochemistry of the coelomocytes. Histochemie 17: 213–231.

———. 1969*c*. The coelomic elements of sea urchins (*Strongylocentrotus*). III. *In vitro* reaction to bacteria. J. Invertebr. Pathol. 13: 42–62.

———. 1970. The coelomic elements of sea urchins (*Strongylocentrotus* and *Centrostephanus*). VI. Cellulose-acetate electrophoresis. Comp. Biochem. Physiol. 37: 289–300.

———. 1971. Invertebrate Pathology. Vol. 7, pp. 263–269 *in* McGraw-Hill Encyclopedia of Science and Technology. McGraw-Hill Book Co., Inc., New York, Toronto, London. 15 vols.

Johnson, P. T., and R. J. Beeson. 1966. *In vitro* studies on *Patiria miniata* (Brandt) coelomocytes, with remarks on revolving cysts. Life Sciences 5: 1641–1666.

Johnson, P. T., and F. A. Chapman. 1969. An annotated bibliography of pathology in invertebrates other than insects, supplement. Center for Pathobiology, Misc. Publ. No. 1. 76 pp.

———. 1970*a*. Comparative studies on the *in vitro* response of bacteria to invertebrate body fluids. I. *Dendrostomum zostericolum*, a sipunculid worm. J. Invertebr. Pathol. 16: 127–138.

———. 1970*b*. Infection with diatoms and other microorganisms in sea-urchin spines (*Strongylocentrotus franciscanus*). J. Invertebr. Pathol. 16: 268–276.

———. 1970*c*. Comparative studies on the *in vitro* response of bacteria to invertebrate body fluids. II. *Aplysia californica* (sea hare) and *Ciona intestinalis* (tunicate). J. Invertebr. Pathol. 16: 259–267.

———. 1971. Comparative studies on the *in vitro* response of bacteria to invertebrate body fluids. III. *Stichopus tremulus* (sea cucumber) and *Dendraster excentricus* (sand dollar). J. Invertebr. Pathol. 17: 94–106.

Jolly, M. S. 1968. Tasar Research Sci. Brochure No. 4, Central Tasar Research Station, Central Silk Board, Govern. of India, Ministry of Commerce. 44 pp.

Josselyn, J. 1865. Voyage in New England. Rpt. W. Veazie, Boston. 211 pp.

Kamm, M. W. 1922. Studies on gregarines. II. Synopsis of the polycystid gregarines of the world, excluding those from the Myriapoda, Orthoptera, and Coleoptera. Illinois Biol. Monogr. 7: 7–104.

Karling, J. S. 1948. Chytridiosis of scale insects. Am. J. Botany 35: 246–254.

Keilin, D. 1921. On a new type of fungus: *Coelomomyces stegomyiae*, n. g., n. sp., parasitic in the body-cavity of the larva of *Stegomyia scutellaris* Walker (Diptera, Nematocera, Culicidae). Parasitology 13: 225–234.

Kellogg, V. L. 1894. A European experiment with insect diseases. Pp. 227–239. *in* Contagious diseases of the chinch bug. 3rd Ann. Rept. Director (F. H. Snow) for the year 1893, Kansas Exp. Sta.

Kelly, E. O. G., and T. H. Parks. 1911. Chinch bug investigations west of the Mississippi River. U.S. Dep. Agr., Div. Entomol. Bull. 95, part 3. 52 pp.

King, K. M., chairman. 1946. Report of committee on grasshopper research. J. Econ. Entomol. 39: 111–113.

Kirby, W. 1826. Diseases of insects. Pp. 197–232 *in* W. Kirby and W. Spence, An introduction to entomology or elements of the natural history of insects, vol. 4. Longman, Rees, Orme, Brown, and Green, London. 634 pp.

Koch, A. 1951. Paul Buchner: Leben und Werk. Dem Freunde und Lehrmeister Gewidmet von Seinen Schülern. 21 pp.

———. 1963. The role of the symbionts in wood-destroying insects. Pp. 151–161 *in* Recent progress in microbiology. Symposia 8. International Congress Microbiology, Montreal, 1962, vol. 8.

———. 1966. Ein Leben für die Symbioseforschung. Paul Buchner zu seinem 80. Geburstag am 12. April 1966. Zool. Beitr. 12: 1–16.

Kodaira, Y. 1961. Biochemical studies on the muscardine fungi in the silkworms, *Bombyx mori.* J. Facult, Text Sci., Technol., Shinshu Univ. No. 29, Ser. E., 1–68.

Koebele, A. 1898. Report of Prof. Albert Koebele, entomologist of Hawaiian government. Hawaiian Planter's Monthly 17: 258–269.

———. 1900. [No title] Rpt. Bd. Comm. Agr. and Forest. (Hawaii) for 1900. pp. 18 and 44.

Kohnke, A. R. 1882. Foul brood, its origin, development, and cure. Youngstown Publ. Co. 13 pp.

Köppen, F. T. 1865. Über die Heuschrecken in Südrussland, nebst einem Anghange über einige andere daselbst vorkommende schädliche Insekten. Soc. Entomol. Rossicae 3: 81–294.

Korringa, P. 1948. Shell disease in *Ostrea edulis*, its dangers, its cause, its control. Nat. Shellf. Assoc. Convention Addresses, 1948, pp. 86–94.

Kramer, J. P. 1963. The entomophilic microsporidians. Pp. 175–182 *in* Recent progress in microbiology. Symposia 8. International Congress of Microbiology, Montreal, 1962, vol. 8.

Krassilstschik, I. M. 1888. La production industrielle des parasites végétaux pour la destruction des insectes nuisibles. Bull. Sci. France et Belg. 19: 461–472.

———. 1896. Sur les microbes de la flacherie et de la grasserie des vers à soie. [Paris] C. R. Seances Hebd. Acad. Sci. 123: 427–429.

Krieg, A. 1961. Grundlagen der insektenpathologie. Viren-, Rickettesien-, und Bakterien-Infektionen. Dietrich Steinkopff, Darmstadt. 304 pp.

———. 1963*a*. Crystalliforous bacteria. Pp. 134–140 *in* Recent progress in microbiology. Symposia 8. International Congress Microbiology, Montreal, 1962, vol. 8.

———. 1963*b*. Rickettsiae and Rickettsioses. Pp. 577–617 *in* E. A. Steinhaus, ed., Insect pathology, vol. 1. Academic Press, New York and London. 661 pp.

Krumbhaar, E. B. 1962. Clio Medica. A series of primers on the history of medicine. XIX. Pathology. Rpt. (1st ed. 1937), Hafner Publishing Co., New York. 206 pp.

Künckel d'Herculais, J., and C. Langlois. 1891. Les champignons parasites des acridiens. C. R. Seances Hebd. Acad. Sci. 112: 1465–1468.

Labbé, A. 1899. Sporozoa. Das Tierreich, Lief. 5: 180 pp.

Lacaze-Duthiers, H. de. 1854. Memoire sur le bucéphale Haime (*Bucephalus haimeanus*) helminthe parasite des huitres et des buscardes. Ann. Sci. Nat. Zool. [Paris] 4s, 1: 249–302.

Laird, M. 1960. *Coelomomyces*, and the biological control of mosquitoes. Conference on Biological Control of Medical Importance. Tech. Rpt., pp. 84–93. American Institute of Biological Science. 144 pp.

———. 1962. Scientific group on the utilization of biotic factors in vector control. Rpt. to Director-General. Bull. Organ. Mondiale Santé (Bull. World Health Organization). 23 pp.

———. 1963. Vector ecology and integrated control procedures. Bull. Organ. Mondiale Santé (Bull. World Health Organization), 29 (suppl.): 147–151.

Langstroth, L. L. 1853. Hive and the honey-bee. Hopkins, Bridgman, and Co., Northhampton, Mass. 384 pp. [Revised by C. Dadant in 1888.]

Lankester, E. R. 1863. On our present knowledge of the Gregarinidae. Quart. J. Microsc. Sci. 3: 83–96.

———. 1880. The destruction of insect pests, an unforeseen application of the results of biological investigation. Nature 21: 447–448.

Lapicque, L. 1945. Notice nécrologique sur André Paillot. C. R. Seances Hebd. Acad. Sci. 220: 205–206.

Latreille, P. A. 1805. Histoire naturelle, générale et particulière des crustacés et des insectes. Vol. 14, F. Dufart, Paris. 432 pp.

Leader, R. W., and I. Leader. 1971. Dictionary of comparative pathology and experimental biology. W. B. Saunders Co., Philadelphia, London, Toronto. 239 pp.

Lebert, H. 1858. Ueber die gegenwärtig herrschende Krankheit des Insekts der Seide. Berliner Entomol. Z. 2: 149–186.

Lebert, H., and H. Frey. 1856. Beobachtungen über die gegenwärtig in Mailändischen herrschende Krankheit der Seidenraupe, der Puppe und Schmetterlings. Vierteljahresschr. Zürich 1: 374–389.

LeConte, J. L. 1874. Hints for the promotion of economic entomology. Am. Assoc. Adv. Sci. Proc. 22: 10–22.

Ledermüller, M. F. 1763. Mikroscopische Gemüths- und Augen- Ergössunfen. C. de Launoy, Nürnberg. 204 pp.

Lefebvre, C. L. 1931*a*. A destructive fungous disease of the corn borer. Phytopathology 21: 124–125.

———. 1931*b*. Preliminary observations on two species of *Beauveria* attacking the corn borer, *Pyrausta nubilalis* Huber. Phytopathology 21: 1115–1118.

———. 1934. Penetration and development of the fungus *Beauveria bassiana*, in the tissues of the corn borer. Ann. Bot. London 48: 441–452.

Legér, L. 1897. Le cycle évolutif des coccidies chez les arthropodes. C. R. Seances Hebd. Acad. Sci. 124: 966–969.

———. 1892. Recherches sur les grégarines. Tabl. Zool. 3: 1–182.

Leidy, J. 1850. [Remarks on isolation of Entophyta and his disbelief in spontaneous generation.] Acad. Natur. Sci. Proc. Philad. 5: 7–9.

———. 1851*a*. [Remarks on the parasites of insects.] Acad. Natur. Sci. Proc. Philad. 5: 204, 210, 211.

———. 1851*b*. [Fungi of *Cicada septendecim*, lemellicorn larvae, and mole cricket.] Acad. Natur. Sci. Proc. Philad. 5: 235.

———. 1853. A flora and fauna within living animals. Smithsonian Contributions to Knowledge 5 (2). 67 pp.

———. 1879. On *Amoeba blattae*. Acad. Natur. Sci. Proc. Philad. 31: 204–205.

———. 1884. Ant infested with a fungus. Acad. Natur. Sci. Proc. Philad., Jan. 1, p. 9.

———. 1889. On several gregarines and a singular mode of conjugation of one of them. Acad. Natur. Sci. Proc. Philad. 41: 9–11.

Lenssen, A. 1897. Sur la présence de Sporozoaires chez un Rotateur. Zool. Anz. 20: 330–333.

Lépine, F. 1949. Félix d'Herelle. Ann. Inst. Pasteur 76: 457–460.

Lesley, J. P. 1880. Fungus inoculation of insects. Nature 22: 31.

Leuckart. 1860. Zur Naturgeschichte der Bienen. 3. Zur Kenntniss der Faulbrut und der Pilzkrankheiten bei den Beinen. Eichstadt Bienenzeitung, 16 Jahrg., Nro. 20, pp. 232–233. [Cited in Phillips and White, 1912.]

Leutenegger, R. 1964. Development of an icosahedral virus in hemocytes of *Galleria mellonella* (L.). Virology 24: 200–204.

Leydig, F. 1854. Zur Anatomie von *Coccus hesperidum*. Z. Wiss. Zool. 5: 1–12. [See p. 11.]

———. 1863. Der Parasit in der neuren Krankheit der Seidenpaupe noch einmal. Arch. Anat. Physiol., Wiss. Med. Jahrgang, pp. 186–192.

Lieberkühn, N. 1853. Evolution des gregarines. Mém. Cour. et Mém. des Savants Étrangers, Acad. Roy. Belg. 26: 3–40.

Lipa, J. J., and M. E. Martignoni. 1960. *Nosema phryganidiae*, n. sp., a microsporidian parasite of *Phryganidia californica* Packard. J. Insect Pathol. 2: 396–410.

Lipa, J. J., and E. A. Steinhaus. 1959. *Nosema hippodamiae* n. sp., a microsporidian parasite of *Hippodamie convergens* Guérin (Coleoptera, Coccinellidae). J. Insect Pathol. 1: 304–308.

Liu, G. K. C. 1952. The silkworm and Chinese culture. Osiris 10: 129–193.

Long, E. R. 1965. A history of pathology. Dover Publications, New York. 199 pp.

Lortet. 1890. La bacterie loqueuse. Traitement de la loque par le napthol B [Beta]. Revue Internationale d'Apiculture 12 (2): 50–54. [Cited in Phillips and White, 1912.]

Lotmar, R. 1941. Die Polyederkrankheit der Kleidermotte (*Tineola biselliella*). Mitt. Schweiz, Entomol. Ges. 18: 372–373.

Lugger, O. 1888. Fungi which kill insects. Univ. Minnesota Agr. Exp. Sta. Tech. Bull. 4. 37 pp.

McCoy, E. E., and C. W. Carver. 1941. A method of obtaining spores of the fungus *Beauveria bassiana* (Bals.) Vuill. in quantity. J. N. Y. Entomol. Soc. 49: 205–210.

McCrady, J. 1873. Observations on the food and reproductive organs of *Ostrea virginiana*, with some account of *Bucephalus cuculus*, nov. spec. Proc. Boston Soc. Natur. Hist. 16: 170–182.

McEwen, F. L. 1963. *Cordyceps* infections. Pp. 273–290 *in* E. A. Steinhaus, ed., Insect pathology: an advanced treatise, vol. 2. Academic Press, New York and London. 689 pp.

McLain, N. W. 1887. Report on experiments in apiculture. Rpt. Comm. Agr. 1886, pp. 583–591. Washington Printing Office.

MacKenzie, J. J. 1892. The foul brood bacillus (*B. alvei*); its vitality and development. 18th Annu. Rpt. Ontario Agr. Coll. Exp. Sta., pp. 267–273. [Cited in Phillips and White, 1912.]

MacLeod, D. M. 1954. Investigations on the genera *Beauveria* Vuill. and *Tritirachium* Limber. Can. J. Botany 32: 818–890.

MacLeod, D. M., and T. C. Loughheed. 1963. Entomogenous fungi. Pp. 141–150 *in* Recent progress in microbiology. Symposia 8. International Congress Microbiology Montreal, 1962, vol. 8.

Maestri, A. 1856. Frammenti anatomiei, fisiologiei e patologici sul baco da seta (*Bombyx mori* Linn.). Fratelli Fusi, Pavia. 172 pp.

Magnus, A. 1270. De animalibus libri vigintiser nouissume impressi, I Calaphon: Venetija impensa heredu Octaviani Scoti, 1519. 105 pp.

Mains, E. B. 1948. Entomogenous fungi. Mycologia 40: 402–416.

Major, R. H. 1944. Agnostino Bassi and the parasitic theory of disease. Bull. Hist. Med. 16: 97–107.

Martignoni, M. E. 1957. Contributo alla conoscenze di una granulosi di *Eucosma griseana* (Hüber) (Tortricidae, Lepidoptera) quale fattore limitante il pullulamento dell insetto nella Engadina alta. Swiss Forest Res. Inst. Memoirs 32: 371–418.

———. 1958. Exploring new methods for mass production of insect viruses. Pest Control Review, Univ. Calif. Agr. Extension Serv., July issue, p. 8.

———. 1964*a*. Mass production of insect pathogens. Pp. 579–609 *in* Paul Debach and E. I. Schlinger, eds., Biological control of insect pests and weeds. Chapman and Hall, London, 844 pp.

———. 1964*b*. Progressive nucleopolyhedrosis in adults of *Peridroma saucia* (Hübner). J. Insect Pathol. 6: 368–372.

———. 1964*c*. Pathophysiology in the insects. Annu. Rev. Entomol. 9: 179–206.

Martignoni, M. E., and C. Auer. 1957. Bekämpfungsversuch gegen *Eucosma griseana* (Hübner) (Lepidoptera, Tortricidae) mit einem granulosis-virus. Mitt. schweiz, Anst. forstl. Versuchsw. 33: 73–93.

Martignoni, M. E., and R. L. Langston. 1960. Supplement to an annotated list and bibliography of insects reported to have virus diseases. Hilgardia 30: 1–40.

Martignoni, M. E., and J. E. Milstead. 1962. Transovum transmission of the nuclear polyhedrosis virus of *Colias eurytheme* Boisduval through contamination of the female genitalia. J. Insect Pathol 4: 113–121.

———. 1964. Hypoproteinemia in a noctuid larva during the course of nucleopolyhedrosis. J. Insect Pathol. 6: 517–531.

———. 1966*a*. Hypoproteinemia in a noctuid larva during the course of a granulosis. J. Invertebr. Path. 8: 261–263.

———. 1966*b*. Glutamate-aspartate transaminase activity in the blood plasma of an insect during the course of two viral diseases. Ann. Entomol. Soc. Am. 60: 428–430.

Martignoni, M. E., and R. J. Scallion. 1961*a*. Multiplication *in vitro* of a nuclear-polyhedrosis virus in insect amoebocytes. Nature 190: 1133–1134.

———. 1961*b*. Preparation and uses of insect hemocyte monolayers *in vitro*. Biol. Bull. 121: 507–520.

Martignoni, M. E., and P. Schmid. 1961. Studies on the resistance to virus infections in natural populations of Lepidoptera. J. Insect Pathol. 3: 62–74.

Martignoni, M. E., E. M. Zitcer, and R. P. Wagner. 1958. Preparation of cell suspensions from insect tissues for *in vitro* cultivation. Science 128: 360–361.

Masera, E. 1936. Le malattie infettive degli insetti. E Loro indice bibliografico. Licinio Cappelli, Bologna. 343 pp.

———. 1956*a*. Nel centenario della moeta Agostino Bassi fondatore della parassitologia. L'Allevatore 12: 6, No. 21.

———. 1956*b*. Scoperta della genesi del calcino. L'Allevatore 12: 7, No. 25.

Massee, G. 1895. A revision of the genus *Cordyceps*. Ann. Bot. 9: 1–44.

Mattes, O. 1927. Parasitäre Krankheiten der Mehlmottenlarven und Versuche über ihre Verwenlbarkeit als biologische Bekämpfungsmittel. (Zugleichein Beitrag zur Zytologie der Bakterien.) Gesell. f. Beförd. Gesam. Naturw. Sitzber. Marburg 62: 381–417.

Mechalas, B. J., and P. H. Dunn. 1964. Bioassay of *Bacillus thuringiensis* Berliner-based microbial insecticides. 1. Bioassay procedures. J. Insect Pathol. 6: 214–217.

Meckel. Cited by Lieberkühn, 1853.

Merian [Graf], Maria Sibilla. 1679–83. Der Raupen wunderbare Verwandelung und sonderbare Blume-Nahrung, worinnen—der Raupen, Würmer—und anderer dergleichen Thierlein, Ursprung, Speisen und Veränderungen—beschrieben—und—verlegt von M.S.M. 2 The. Nürnberg. Frankfurt am Mayn.

Metalnikov, S., and V. Chorine. 1928. Maladies bacteriennes chez les chenilles de la pyrale du mais (*Pyrausta nubilalis* Hbn.). C. R. Seances Hebd. Acad. Sci. 186: 546–549.

———. 1929. On the infection of gypsy moth and certain other insects with *Bacterium thuringiensis*. A preliminary report. Pp. 60–61 *in* T. Ellinger, ed., International corn borer investigations, vol. 2. International Livestock Exposition, Union Stock Yards, Chicago. 183 pp.

Metalnikov, S., and C. Toumanoff. 1928. Experimental researches on the infection of *Pyrausta nubilalis* by entomophytic fungi. Pp. 72–73 *in* T. Ellinger, ed., International corn borer investigations, vol. 1. International Livestock Exposition, Union Stock Yards, Chicago. 236 pp.

Metchnikoff, E. 1879. O boleznach litchinok khlebnogo zhuka. Zapiski Imperatorskogo Obschestva sel' skogo khoziaistva Iuzhnoi Rossi. Odessa. Pp. 21–50. Also published as reprint-pamphlet: O boleznach litchinok chlebnogo shuka of a series, O vrednych dla zemledelija nasekomych. Vyp. 3, Odessa. 32 pp.

———. 1880. Zur Lehre über Insectenkrankheiten. Zool. Anz. 3: 44–47.

———. 1882. Vergleichend-embryologische Studien: 3. Über die Gastrula einiger Metazoen. Zeit. Wissen. Zool. 37: 286.

———. 1883. Untersuchungen über die mesodermalen Phagocyten einiger Wirbeltiere. Biol. Centralbl. 3: 560–565.

———. 1884*a*. Untersuchungen über die intracelluläre Verdauung bei wirbellosen Thieren. Arb. zool. Inst. Univ. Wien, zool. Stat. Triest 5: 141–168.

———. 1884*b*. Ueber eine Sprosspilzkrankheit der Daphnien. Beitrag zur Lehre über den Kampf der Phagocyten gegan Krankheitserreger. Arch. Pathol. Anat. Physiol. Klin. Med. 96: 177–195.

———. 1884*c*. Ueber die Beziehungen der Phagocyten zu Milzbrandbacillen. Arch. Pathol. Anat. Physiol. Klin. Med. 97: 502–526.

———. 1892. Leçons sur la pathologie camparée de l'inflammation. G. Masson, Paris. 235 pp. [English trans. 1893 by F. A. and E. H. Starling, published by Kegan Paul, Trench, Trübner, and Co., Ltd., London. 197 pp. Dover edition, with a new introduction by A. M. Silverstein, 1968. Dover Publications, New York. 224 pp.]

———. 1905. L'Immunité dans les maladies infectieuses. Ann. Inst. Pasteur, 19 (11). [English trans. by F. G. Binnie, 1907. University Press, Cambridge. 591 pp.]

Metchnikoff, O. 1921. Life of Elie Metchnikoff, 1845–1916. Constable and Co., Ltd., London. 297 pp. [See p. 111.]

Michelbacher, A. E., and R. F. Smith. 1943. Some natural factors limiting the abundance of the alfalfa butterfly. Hilgardia 15: 369–397.

Michelbacher, A. E., and W. W. Middlekauff. 1952. Fungus on codling moth. Calif. Agr., March, pp. 13–14.

Mitani, K. 1929. Disease of silkworms by protozoa, pp. 2–285, and Disease of silkworms by fungi, pp. 291–476 [in Japanese]. *in* Silkworm pathology, vol. 2. Ind. ed. Meibundo, Tokyo. 478 pp.

Mitchell, S. L. 1827. Views of the process in nature by which, under particular circumstances, vegetables grow on bodies of living animals. Am. J. Sci. Arts 12: 21–28.

Molitor-Mühlfeld. 1868. Die Faulbrut, ihre Entstehung, Fortpflanzung und Heilung. Eichstädt Bienenzeitung 24 (8): 93–97. [Cited in Phillips and White, 1912.]

Montagne, J. F. C. 1836. Experiences et observations sur le champignon entomoctone, ou histoire botanique de la muscardine. C. R. Seances Hebd. Acad. Sci. 3: 166–170.

Morrill, A. W., and E. A. Back. 1912. Natural control of white flies in Florida. U. S. Dep. Agr., Bureau of Entomol. Bull. 102. 73 pp.

Müller-Kögler, E. 1941. Beobachtungen über das Verpilzen von Forleulenraupen durch *Empusa aulicae* Reich. A. Pflanzenkr. Pflanzenschutz 51: 124–135.

———. 1965. Pilzkrankheiten bei Insekten. Paul Parey, Berlin and Hamburg. 444 pp.

Muma, M. H. 1955. Factors contributing to the natural control of citrus insects and mites in Florida. J. Econ. Entomol. 48: 432–438.

———. 1956. Predators and parasites of citrus mites in Florida. Pp. 633–647 *in* Proc. Int. Congr. Entomol., 10th, Montreal 4.

Muma, M. H., and D. W. Clancy. 1961. Parasitism of purple scale in Florida Citrus groves. Florida Entomologist 44: 159–165.

Munn, W. A. 1870. Bevan on the honey bee. The honey bee, its natural history, physiology and management by Edward Bevan. Revised and enlarged and illustrated by William Augustus Munn. 3rd. edition. John Van Voorst. London. 384 pp. [Bevan edition published first 1838.]

Naegeli, K. W. 1857. Über die neue Krankheit der Seidenraupe und verwandte Organismen. Beitr. Botan. Z. 15: 760–761.

Needham, J. 1961. Science and civilization in China. Vol. 1. Cambridge University Press, Cambridge, England. 318 pp.

Nelson, E. V. 1967. History of beekeeping in the United States. Pp. 2–4 *in* Beekeeping in the United States. U. S. Dep. Agr. Handbook No. 335. 147 pp.

Nicolle, J. 1961. Louis Pasteur. The story of his major discoveries. Trans. by Hutchinson and Co., Fawcett Publications, Inc., Greenwich, Conn. 192 pp.

Nowakowski, L. 1883. Entomophthoreae. Pro. Acad. Sci. Krakow. 34 pp.

Nysten, P. H. 1808. Recherches sur les maladies des vers à soie et les moyens de les prévenir. Impr. Impéarile, Paris. 188 pp.

Omori, J. 1899. Modern treatise on Japanese silkworm disease [in Japanese]. Sangyo no Tomoshibi Sha, Ibaragi Perfecture, Japan. 340 pp.

Osborn, H. A. 1913. Biographical memoir of Joseph Leidy, 1823–1891. Nat. Acad. Sci. Biographical Memoirs 7: 335–396.

Osburn, M. R., and H. Spencer. 1938. Effect of spray on scale insect populations. J. Econ. Entomol. 31: 731–732.

Osimo, M. 1859 Richerche e considerazioni ulteriori sull'attuale malatia dei bachi. Riv. Period. Lavori I. R. Accad. Sci., Lett. and Arti, Padova 6: 157–184.

Packard, C. M., and C. Benton. 1937. How to fight the chinch bug. U. S. Dep. Agr. Farmer's Bull. 1780. 21 pp.

Paillot, A. 1928. Les maladies du ver à soie grasserie et dysenteries. Editions du service Photographique, de l'Université, Lyon. 328 pp.

———. 1930. Traite des maladies du ver à soie. G. Doin et Cie, Paris. 279 pp.

———. 1933. L'Infection chez les insectes. G. Patissier, Trévoux. 535 pp.

Panebianco, R. 1895. Asservazione sui granuli d. giallume. Boll. Mens. di Bachicolt 10, pp. 145–160.

Parkin, J. 1906. Fungi parasitic on scale insects. Ann. Royal Botan. Gardens, Peradeniya, 3, part 1. [Cited by Fawcett, 1908.]

Paroletti, [Sig.] 1803. Sull'uso dei suffumigi d'acido muriatico ossigenato per disinfettare l'aria delle stanze dove si allevano i bachi de seta. Biblioteca Italiana, October 30. [See Cutbush, 1808.]

Pasteur, L. 1870. Études sur la maladie des vers à soie. Gauthier-Villars, Paris. Vol. 1, 330 pp.; vol. 2, 327 pp.

———. 1874. [On the use of fungi against Phylloxera.] C. R. Seances Hebd. Acad. Sci. 79: 1233–1234.

———. 1888. Sur la destruction des lapin en Australie et dans la Nouvelle-Zélande. Ann. Inst. Pasteur 2: 1–8.

Patterson, R. 1956. Spinning and weaving. Pp. 197–198 *in* C. J. Singer et al., eds., A history of technology, vol. 2. The Clarendon Press, Oxford. 802 pp.

Pavlovsky, E. N. 1952. The place of insect pathology and entomology in the development of Soviet science. Pp. 1–15 *in* Russian edition of Principles of insect pathology. E. A. Steinhuas. McGraw-Hill, New York. 757 pp. Publications of Foreign Literature, Moscow.

Peck, W. D. 1795. The description and history of the canker worm. Massachusetts Magazine 7: 323–327; 415–416.

———. 1796. Natural history of the canker worm. Rules and Regulations of the Massachusetts Society for Promoting Agriculture, pp. 34–45.

———. 1827. Natural history of the canker worm. New England Farmer 5: 393–394.

Pellet, F. C. 1938. Efforts toward disease control. Pp. 186–197 *in* History of American beekeeping. Collegiate Press, Inc., Ames, Iowa. 213 pp.

Pérez, C. 1899. Sur une coccidie nouvelle *Adelea mesnili* (n. sp.) parasite coelomique d'un Lépidoptère. Mem. Soc. Biol. 51: 694–696.

Persoon, D. C. H. 1801. Synopsis methodica fungorum. H. Dietrich, Göttingen. 706 pp. [See p. 687.]

Petch, T. 1921*a*. Fungi parasitic on scale insects. Trans. Brit. Mycol. Soc. 7: 18–40.

———. 1921*b*. Studies on entomogenous fungi. The *Nectriae* parasitic on scale insects. Trans. Brit. Mycol. Soc. 7: 89–166.

———. 1939. Notes on entomogenous fungi. Trans. Brit. Mycol. Soc. 23: 127.

———. 1940*a*. *Myrophagus ucrainicus* (Wize) Sparrow, a fungus new to Britain. The Naturalist, 1940, p. 68.

———. 1940*b*. An *Empusa* on a mite. Proc. Linn. Soc. New South Wales 65: 259–260.

Pettit, R. H. 1895. Studies in artificial cultures of entomogenous fungi. Cornell University Agr. Exp. Sta. Bull. No. 97, pp. 339–378.

Peyritsch, J. 1873. Beiträge zur Kenntniss der Laboulbenien. Sitzunsh. Kaiserl. Akad. Wiss., Wien 68: 227–254.

Phillips, E. F. 1918. The Control of European Foulbrood. U. S. Dep. Agr. Farmer's Bulletin 975. Government Printing Office, Washington, D. C. 16 pp.

Phillips, E. F., and G. F. White. 1912. Historical notes on the causes of bee diseases. U. S. Dep. Agr., Bureau Entomol. Bull. 98. Government Printing Office, Washington, D. C. 96 pp.

Picou, R., and E. Ramond. 1899. Action bactéricide de l'extrait de toenia inerme. C. R. Seances Soc. Biol. 51: 176–177.

Plenciz, M. A. 1762. Cited by A. Berg, 1963.

Pliny (Plinius Secundus, Caius). A.D. 77. The natural history of Pliny. Vol. 3, books 11–17 Trans. by John Bostock and H. T. Riley. 1855. H. G. Bohn, London. 536 pp. [Chap. 20 titled "The Diseases of Bees"; chap. 21 titled "Things that are Noxious to Bees."]

Pomier, L. 1754. L'art decultiver les muriers-blancs, délever les vers à soye, et de tirer les soye cocons. *Half title*: Traité des muriers-blancs et des vers à soye. Lottin and Butard, Paris. 234 pp.

———. 1763. Traité sur la culture des muriers blancs, la manière d'élever les vers à soie, et l'usage qu'on doit faire des cocons. [Not seen by the author, but apparently a revision of the 1754 work.]

Pramer, D. 1965. Symposium on microbial insecticides. 3. Fungal parasites of insects and nematodes. Bacteriol. Rev. 29: 382–387.

Prentiss, A. N. 1880. Destruction of obnoxious insects by means of fungoid growths. Am. Naturalist 14: 575–581; 630–635.

Preston, J. W., and R. L. Taylor. 1970. Observations on the phenoloxidase system in the hemolymph of the cockroach, *Leucophaea maderae*. J. Insect Physiol. 16: 192.

Preuss, 1868. Das Wesen der bösartigen Faulbrut besteht in einem mikroskopischen Pilze, *Cryptococcus alvearis*. Sie kann verhutet und geheilt werden. Eichstädt Bienenzeitung, 24 (19, 20): 225–228. [Cited by Phillips and White, 1912.]

Prevost, L. 1867. California silk grower's manual. H. H. Bancroft & Co., San Francisco. 246 pp. [Entered according to act of Congress in 1866].

Prillieux, E., and G. Delacroix. 1891. Le champignon parasite de la larve du hanneton. C. R. Seances Hebd. Acad. Sci. 112: 1079–1081.

Prowazek, S. von. 1907. Chlamydozoa. 2. Gelbsucht der Seidenraupen. Arch. Protistenk. 10: 358–364.

———. 1912. Untersuchungen über die Gelbsucht der Seidenraupen. Zentrabl. Bakteriol. Parasitenk. Infektionskr., 1 Orig. 67: 268–284.

Quatrefages, A. de. 1859. Études sur les maladies actuelles du ver à soie. Victor Masson, Paris. 382 pp. Supplement: 1860. Nouvelles recherches faites en 1859 sur les maladies actuelles des vers à soie. 118 pp.

Quinby, M. 1853. Mysteries of bee-keeping explained. A. O. Moore, Agricultural Book Publ., New York. 384 pp. 8th ed. 1859.

Rabb, R. L., E. A. Steinhaus, and F. E. Guthrie. 1957. Preliminary tests using *Bacillus thuringiensis* Berliner against hornworms. J. Econ. Entomol. 50: 259–262.

Rademacher, B. 1958. Prof. Dr. h. c. Hans Blunck. Z. Pflanzenkr. Pflanzenpathol. Pflanzenschutz 65: 1–10.

Ratzeburg, J. T. C. 1869. [Fungoid parasitism of insects. "Mycetinosis."] Boston Soc. Natur. Hist. Proc. 12: 281.

Réamur, R. A. F. de. 1726. Remarques sur la plante appellée à la Chine Hia Tsao Tom Tchom, ou plante ver. [Paris] Acad. Roy. Sci. Mém., pp. 302–305.

Rex, E. G. 1940. A promising fungus pathogen of adult Japanese beetles (*Popillia japonica*). J. N.Y. Entomol. Soc. 48: 401–403.

Rhodes, R. A. 1965. Symposium on microbial insecticides. 2. Milky disease of the Japanese beetle. Bacteriol. Rev. 29: 373–381.

Richards, A. G. 1953. The penetration of substances through the cuticle. Pp. 42–54 *in* K. D. Roeder, ed., Insect physiology. John Wiley and Son, New York. 1100 pp.

Rienzi, L. 1886. General instructions for rearing silkworms, with a treatise on securing healthy silkworm eggs. Also, a sketch of the habits and structure of the silkworm.

Calif. State Board of Silk Culture. Separate publication. State Printing Office, Sacramento. 21 pp.

Riley, C. V. 1879. The silkworm; being a brief manual of instructions for the production of silk. U. S. Dep. Agr., Special Rept. No. 11. 31 pp.

———. 1883. The use of contagious germs as insecticides. Am. Naturalist 17: 1169–1170.

———. 1885. [Untitled circular,] No. 9, issued by the Division of Entomology of the U. S. Dep. Agr., May 1, 1 p.

Roberts, D. W. 1966*a*. Toxins from the entomogenous fungus *Metarrhizium anisopliae.* I. Production in submerged and surface cultures, and in inorganic and organic nitrogen media. J. Invertebr. Pathol. 8: 212–221.

———. 1966*b*. Toxins from the entomogenous fungus *Metarrhizium anisopliae.* II. Symptoms and detection in moribund hosts. J. Invertebr. Pathol. 8: 222–227.

———. 1969. Toxins from the entomogenous fungus *Metarrhizium anisopliae*: isolation of destruxins from submerged cultures. J. Invertebr. Pathol. 14: 82–88.

Robin, C. 1847. Des végétaux qui croissent sur l'homme et sur les animaux vivants. J.-B. Baillière, Paris. 120 pp.

———. 1853. Histoire naturelle des végétaux parasites qui croissent sur l'homme et sur les animaux vivants. J.-B. Baillière, Paris. 702 pp.

Robinet, S. 1845. La muscardine; des causes de cette maladie, et des moyens d'en preserver les vers à soie. Millet and Robinet, Paris. 288 pp. 2nd ed. [1st ed. published in 1843.]

Rockwood, L. P. 1951. Some hyphomycetous fungi found on insects in the Pacific Northwest. J. Econ. Entomol. 44: 215–217.

Rolfs, P. H. 1897. A fungus disease of the San Jose scale (*Sphaerostilbe coccophila* Tul.). Florida Agr. Exp. Sta. Bull. 41: 513–543.

Rolfs, P. H., and H. S. Fawcett. 1908. Fungous diseases of scale insects and whitefly. Florida Agr. Exp. Sta. Bull. 94. 17 pp.

———. 1913. Fungous diseases of scale insects and whitefly. Revised by P. H. Rolfs. Florida Agr. Exp. Sta. Bull. 119 pp. 71–82.

Root, A. I. 1877. The ABC of bee culture. A. I. Root, Medina, Ohio. 340 pp.

Rorer, J. B. 1910. The green muscardine of froghoppers. Proc. Agr. Soc. Trinidad Tobago 10: 467–482. Society paper 442.

———. 1913. The use of the green muscardine in the control of some sugarcane pests. Phytopathology 3 (2): 88–92.

Rosenau, M. J. 1901. An investigation of a pathogenic microbe (*B. typhimurium* Danysz) applied to the destruction of rats. U. S. Marine Hospital Service Bull. 5 of the Hygiene Laboratory. 11 pp.

Ross, R. 1895. The crescent-sphere-flagella metamorphosis of the malarial parasite in the mosquito. Trans. S. Indian Branch Brit. Med. Assoc. 6: 334–338.

———. 1897. On some peculiar pigmented cells found in two mosquitoes fed on malarial blood. Brit. Med. J., Dec. 18, pp. 1786–1788. [See also Brit. Med. J., Feb. 26, 1898, pp. 550–551.]

Rouget, A. 1850. Notice sur une production parasite sur le *Brachinus crepitans.* Ann. Soc. Entomol. France 8: 21–24.

Rubtzov, I. A. 1948. Biological methods of combat against harmful insects [in Russian]. Moscow. [Quoted by Pavlovsky, 1952.]

Rudolphi, K. A. 1819. Entozoorum synopsis, ciu accedunt mantissa duplex et indices locupletissimi Berolini, Rucker, 8. 811 pp. (Entozoen in Insecten, pp. 786–788.)

Ryder, J. A. 1887. On a tumor in the oyster. Acad. Natur. Sci. Proc. Philad. 39: 25–27.

Sager, S. M. 1960. On the transtadial transmission of insect viruses. J. Insect Pathol. 3: 307–309.

Sano, E. 1898. Application of Japanese sericulture [in Japanese]. Editorial Committee on Japanese Sericulture. Tokyo. 146 pp.

Sarton, G. 1959. A history of science. Harvard University Press, Cambridge. 554 pp.

Sasaki, C. 1900. Study book of Japanese silkworm pébrine [in Japanese]. Keigyosha, Tokyo. 246 pp.

Sauvageau, C., and J. Perraud. 1893. Sur un champignon parasite de la cochylis. C. R. Seances Hebd. Acad. Sci. 117: 187–191.

Schierbeek, A. 1959. Measuring the invisible world: the life and works of Antoni van Leeuwenhoek FRS. Abelard-Schuman, London and New York. 223 pp.

Schirack, A. G. 1771. Histoire naturelle de la Reine des Abeilles, avec l'art de former des essaims. F. Staatman, La Haye. 269 pp.

Schneider, A. C. J. 1885. Coccidies nouvelles on peu connues. Tablettes Zool. 1: 4–9.

Schönfeld 1873. Faulbrut-studien, Pt. I. Eichstädt Bienenzeitung, 29 Jahrg., Nro. 21, 250-254. Cited in Phillipps and White, 1912.

———. 1874. Faulbrut-studien. Pt. II. Eichstädt Bienenzeitung. 30 Jahrg., Nro. 1, 3-5. Cited in Phillipps and White, 1912.

Schönlein, J. L. 1839. Zur Pathogenie der Impetigines. Müller's Arch. (Arch. Anat. Physiol. wiss. Med.), p. 82.

Schwabe, C. W. 1969. Veterinary medicine and human health. 2nd ed. Williams and Wilkins Co., Baltimore. 713 pp.

Schwartz, W. 1966. Paul Buchner 80 Jahre. Z. Allg. Mikrobiol. 6: 93.

Selhime, A. G., and M. H. Muma. 1966. Biology of *Entomophthora floridana* attacking *Eutetranychus banksi*. Florida Entomologist 49: 161–168.

Shimer, H. 1867. Notes on *Micropus* (*Lygarus*) *leucopterus* Say ("the chinch bug"). With an account of the great epidemic disease of 1865 among insects. Acad. Natur. Sci. Proc. Philad. 19: 75–80.

Shimizu, K. 1844-47. Cited by Sano, 1898.

Siebold, C. T. E. von. 1839. Beiträge zur Naturgeschichte der wirbellosen Thiere. Neuest. Schrift. Naturf. Gesell. Danzig 3: 56–71.

Sigerist, H. E. 1960. On the history of medicine. Edited and with an introduction by F. Marti-Ibañez. M. D. Publications, Inc., New York. 313 pp.

Silvestri, F. 1909. A survey of the actual state of agricultural entomology in the United States of North America. The Hawaiian Forester and Agriculturalist 6: 287–336.

Singer, C. 1959. A short history of scientific ideas to 1900. Oxford Univ. Press, London. 525 pp.

Singer, C., and E. A. Underwood. 1962. A Short History of Medicine. Oxford Univ. Press, Oxford. 854 pp.

Smith, A. C., and R. L. Taylor. 1968*a*. Tumefactions (tumorlike swellings) on the foot of the moon snail, *Polinices lewisii*. J. Invertebr. Pathol. 10: 263–268.

———. 1968*b*. Digestive gland and integument lesions associated with malnutrition in a ghost shrimp, *Callianssa affinis*. J. Invertebr. Pathol. 12: 1–6.

Smith, O. J., K. M. Hughes, P. H. Dunn, and I. M. Hall. 1956. A granulosis-virus disease of the western grape leaf skeletonizer and its transmission. Can. Entomologist, 88: 507–515.

Smith, T., and F. E. Kilbourne. 1893. Investigations into the nature, causation, and prevention of Texas or southern cattle fever. U. S. Dep. Agr., Bur. Anim. Ind. Bull. No. 1, 177 pp.

Smyth, E. G. 1916. *Lachnosterna grande.* 3rd Annu. Rpt. Board of Comm. Agr. Porto Rico, 42–47.

Snow, F. H. 1890. Experiments for the destruction of chinch bugs. 21st Annu. Rpt. Entomol. Soc., Ontario. 93–97. [Published in 1891.]

———. 1891. Chinch-bugs. Experiments in 1890 for their destruction in the field by the artificial introduction of contagious diseases. 7th Bienn. Rpt. Kansas State Board Agr. 12: 184–188.

———. 1894. Contagious diseases of the chinch bug. 3rd Annu. Rpt. of the Director Kansas Agr. Exp. Sta. for the year 1893. 247 pp.

———. 1895. Contagious diseases of the chinch bug. 4th Annu. Rpt. of the Director Kansas Agr. Exp. Sta., for the year 1894. 46 pp.

———. 1896. Contagious diseases of the chinch-bug. 5th Annu. Rpt. of the Director Kansas Agr. Exp. Sta., for the year 1895. 55 pp.

Sokoloff, V. P., and L. J. Klotz. 1941. A bacterial pathogen of the citrus red scale. Science 94: 40–41.

———. 1942. Mortality of the red scale on citrus through infection with a spore-forming bacterium. Phytopathology, 32: 187–198.

South, F. W. 1910. Further notes on the fungus parasites of scale insects. West Indian Bull. 12: 403.

Sparrow, F. K., Jr. 1939. The entomogenous chytrid *Myrophagus* Thaxter. Mycologia 31: 439–444.

Speare, A. T. 1912. Fungi parasitic upon insects injurious to sugar cane. Hawaii Sugar Planters' Assoc. Exp. Sta., Div. Pathol. and Physiol. Bull. 12: 1–62.

———. 1922. Natural control of the citrus mealybug in Florida. U.S. Dep. Agr. Bull., 1117. Government Printing Office, Washington, D.C. 19 pp.

Speare, A. T., and R. H. Colley. 1912. The artificial use of the brown-tail fungus in Massachusetts, with practical suggestions for private experiment, and a brief note on a fungus disease of the gypsy caterpillar. Wright and Potter, Boston, Mass. 29 pp.

Speare, A. T., and W. W. Yothers. 1924. Is there an entomogenous fungus attacking the citrus rust mite in Florida? Science 60: 41–42.

Steinhaus, E. A. 1939. Studies on the life and death of bacteria. I. The senescent phase in aging cultures and the probably mechanisms involved. J. Bacteriol. 38: 249–261.

———. 1940. The microbiology of insects with special reference to the biologic relationships between bacteria and insects. Bacteriol. Rev. 4: 17–57.

———. 1941. A study of the bacteria associated with thirty species of insects. J. Bacteriol. 42: 757–790.

———. 1942. Catalog of bacteria associated extracellularly with insects and ticks. Burgess Publishing Co., Minneapolis, Minn. 206 pp.

———. 1945*a*. Insect pathology and biological control. J. Econ. Entomol. 38: 591–596.

———. 1945*b*. Bacterial infections of potato tuber moth larvae in an insectary. J. Econ. Entomol. 38: 718–719.

———. 1946. Insect microbiology. Comstock Pub. Co., Inc., Ithaca, New York. 757 pp.

———. 1947*a*. A coccidian parasite of *Ephestia kühniella* Zeller and of *Plodia interpunctella* (Hbn.) (Lepidoptera, Phycitidae). J. Parasitol. 33: 29–32.

———. 1947*b*. Control of insect pests by means of disease agents. Calif. Agr. 1: no. 6, 2.

———. 1947*c*. A new disease of the variegated cutworm, *Peridroma margaritose* (Haw.). Science 106: 323.

———. 1948. Polyhedrosis ("wilt disease") of the alfalfa caterpillar. J. Econ. Entomol. 41: 859–865.

———. 1949*a*. Principles of insect pathology. McGraw-Hill Book Co., Inc., New York. 757 pp.

———. 1949*b*. Nomenclature and classification of insect viruses. Bacteriol. Rev. 13: 203–223.

———. 1951*a*. Possible use of *Bacillus thuringiensis* Berliner as an aid in the biological control of the alfalfa caterpillar. Hilgardia 20: 359–381.

———. 1951*b*. Report on diagnoses of diseases insects. 1944–1950. Hilgardia 20: 629–678.

———. 1952*a*. Microbial infections in European corn borer larvae held in the laboratory. J. Econ. Entomol. 45: 48–51.

———. 1952*b*. The susceptibility of two species of *Colias* to the same virus. J. Econ. Entomol. 45: 897–899.

———. 1953 Taxonomy of insect viruses. Ann. N. Y. Acad. Sci. 56: 517–537.

———. 1954*a*. Duration of infectivity of the virus of silkworm jaundice. Science 120: 186–187.

———. 1954*b*. The effects of disease on insect populations. Hilgardia 23: 197–261.

———. 1955. The diagnoses of insect diseases including *instructions for submitting specimens* to the Laboratory of Insect Pathology. Department of Biological Control, University of California, Berkeley. Mimeographed. 4 pp.

———. 1956*a*. Living insecticides. Sci. Am. 195: 96–103.

———. 1956*b*. Potentialities for microbial control of insects. Agr. Food Chem. 4: 676–680.

———. 1956*c*. Microbial control—the emergence of an idea. A brief history of insect pathology through the nineteenth century. Hilgardia 26: 107–160.

———. 1957*a*. New records of insect-virus diseases. Hilgardia, 26: 417–430.

———. 1957*b*. Microbial diseases of insects. Annu. Rev. Microbiol. 11: 165–182.

———. 1957*c*. Concerning the harmlessness of insect pathogens and the standardization of microbial control products. J. Econ. Entomol. 50: 715–720.

———. 1958. Stress as a factor in insect disease. Ecology 39: 113–121.

———. 1959*a*. Insect pathology and microbial control. Pest Control Rev., University of California Ext. Serv., Feb., 1959, 1–3.

———. 1959*b*. *Serratia marcescens* Bizio as an insect pathogen. Hilgardia 28: 351–380.

———. 1959*c*. Granuloses in two Alaskan insects. J. Econ. Entomol. 52: 350–380.

———. 1959*d*. On the improbability of *Bacillus thuringiensis* Berliner mutating to forms pathogenic for vertebrates. J. Econ. Entomol. 52: 506–508.

———. 1960*a*. Insect pathology; challenge, achievement, and promise. Bull. Entomol. Soc. Am. 6: 9–16.

———. 1960*b*. The duration of viability and infectivity of certain insect pathogens. J. Insect Pathol. 2: 225–229.

———. 1960*c*. Notes on polyhedroses in *Peridroma, Prodenia, Colias, Heliothis*, and other Lepidoptera. J. Insect Pathol. 2: 327–333.

———. 1960*d*. Bacterial and viral diseases of insects of medical importance (and other excerpts from Report of Conference on Biological Control of Insects of Medical Importance, held in Washington, D. C., February 3-4, 1960). Technical Rpt., published by Am. Inst. Bio. Sci., 144 pp.

———. 1960*e*. The importance of environmental factors in the insect-microbe ecosystem. Bacteriol. Rev. 24: 365–373.

———. 1961. On the correct author of *Bacillus sotto*. J. Insect Pathol. 3: 97–100.

———. 1963. Background for the diagnosis of insect diseases. Pp. 549–589 *in* E. A. Steinhaus, ed., Insect pathology: an advanced treatise, vol. 2. Academic Press, New York. 689 pp.

———. 1964. "The day is at hand . . . " (Entomol. Soc. Am. presidential address.) Bull. Entomol. Soc. Am. 10: 3–7.

———. 1965. Symposium on microbial insecticides. IV. Diseases of invertebrates other than insects. Bacteriol. Rev. 29: 388–396.

———. 1967*a*. On the importance of invertebrate pathology in comparative pathology. Rev. Pathol. Comparée 67: 139–142. Also J. Invertebr. Pathol. 9: i-v.

———. 1967*b*. Microbial control—a comment on its present status in the United States. Bull. Entomol. Soc. Am. 13: 104–108.

———. 1967*c*. Microbial control is not all. P. 58. *in* Abstracts of papers presented at the Joint United States-Japan Seminar on Microbial Control of Insect Pests. Fukuoka, 61 pp.

———. 1968. Centers for pathobiology. J. Invertebr. Pathol. 11: i-iv.

———. 1969. Invertebrates as models for the study of diseases of man. Federation Proc. 28: 1801–1804.

Steinhaus, E. A., M. M. Batey, and C. L. Boerke. 1956. Bacterial symbiotes from the caeca of certain Heteroptera. Hilgardia 24: 495–518.

Steinhaus, E. A., and C. R. Bell. 1953. The effect of certain microorganisms and antibiotics on stored-grain insects. J. Econ. Entomol. 46: 582–598.

Steinhaus, E. A., and J. M. Birkeland. 1938. "Cannibalism" among bacteria. J. Bacteriol. (Abstract) 36: 216–217.

———. 1939. Studies on the life and death of bacteria. I. The senescent phase in aging cultures and the probable mechanisms involved. J. Bacteriol. 38: 249–261.

Steinhaus, E. A., and F. J. Brinley. 1957. Some relationships between bacteria and certain sewage-inhabiting insects. Mosquito News, 17: 299–302.

Steinhaus, E. A., and J. P. Dineen. 1959. A cytoplasmic polyhedrosis of the alfalfa caterpillar. J. Insect Pathol. 1: 171–183.

———. 1960. Observations on the role of stress in a granulosis of the variegated cutworm. J. Insect Pathol. 2: 55–65.

Steinhaus, E. A. and K. M. Hughes. 1949. Two newly described species of Microsporidia from the potato tuberworm, *Gnorimoschema operculella* (Zeller) (Lepidoptera, Gelechiidae). J. Parasitol. 35: 67–75.

———. 1952. A granulosis of the western grape leaf skeletonizer. J. Econ. Entomol. 45: 744–745.

Steinhaus, E. A., K. M. Hughes, and H. B. Wasser. 1949. Demonstration of the granulosis virus of the variegated cutworm. J. Bacteriol. 57: 219–224.

Steinhaus, E. A., and E. A. Jerrel. 1954. Further observations of *Bacillus thuringiensis* Berliner and other sporeforming bacteria. Hilgardia 23: 1–23.

Steinhaus, E. A., and R. Leutenegger. 1963. Icosahedral virus from a scarab (*Sericesthis*) . J. Insect Pathol. 5: 266–270.

Steinhaus, E. A., and G. A. Marsh. 1960. Granulosis of the granulate cutworm. J. Insect Pathol. 2: 115–117.

———. 1962. Report on diagnoses of diseased insects. 1951-1961. Hilgardia 33: 349–490.

———. 1967. Previously unreported accessions for diagnosis and new records. J. Invertebr. Pathol. 9: 436–438.

Steinhaus, E. A., and M. E. Martignoni. 1967. An abridged glossary of terms used in invertebrate pathology. Pacific Northwest Forest and Range Experiment Station. U. S. Dep. Agr., Forest Service. 22 pp. [2nd edition, 1970, 38 pp.]

Steinhaus, E. A., and C. G. Thompson. 1949*a*. Preliminary field tests using a polyhedrosis virus to control the alfalfa caterpillar. J. Econ. Entomol. 42: 301–305.

———. 1949*b*. Alfalfa caterpillar tests; biological control by artificial spread of virus disease studied. Calif. Agr. 3: March, 5–6.

———. 1949*c*. Granulosis disease in the buckeye caterpillar, *Junonia coenia* Hübner. Science 110: 276–278.

Steinhaus, E. A., and R. D. Zeikus. 1968*a*. Teratology of the beetle *Tenebrio molitor*. I. Gross morphology of certain abnormality types. J. Invertebr. Pathol. 10: 190–210.

———. 1968*b*. Teratology of the beetle *Tenebrio molitor*. III. Ultrastructural alterations in the flight musculature of the pupal-winged adult. J. Invertebr. Pathol. 12: 40–52.

———. 1969*a*. Teratology of the beetle *Tenebrio molitor*. IV. Ultrastructure of the necrotic fat body and foregut associated with the pupal-winged adult. J. Invertebr. Pathol 13: 337–344.

———. 1969*b*. Teratology of the beetle *Tenebrio molitor*. V. Ultrastructural changes and viruslike particles in the foregut epithelium of pupal-winged adults. J. Invertebr. Pathol. 14: 115–121.

Stempell, W. 1909. Über *Nosema bombycis* Nägeli. Arch. Protistenk. 16: 281–358.

Stevenson, J. A. 1916. Work with the green muscardine. 4th Rpt. Board Comm. Agr. Porto Rico, 34–35.

———. 1918. The green muscardine fungus in Porto Rico. (*Metarrhizium anisopliae* [Metsch.] Sorokin). J. Agr. Porto Rico. pp. 19–32.

Stirrett, G. M., G. Beall, and M. I. Timonin. 1937. A field experiment on the control of European corn borer, *Pyrausta nubilalis* Hubn., by *Beauveria bassiana* Vuill. Sci. Agr., Ottawa, 17: 587–591.

Stubbs, A. C. 1898. Observations on abnormal specimens of *Planorbis spirorbis* and other freshwater shells at Tenby. J. Conchol. 9: 106–108.

Sweetman, H. L. 1958. The Principles of Biological Control. Wm. C. Brown Co., Dubuque, Iowa. 560 pp.

Swenk, M. H. 1925. The chinch bug and its control. Nebraska Agr. Exp. Sta. Circular 28, 34 pp.

Swift, J. 1773. Verse 1. *In* On Poetry: a Rhapsody.

Tamura, S., S. Kuyama, Y. Kodaira, and S. Higashikawa. 1964. The structure of destruxin B, a toxic metabolite of *Oospora destructor* [insect pathogen]. Agr. Biol. Chem. 28: 137–138.

———. 1965. Destruxin B, an insecticidal metabolite of *Oospora destructor*. Pp. 749–750 *in* Proc. Int. Congr. Entomol. 12th, London, 1964.

Tanada, Y. 1953. Description and characteristics of a granulosis virus of the imported cabbageworm. Proc. Hawaiian Entomol. Soc. 15: 235–260.

———. 1954. A polyhedrosis virus of the imported cabbageworm and its relation to a polyhedrosis virus of the alfalfa caterpillar. Ann. Entomol. Soc. Am. 47: 553–574.

———. 1955. Field observations on a microsporidian parasite of *Pieris rapae* (L.) and *Apanteles glomeratus* (L.). Proc. Hawaiian Entomol. Soc. 15: 609–616.

———. 1956. Some factors affecting the susceptibility of the armyworm to virus infections. J. Econ. Entomol. 49: 52–57.

———. 1957. Probable origin and dissemination of a polyhedrosis virus of an armyworm in Hawaii. J. Econ. Entomol. 50: 118–120.

———. 1959*a*. Synergism between two viruses of the armyworm, *Pseudaletia unipuncta* (Haworth) (Lepidoptera, Noctuidae). J. Insect Pathol. 1: 215–231.

———. 1959*b*. Descriptions and characteristics of a nuclear polyhedrosis virus and a granulosis virus of the armyworm, *Pseudaletia unipuncta* (Haworth) (Lepidoptera, Noctuidae). J. Insect Pathol. 1: 197–214.

———. 1960*a*. A nuclear-polyhedrosis virus of the lawn armyworm, *Spodoptera mauritia* (Boisduval) (Lepidoptera: Noctuidae). Proc. Hawaiian Entomol. Soc. 17: 304–308.

———. 1960*b*. Epizootiology of the virus diseases of the armyworm, *Pseudaletia unipuncta* (Haworth). *In* Intern. Kongr. Entomol. Verhandl. 11th. K. Vienna, 1960. [Abstract]

———. 1961. The epizootiology of virus disease in field populations of the armyworm, *Pseudaletia unipuncta* (Haworth). J. Insect Pathol. 3: 310–323.

———. 1964*a*. Incidence of microsporidiosis in field population of the armyworm, *Pseudaletia unipuncta* (Haworth). Proc. Hawaiian Entomol. Soc. 18: 435–436.

———. 1964*b*. A granulosis virus of the codling moth, *Carpocapsa pomonella* (Linnaeus) (Olethreutidae, Lepidoptera). J. Insect Pathol. 6: 378–380.

———. 1965. Factors affecting the susceptibility of insects to viruses. Entomophaga 10: 139–150.

———. 1966. Field observation of the biotic factors regulating the population of the armyworm *Pseudaletia unipuncta* (Haworth). Proc. Hawaiian Entomol. Soc. 19: 302–308.

———. 1967. The role of viruses in the regulation of the population of the armyworm, *Pseudaletia unipuncta* (Haworth). Pp. 15–16 *in* Abstracts of papers presented at the Joint United States-Japan Seminar on Microbial Control of Insect Pests. Fukuoka, 61 pp.

Tanada, Y., and J. W. Beardsley. 1957. Probable origin and dissemination of a polyhedrosis virus of an armyworm in Hawaii. J. Econ. Entomol. 50: 118–120.

Tanada, Y., and G. Y. Chang. 1960. A cytoplasmic polyhedrosis of the armyworm, *Pseudaletia unipuncta* (Haworth) (Lepidoptera, Noctuidae). J. Insect Pathol. 2: 201–208.

———. 1962*a*. An epizootic resulting from a microsporidian and two virus infections in the armyworm, *Pseudaletia unipuncta* (Haworth). J. Insect Pathol. 4: 129–131.

———. 1962*b*. Cross-transmission studies with three cytoplasmic-polyhedrosis viruses. J. Insect Pathol. 4: 361–370.

———. 1964. Interactions of two cytoplasmic-polyhedrosis viruses in three insect species. J. Insect Pathol. 6: 500–516.

———. 1968. Resistance of the alfalfa caterpillar, *Colias eurytheme*, at high temperatures to a cytoplasmic-polyhedrosis virus and thermal inactivation point of the virus. J. Invertebr. Pathol. 10: 79–83.

Tanada, Y., and R. Leutenegger. 1965. A cytoplasmic-polyhedrosis virus in field populations of the armyworm, *Pseudaletia unipuncta* (Haworth) in Hawaii. J. Invertebr. Pathol. 7: 517–519.

Tanada, Y., and C. Reiner. 1960. Microbial control of the artichoke plume moth, *Platyptilia carduidactyla* (Riley) (Pterophoridae, Lepidoptera). J. Insect Pathol. 2: 230–246.

———. 1962. The use of pathogens in the control of the corn earworm, *Heliothis zea* (Boddie). J. Insect Pathol. 4: 139–154.

Tanada, Y., and A. M. Tanabe. 1964. Response of the adult of the armyworm, *Pseudaletia unipuncta* (Haworth) to cytoplasmic-polyhedrosis-virus infection in the larval stage. J. Insect Pathol. 6: 486–490.

———. 1965. Resistance of *Galleria mellonella* (Linnaeus) to the *Tipula* iridescent virus at high temperatures. J. Invertebr. Pathol. 7: 184–188.

Tanada, Y., A. M. Tanabe, and C. E. Reiner. 1964. Survey of the presence of a cytoplasmic-polyhedrosis virus in field populations of the alfalfa caterpillar, *Colias eurytheme* Boisduval, in California. J. Insect Pathol. 6: 439–447.

Tangl, F. 1893. Bakteriologischer Beitrag zur Nonnenraupenfrage. Forstwiss. Centralbl. 15: 209–230.

Tasnádi-Kubacska, A. 1962. Pathologie der Vorqeitlichen Tiere. pp. 43–46. *In* Paläopathologie. Veb Gustav Fischer Verlag, Jena. 2 Vols.

Taylor, R. L. 1965. Partial-pupae of the greater wax moth, *Galleria mellonella* (Linnaeus). J. Invertebr. Pathol. 7: 489–492.

———. 1966. *Haplosporidium tumefacientis* sp. n., the etiologic agent of a disease of the California sea mussel, *Mytilus californianus* Conrad. J. Invertebr. Pathol. 8: 109–121.

———. 1967. A fibrous banded structure in a crop of the cockroach, *Leucophaea maderae*. J. Ultrastruct. Res. 19: 130–141.

———. 1968. Tissue damage induced by an oxyuroid nematode, *Leidynema* sp. in the hindgut of the cockroach, *Leucophaea maderae*. J. Invertebr. Pathol. 11: 214–218.

———. 1969*a*. Formation of tumorlike lesions in the cockroach *Leucophaea maderae* after decapitation. National Cancer Institute Monograph, 31 pp.

———. 1969*b*. Formation of tumorlike lesions in the cockroach *Leucophaea maderae* after nerve severance. J. Invertebr. Pathol. 13: 167–187.

———. 1969*c*. Formation of tumorlike lesions in the cockroach *Leucophaea maderae* after anal blockage. J. Invertebr. Pathol. 13: 416–422.

———. 1969*d*. A suggested role for the polyphenol-phenoloxidase system in invertebrate immunity. J. Invertebr. Pathol. 14: 427–428.

Taylor, R. L., and W. C. Freckleton, Jr. 1968. The use of the tissue adhesive, methly-2-cyanoacrylate monomer, in invertebrate surgery. J. Invertebr. Pathol. 12: 412–414.

———. 1969. Sex differences in longevity of starved cockroaches *Leucophaea maderae*. J. Invertebr. Pathol. 13: 68–73.

Taylor, R. L., and J. W. Preston. 1969. The inflammatory response of the cockroach *Leucophaea maderae* to injections of normal and abnormal tissues. J. Invertebr. Pathol. 14: 150–157.

Taylor, R. L., and A. C. Smith. 1966. Polypoid and papillary lesions in the foot of the gaper clam, *Tresus nuttalli*. J. Invertebr. Pathol. 8: 264–266.

Thaxter, R. 1888. The Entomophthoreae of the United States. Boston Soc. Natur. Hist. 4 (VI): 133–201.

———. 1896–1931. Contribution towards a monograph of the Laboulbeniaceae. Am. Acad. Arts Sci. Memoirs 12: 187–429, 1896; 13: 217–469, 1908; 14: 309–426, 1924; 15: 427–580, 1926; 16: 1–435, 1931.

Thélohan, P. 1895. Recherches sur les Myxosporidies. Bull. Sci. France et Belg. 26: 100–394.

Thompson, C. G., and E. A. Steinhaus. 1950*a*. Further tests using a polyhedrosis virus to control the alfalfa caterpillar. Hilgardia 19: 411–445.

———. 1950*b*. Alfalfa caterpillar control: treatment of fields by airplane application of spray advances destruction of pests. California Agr. 4(5): 8, 16.

Thompson, W. L. 1939. Cultural practices and their influence upon citrus pests. J. Econ. Entomol. 32: 782–789.

Timonin, M. I. 1939. Pathogenicity of *Beauveria bassiana* (Bals.) Vuill. on Colorado potato beetle larvae. Can. J. Res. (D) 17: 103–107.

Tillet. 1755. Cited by E. Müller-Kögler, 1965, *in* Introduction.

Torrubia, J. 1754. Aparto para la historia natural Española. Don Agustin de Gordejuela y Sierra. Madrid, 204 pp.

Toumanoff, C., and C. Vago. 1951. L'agent pathogène de la flacherie des vers à soie endémique dans la region des Cévennes: *Bacillus cereus* var. *alesti* var. nov. C. R. Seances Hebd. Acad. Sci. 233: 1504–1506.

———. 1952*a*. La nature de l'affection des vers à soie due à *Bacillus cereus* var. *alesti* Toum. et Vago, et les modalités d'action de ce bacille. Ann. Inst. Pasteur 83: 421–422.

———. 1952*b*. L'effet de l'alcalinité du milieu de culture sur la virulence de *Bacillus cereus* var. *alesti* Toum. et Vago, pour les vers à soie. C. R. Seances Hebd. Acad. Sci. 235: 1715–1717.

Trabut, L. 1898*a*. Le champignon des altises (*Sporotrichum globuliferum*). C. R. Seances Hebd. Acad. Sci. 75: 359.

———. 1898*b*. Destruction de l'altise de la vigne par un champignon parasite (*Sporotrichum globuliferum* ou *Isaria globulifera*). Gouvern. Génér. de l'Algérie, Serv. Bot. Inform. Agr. Bull. 15.

———. 1899. La destruction des altises en hiver (*Sporotrichum globuliferum*). Bull. Agr. de l'Algérie et de la Tunisie, Oct. 15, 1899.

Tubeuf, C. von. 1892. Die Krankheiten der Nonne (*Liparis monacha*). Forstl. Naturw. Z. 1: 34–57; 62–79; 277.

Tulasne, L. R., and C. Tulasne. 1863-1865. Selecta fungorum carpologia. Fussu, Paris. Vol. 2, 221 pp.; Vol. 3, 319 pp.

Turpin, P. J. F. 1836. Observations sur le *Botrytis* de la muscardine. C. R. Seances Hebd. Acad. Sci. 3: 170–173.

Urich, F. W. 1913. Notes on some Mexican sugarcane insects. J. Econ. Entomol. 6: 247.

Vago, C. 1963. La pathogénèse des viroses d'insects. pp. 162–174. *In* Recent Progress in Microbiology. Symposia VIII. International Congress Microbiology. Montreal, 1962.

Vallery-Radot, R. 1937. The life of Pasteur. [Trans. from French by Mrs. R. L. Devonshire. The Sun Dial Press, Inc. Garden City, New York. 484 pp.]

Van der Laan, P. A., ed. 1967. Insect Pathology and Microbial Control. Proceedings of the International Colloquium on Insect Pathology and Microbial Control. Wageningen, The Netherlands, September 5–10, 1966. North-Holland Publishing Co., Amsterdam. 360 pp.

Van Dine, D. L. 1911. Progress report on introduction to beneficial parasites into Porto Rico. 1st Rpt. Board Comm. Agr. Porto Rico, 34–35.

Vida, Marcus Girolamo. 1527. The Silkworm (Bombycum): a poem in two books. Trans. into English verse by the Reverend Samuel Pullein of Trinity College, Dublin, Bi-lingual edition: English translation facing original Latin text. Printed by S. Powell, 1750, 141 pp. (Book II, pp. 81–93).

Virgil (Publius Virgilius Maro). 37–29 B.C. The Georgics of Virgil, Book IV. [Trans. by C. Day Lewis, 1941. Oxford University Press, 83 pp.]

Vittadini, C. 1859. Sul mado di distinguere nei bachi da seta la semente infetta dalla sana Milan, Atti Inst. Lomb (1858) I, 360–363.

Vouk, V. 1931. Four years of international corn borer investigations. Pp. 92–96 *in* T. Ellinger, ed., International corn borer investigations, vol. 4. International Livestock Exposition, Union Stock Yards, Chicago. 96 pp.

———. 1932. Rad botaničkog instituta univerziteta u Zagrebu na izucavanju kukuruznog crva. Acta Botanica, Inst. Botan. Univ. Zagrebensis 7: 129–144.

Vouk, V., and Z. Klas. 1931. Conditions influencing the growth of the insecticidal fungus *Metarrhizium anisopliae* (Metsch.) Sor. Pp. 24–45 *in* T. Ellinger, ed., International corn borer investigations, vol. 4. International Livestock Exposition, Union Stock Yards, Chicago. 96 pp.

———. 1932. Über einige Kulturbedingungen des insektentötenden Pilzes *Metarrhizium anisopliae* (Metsch.) Sor. Acta Botanica, Inst. Botan. Univ. Zagrebensis 7: 35–58.

Vuillemin, P. 1912. *Beauveria*, nouveau genre de Verticilliacées. Soc. Bot. de France [Paris] Bull. 59: 34–40.

Walker, A. J. 1938. Fungal infections of mosquitoes, especially of *Anopheles costalis* Ann. Trop. Med. Parasitol. 32: 231–244.

Walker, J. C. 1969. Plant Pathology. 3rd ed., McGraw-Hill Book Co., 819 pp.

Wallengren, H. 1930. On the infection of *Pyrausta nubitalis* Hb. by *Metarrhizium anisopliae* (Metsch.) Sor. II. Pp. 64-73 *in* T. Ellinger, ed., International corn borer investigations, vol. 3. International Livestock Exposition, Union Stock Yards, Chicago. 174 pp.

Wallengren, H., and R. Johannson. 1929. On the infection of *Pyrausta nubitalis* Hb. by *Metarrhizium anisopliae* (Metsch.) Sor. Pp. 131–145 *in* T. Ellinger, ed., International corn borer icvestigations, vol. 2. International Livestock Exposition, Union Stock Yards, Chicago. 183 pp.

Ward, H. B. 1930. Stephen Alfred Forbes—a tribute. (Obituary.) Science 71: 378–381.

Wasser, H. B. 1952. Demonstration of a new insect virus not associated with inclusion bodies. J. Bacteriol. 64: 787–792.

Wasser, H. B., and E. A. Steinhaus. 1951. Isolation of a virus causing granulosis in the red-banded leaf roller. Virginia J. Sci. 2: 91–93.

Waterson, J. M. 1947. Report of the plant pathologist, 1946. Bermuda Dept. Agr. Bermuda Press, Ltd., Hamilton, Bermuda. 18 pp.

Watkins, L. H. 1967. Harbison's second importation of bees to California. Am. Bee J. 107 (10): 378–379.

———. 1968*a*. The myth of the Russian bees in California. Am. Bee J. 108 (4): 145–146.

———. 1968*b*. California's first honey bees. Am. Bee J. 108 (5): 190–191.

———. 1968*c*. Mr. Buck's importation of bees to California. Am. Bee J. 108 (6): 232–233.

Watson, J. R. 1914. Whitefly control, 1914. Florida Agr. Exp. Sta. Bull. 123, 23 pp.

———. 1916. Preserving fungus parasites of whitefly. Florida Agr. Exp. Sta. Press Bull. 257. 2 pp. (Pages unnumbered.)

———. 1923. Entomogenous fungi on citrus. Florida Agr. Exp. Sta. Press Bull. 346, 2 pp. (Pages unnumbered.)

Watson, J. R., and E. W. Berger. 1937. Citrus insects and their control. Univ. Florida Ext. Bull. 88, 1–135.

Webber, H. J. 1895. Preliminary notice of a fungus parasite of *Aleurodes citri*. J. Mycology 7: 363.

———. 1897. The new method of warfare against scale insects. Florida State Hort. Soc. Proc. 10: 53–58.

Weibel, A. C. 1952. Two thousand years of textiles. Pantheon Books. New York. 169 pp. plus 330 plates.

Weiser, J., and M. H. Muma. 1966. *Entomophthora floridana* n. sp. (Phycomycetes: Entomophthoraceae), a parasite of the Texas citrus mite, *Eutetranychus banksi.* Florida Entomologist 49: 155–159.

Weiss, H. B. 1936. The pioneer century of American entomology. Publ. by author. New Brunswick, New Jersey. 320 pp. [Insect pathology items on pp. 6, 8-11, 42-43, 50-51, 57, 69, 102, 104-105, 116-117, 134, 145, 176, 185.]

Weston, W. H. 1933. Roland Thaxter. Mycologia 25(2): 69–89.

Wheeler, W. M. 1930. Demons of the dust. W. W. Norton and Co., New York. 378 pp.

White, G. F. 1906. The bacteria of the apiary with special reference to bee diseases. U. S. Dep. Agr., Bur. Entomol., Tech. Bull. 14, 50 pp.

———. 1907. The cause of American foulbrood. U. S. Dep. Agr., Bull. Entomol. Circ. 94. 4 pp.

———. 1912. The cause of European foulbrood. U. S. Dep. Agr., Bur. Entomol. Circ. 157. 15 pp.

———. 1913. Sacbrood, a disease of bees. Dep. Agr., Bur. Entomol., Circ. 0169, 5 pp.

———. 1917. Sacbrood. U. S. Dep. Agr. Bull. 431. 54 pp.

———. 1919. Some observations on Nosema-disease. Am. Bee J. July-Aug.-Sept., 1919. 10 pp.

———. 1920*a*. American foulbrood. U. S. Dep. Agr., Bur. Entomol. Bull. 809. 46 pp.

———. 1920*b*. European foulbrood. U. S. Dep. Agr. Bull. 810. 39 pp.

———. 1923*a*. Hornworm septicemia. J. Agr. Research 26: 477–486.

———. 1923*b*. Cutworm septicemia. J. Agr. Research 26: 487–496.

———. 1927. A protozoan and a bacterial disease of *Ephestia kühniella* Zell. Proc. Entomol. Soc. Wash. 29: 147–148.

———. 1928. Potato beetle septicemia, with the proposal of a new species of bacterium. Proc. Entomol. Soc. Wash. 30: 71–72.

———. 1929–30. Preliminary observations on the polyhedral diseases of insects. J. Parasitol. 16: 107.

———. 1931. Production of sterile maggots for surgical use. J. Parasitol. 18: 133.

———. 1935. Potato beetle septicemia. J. Agr. Research 51: 223–234.

Wickson, E. J. 1879–1905. Pamphlets on silk culture. Wickson Collection. University California Library, Berkeley.

Wildermuth, V. L. 1914. The alfalfa caterpillar. U. S. Dep. Agr. Bull. 124. Bureau of Printing, Washington, D. C.

———. 1922. The alfalfa caterpillar (*Eurymus eurytheme* Boisd.). U. S. Dep. Agr. Bur. Entomol. Circ. 133.

Wildy, P. 1971. Classification and Nomenclature of Monographs in Virology 5: S. Karger, Basel and New York. 81 pp.

Willet, J. E., and A. J. Cook. 1880. Experiments with yeast ferment on various insects. Am. Entomologist 3: 289–290.

Williams, G. 1959. Virus Hunters. Knopf, New York. 503 pp.

Williams, J. W. 1890. A tumor in a fresh-water mussel (*Anodonta cygnaea* Linn.). J. Anat. Physiol. Norm. Pathol. 24: 307–308.

Wilson, W. T. 1970. Inoculation of the pupal honeybee with spores of *Bacillus larvae*. J. Agr. Research 9 (1): 33–37.

———. 1971. Survival of *Bacillus larvae* spores in honeybee faeces exposed to sunlight. J. Agr. Research 10 (3): 149–151.

Wilson, W. T., and L. E. Combs. 1970. Flagellar bundles of the honeybee pathogen, *Bacillus larvae*; their occurrence, size, and development *in vivo* and *in vitro*. Can. J. Microbiol. 16: 521–526.

Winston, J. R., J. J. Bowman, and W. W. Yothers. 1923. Bordeaux-oil emulsion. U. S. Dep. Agr. Tech. Bull. 176.

Wittig, G., E. A. Steinhaus, and J. P. Dineen. 1960. Further studies of the cytoplasmic-polyhedrosis virus of the alfalfa caterpillar. J. Insect Pathol. 2: 334–345.

Wize, C. 1904. Choroby komósnika buraczanego (*Cleonus punctiventris*) powodowane przez grzyby owadobójeze, ze szczégolnen uwzglednieniem gatunków now; ch.— Die durch Pilze hervorgerufenen Krakheiten des Rübenrüsselkafers (*Cleonus punctiventris* Germ.) mit besonderer Berücksichtigung neuer Arten Bull. Internat. Acad. Sci. Cracovie, Classe Sci. Math. Nat. 10: 713–727.

Yamafuji, K. 1958. Role of deoxyribonucleic acid and deoxyribonuclease in induction processes of polyhedrosis virus. Pp. 731–736 *in* Proc. Inter. Cong. Entomol., 10th, Montreal, 4, 1956.

Yamafuji, K., and H. Omura. 1950. Studies on the provoking action of hydroxylamine derivatives for the virus formation. J. Agr. Chem. Soc. Japan 23: 325. [In Japanese. English abstract.]

Yamazaki, H. 1960. Flacherie of the silkworm is infectious. Virus-like particles in the homogenate of the diseased silkworm. Shinano Mainichi [Newspaper], 9-26-1960. [In Japanese.]

Yothers, W. W., and A. C. Mason. 1930. The citrus rust mite and its control. U. S. Dep. Agr. Tech. Bull. 176. 56 pp.

Ziegler, L. W. 1949. The possible truer status of the red-headed scale-fungus. A preliminary report. Florida Entomologist 32: 151–157.

Zeikus, R. D., and E. A. Steinhaus. 1968. Teratology of the beetle *Tenebrio molitor*. II. The Development and gross description of the pupal-winged adult. J. Invertebr. Pathol. 11: 8–24.

Zopf, W. 1890. Die Pilze in morphologischer, physiologischer, biologischer und systematischer Beziehung. Eduard Trewendt. Breslau. 500 pp.

Index to Authors Cited

Subject Index

DATE DUE

DEMCO 38-297